Lecture Notes in Physics

Volume 1042

The series Lecture Notes in Physics (LNP), founded in 1969, reports new developments in physics research and teaching - quickly and informally, but with a high quality and the explicit aim to summarize and communicate current knowledge in an accessible way. Books published in this series are conceived as bridging material between advanced graduate textbooks and the forefront of research and to serve three purposes:

- to be a compact and modern up-to-date source of reference on a well-defined topic;
- to serve as an accessible introduction to the field to postgraduate students and non-specialist researchers from related areas;
- to be a source of advanced teaching material for specialized seminars, courses and schools.

Both monographs and multi-author volumes will be considered for publication. Edited volumes should however consist of a very limited number of contributions only. Proceedings will not be considered for LNP.

Volumes published in LNP are disseminated both in print and in electronic formats, the electronic archive being available at springerlink.com. The series content is indexed, abstracted and referenced by many abstracting and information services, bibliographic networks, subscription agencies, library networks, and consortia.

Proposals should be sent to a member of the Editorial Board, or directly to the responsible editor at Springer:

Dr Lisa Scalone
lisa.scalone@springernature.com

Shinya Uji

Fermiology

Quantum Oscillations and Angle Dependent Magnetoresistance Oscillations

Springer

Shinya Uji
Research Center for Materials
Nanoarchitectonics
National Institute for Materials Science
Tsukuba, Ibaraki, Japan

ISSN 0075-8450 ISSN 1616-6361 (electronic)
Lecture Notes in Physics
ISBN 978-981-95-0819-8 ISBN 978-981-95-0820-4 (eBook)
https://doi.org/10.1007/978-981-95-0820-4

Translation from the Japanese language edition: "Ferumioroji" by Shinya Uji, © Shinya Uji 2020. Published by University of Tsukuba Press. All Rights Reserved.

This Springer imprint is published by the registered company Springer Nature Singapore Pte Ltd.
The registered company address is: 152 Beach Road, #21-01/04 Gateway East, Singapore 189721, Singapore

Preface

Conductors such as metals with high electrical conductivity exhibit large heat transfer, high optical reflectivity at the surface, and temperature-independent magnetism. These properties are predominantly determined by the electronic states on the Fermi surface. The Fermi surfaces in conductors contain a variety of microscopic information on the electronic states that lead to physical properties. Therefore, extracting this information is an important research issue for understanding the origin of the physical properties of conductors.

In general, conductors with energy bands formed by many orbitals of the constituent atoms have complicated Fermi surface structures. By contrast, the Fermi surfaces of organic conductors, often cited in this Lecture Note, have very simple structures. This is because only one orbital of an organic molecule forms an energy band, and the electronic state is highly anisotropic; i.e., the electronic state is spread out only in a certain direction or plane. The physical properties of organic conductors can be easily deduced from the simple Fermi surface structures, and therefore "*Fermiology*" (Fermi surface study) has developed significantly in organic conductors both experimentally and theoretically.

There are two promising methods to determine Fermi surfaces and energy band structures: photoemission spectroscopy and quantum oscillation measurements. In recent years, there have been significant technical developments in the former, known as angle-resolved photoemission spectroscopy. In this method, light is incident on the surface of a sample and the energy and momentum of the electrons emitted from the sample surface are measured. This allows the direct observation of the energy bands. The advantage is that the band structure can be determined over a wide energy range up to the Fermi level due to the high energy of the incident light. However, because the emitted photoelectrons are limited to those near the surface of the sample, this technique provides information on the electronic states only at the sample surface and it means that a clean surface is required. Its energy resolution is also not high. The latter method, quantum oscillation measurements resulting from Landau quantization of electronic states in a magnetic field, can provide valuable information on the electronic states very close to the Fermi level. This method is insensitive to the sample surface condition, as quantum oscillations are observed in bulk properties such as electrical resistance, magnetization, etc. Furthermore, its energy resolution is higher than that of photoelectron spectroscopy, allowing us to probe the detailed structure of the Fermi surfaces. On the other hand, owing to the geometrical structure of the Fermi surface, electrons have characteristic orbits on the Fermi surface in a magnetic field. This orbital motion causes oscillations in the resistance as a function of the magnetic field orientation. This is called the angle-dependent magnetoresistance oscillation, and the measurement has recently been established as a method to determine the Fermi surface structure in detail. The Fermi surface structure is closely related to the mechanisms of various electronic phase transitions such as superconductiviting, density wave, and electron nematic transitions. Therefore, it is very important to know the detailed structure of Fermi surfaces in order to determine the origin of electronic phase transitions. This Lecture Note focuses on these measurements and examines how the detailed structure of Fermi surfaces can be determined.

Most of the content is described for low-dimensional conductors with simple Fermi surface structures. However, the content is also applicable to conductors with complex three-dimensional Fermi surface structures. This Lecture Note is written for students who have studied elementary solid-state physics after mastering quantum mechanics and statistical mechanics. The chapters contain many appendices for a better understanding. If you wish to have a general overview of the content, you can skip the appendices. It is hoped that this Lecture Note will be of assistance to those beginning their studies in this research area.

The author would like to thank Prof. Toshihito Osada (University of Tokyo) for his advice on the content of Chap. 4. The author would also like to thank the members of the University of Tsukuba Press for their advice. I am grateful to Mr. Kyohei Morisada (University of Tsukuba) for careful proofreading.

Tsukuba, Japan
June 2020

Shinya Uji

Preface for the English Edition

This Lecture Note is the English translation of the Japanese edition "*Fermiology*," published in 2020. This English version corrects errors in the Japanese edition and also includes more details on the Berry phase. Although many textbooks on solid-state physics are published worldwide, there are very few textbooks specialized in Fermiology, especially quantum oscillations and angle-dependent magnetoresistance oscillations. It is hoped that this textbook will help in the study of Fermiology.

Tsukuba, Japan
May 2025

Shinya Uji

Competing Interests The author has no competing interests to declare that are relevant to the content of this manuscript.

Contents

Symbols

a, b, c	Lattice constants
A_F	Cross section of Fermi surface
A''_F	Curvature factor of cross section of Fermi surface
A_N	Area of cyclotron orbit with Landau index N in k-space
A_{osc}	Amplitude of quantum oscillation
A_r	Area between two orbits in real space
B	Magnetic flux density (magnetic field as a convention)
B_c, B'_c	Threshold magnetic field of coherence peak for quasi-2D, 1D Fermi surface
B_{int}	Internal magnetic field
B_{MB}	Magnetic breakdown field
C, $\tilde{C}$	Heat capacity of electron, its oscillatory part
C_L, $\tilde{C}_L$	Elastic constant, its oscillatory part
C_T	Total heat capacity of sample and thermometer
$D(E)$, $\tilde{D}$	Density of state at energy E, its oscillatory part
D_N	Degeneracy of Landau level with N
e	Elementary charge
E, $\tilde{E}$	Electric field or energy, oscillatory part of internal energy
ΔE	Zeeman splitting
$E(k)$	Eigenvalues of electronic states or energy band
E_F	Fermi energy
E_g	Energy gap at zone boundary or energy barrier
E_N	Landau level energy with index N
ΔE_N	Energy spacing between Landau levels
$f(E)$	Fermi distribution function
F	Force or free energy
F_Q, F''_Q	Frequency of quantum oscillation, its second derivative
g, g_c	g factor, effective g factor
$\mathbf{G}$, $\mathbf{g}_i$ $(i = 1, 2, 3)$	Reciprocal lattice vector, unit vector of reciprocal lattice
h, $\hbar$	Planck constant, reduced Planck constant $\hbar = h/2\pi$
H	Magnetic field $B = \mu H$

$\mathcal{H}$, $\mathcal{H}_a$	Hamiltonian, Hamiltonian of localized state at lattice point a
I_{QI}	Oscillation amplitude due to quantum interference
j_i $(i = x, y, z)$	Current density
$J_i(z)$	i-th order Bessel function
k, k_F	Wave number, Fermi wave number
k_B	Boltzmann constant
$k_\perp$	Radius of cyclotron orbit or wave number perpendicular to magnetic field
$k_\parallel$	Wave number parallel to magnetic field
K	Constant in reduction factors, $K = 2\pi^2 m_0 k_B/e\hbar = 14.7$ [T/K]
l_B	Magnetic length, $l_B^2 = \hbar/eB$
m_0, m_b, m_c	Free electron mass, band mass, (cyclotron) effective mass
m_{cyc}	Effective mass obtained from cyclotron resonance experiment
m_{ij}	Mass tensor
M, $\tilde{M}$	Magnetization, its oscillatory part
$\tilde{M}_\parallel$, $\tilde{M}_\perp$	Oscillatory part of parallel, perpendicular magnetization
$n(E)$	Number of electronic state up to E
N	Landau index
N_0, N_L	Number of electrons with $k_z = 0$ in zero magnetic field, finite magnetic field
N_e	Number of electrons per unit volume
N_l	Number of lattice points per unit volume
$\boldsymbol{p}$	Momentum
P, p_T	Probability, transition amplitude of magnetic breakdown
Q, q_T	Probability, transition amplitude of Bragg reflection
$\boldsymbol{R}$, $\boldsymbol{a}_i$ $(i = 1, 2, 3)$	Lattice vector, unit lattice vector
R, R_0	Electric resistance, its non-oscillatory part
R_T, R_D, R_S	Temperature, Dingle, spin splitting reduction factors of quantum oscillation
s, s_z, $\boldsymbol{s}_i$	Electron spin, its z-component, spin operator
S, S_m	Entropy, magnetic entropy
S_{loc}	Local spin
S_N	Area of cyclotron orbit with Landau index N in real space
S_{spin}	Spin splitting factor $S_{\mathrm{spin}} = g_c\mu_c/2$
t	Time
t_i $(i = a, b, c)$	Transfer integral along i-axis
T	Temperature
T_c	Superconducting transition temperature
ΔT, $\Delta\tilde{T}$	Magnetocaloric effect, its oscillatory part
T_{bath}, T_{sample}	Thermal bath temperature, sample temperature
T_D	Dingle temperature $T_D = \hbar/2\pi k_B\tau$
U_j	Spin rotation operator around j-axis

$v,\ v_F$	Velocity of electron, Fermi velocity
$v_\perp$	Velocity perpendicular to magnetic field
$V(r)$	Lattice potential
z	Parameter in temperature reduction factor, $z = Kp\mu_c T/B$
ε	Constant term of energy band
$\Delta\varepsilon$	Uncertainty of energy
θ	Magnetic field angle (mainly elevation angle)
θ_c	Critical angle
κ	Thermal conductivity
λ_{ee}	Mass enhancement factor by electron-electron interaction
λ_{ep}	Mass enhancement factor by electron-phonon interaction
μ	Chemical potential or permeability
μ_0	Permeability of vacuum
μ_B	Bohr magneton $\mu_B = e\hbar/2m_0$
μ_c	Effective mass ratio $\mu_c = m_c/m_0$
ξ_{loc}	Weak localization length
$\rho,\ \rho_0,\ \tilde{\rho}$	Electrical resistivity, its non-oscillatory part, its oscillatory part
$\rho_{ij}\ (i,\ j = x,\ y,\ z)$	Resistivity tensor
$\sigma,\ \sigma_0,\ \tilde{\sigma}$	Electrical conductivity, its non-oscillatory part, its oscillatory part
$\boldsymbol{\sigma}_i\ (i = x,\ y,\ z)$	Pauli matrix
$\sigma_{ij}\ (i,\ j = x,\ y,\ z)$	Conductivity tensor
τ	Scattering time
τ_r	Thermal relaxation time
τ_{torque}	Magnetic torque
Φ	Total magnetic flux
Φ_0	Flux quantum $\Phi_0 = h/e$
φ	Magnetic field angle (mainly azimuthal angle)
φ_0	Initial angle of electron orbit, phase of quantum oscillation
φ_a	Wave function localized at lattice point a
φ_c	Critical angle
$\chi_L,\ \chi_{Pauli}$	Landau diamagnetism, Pauli paramagnetism
Ψ_B	Berry phase
ω_c	Cyclotron (angular) frequency
ω_c^{QI}	Angular frequency of quantum interference
$\omega_y,\ \omega_z$	Angular frequency of periodic motion on 1D Fermi surface
ω_D	Debye frequency
$\Omega,\ \tilde{\Omega}$	Thermodynamic grand potential, its oscillatory part
$\tilde{\Omega}_{2D}$	Oscillatory part of thermodynamic grand potential in 2D electron system
Ω_c	Angular frequency of closed orbit on side of quasi-2D Fermi surface
Ω_c'	Angular frequency of closed orbit on side of quasi-1D Fermi surface

Electronic States and Energy Bands in Crystals

1

Abstract

In crystal potential, the electronic states form energy bands. In conductors such as metals, the electrons occupy the energy bands up to a certain level. The energy bands and level are called the conduction bands and Fermi level, respectively. The electrons near the Fermi level, which have the highest kinetic energy, dominate the physical properties of the conductors. It is therefore important to know the electronic states in the conduction bands near the Fermi level in order to understand the physical properties of the conductors. The theory to calculate the energy bands in crystals is known as band theory. The calculation methods of the energy bands have advanced considerably with the development of computers. In particular, the methods of calculating energy bands without empirical parameters, first-principles calculations, have now been established as a highly effective method for understanding physical properties. In order to provide a straightforward insight into the energy bands, we do not delve into complicated calculations, but instead provide an overview of the band calculations in two limits: the free electron and the tight binding approximations. This approach allows for a more comprehensive understanding of how energy bands and the resulting Fermi surfaces are formed.

1.1 Free Electron Approximation

1.1.1 Wave Functions and Density of States

The most basic model, the free electron approximation, assumes that the periodic potential formed by the crystal lattice is sufficiently small compared to the kinetic energy of the electrons. This assumption is reasonable for alkali and alkaline earth metals, where the atomic orbitals forming the conduction bands are sufficiently broad, such as the *s*-orbitals. In this approximation, the electronic

S. Uji, *Fermiology*, Lecture Notes in Physics 1042,
https://doi.org/10.1007/978-981-95-0820-4_1

states forming the conduction bands are assumed to be equivalent to those in vacuum (free electrons). The electronic states in vacuum can be obtained by solving the quantum mechanical equation of motion, the Schrödinger equation $\mathcal{H}\varphi(\boldsymbol{r}) = E(\boldsymbol{k})\varphi(\boldsymbol{r})$. Using the reduced Planck constant $\hbar$ and the mass m of the electron, the Hamiltonian is expressed as

$$\mathcal{H} = -\frac{\hbar^2}{2m}\nabla^2. \tag{1.1}$$

Taking the volume of space in which the electrons exist as V, the eigenstate $\varphi(\boldsymbol{r})$ of the Schrödinger equation is a plane wave,

$$\varphi(\boldsymbol{r}) = \frac{1}{\sqrt{V}}e^{i\boldsymbol{k}\cdot\boldsymbol{r}}. \tag{1.2}$$

This eigenstate, the wave function of the electron, is normalized as

$$\langle\varphi(\boldsymbol{r})|\varphi(\boldsymbol{r})\rangle = \int_V \varphi(\boldsymbol{r})^*\varphi(\boldsymbol{r})\mathrm{d}\boldsymbol{r} = \int_V |\varphi(\boldsymbol{r})|^2\mathrm{d}\boldsymbol{r} = 1.$$

The eigenvalue, the kinetic energy of the electron, $E(\boldsymbol{k})$ is written as

$$E(\boldsymbol{k}) = \frac{\hbar^2\boldsymbol{k}^2}{2m} = \frac{\hbar^2\left(k_x^2 + k_y^2 + k_z^2\right)}{2m}. \tag{1.3}$$

The energy is proportional to the square of the wave number k.

We restrict the space in which the electrons exist to a cube with side length L. The periodic boundary conditions $\varphi(r) = \varphi(r + L)$, $(r = x, y, z)$ give the allowed wave numbers of the electronic states,

$$k_x = \frac{2\pi}{L}n_x,\ k_y = \frac{2\pi}{L}n_y,\ k_z = \frac{2\pi}{L}n_z,\quad n_x,\ n_y,\ n_z : \text{integer}. \tag{1.4}$$

In the crystal, electrons occupy states from the lowest energy level with $n_i = 0 (i = x, y, z)$. In accordance with the Pauli exclusion principle, a single electron is allowed to occupy a single state if the spin degree of freedom of the electron is disregarded. Given a macroscopic number of electrons in the crystal, states are occupied up to a certain energy. The wave number of this energy state is designated as the Fermi wave number k_F, and its energy is identified as the Fermi energy or Fermi level E_F. The distribution of k_F forms a sphere in k-space as illustrated in Fig. 1.1. All the k states inside this sphere ($k \le k_F$) are occupied by the electrons, and this sphere is called the Fermi surface. The electrons in the states on the Fermi surface have the highest kinetic energy. From the periodic boundary conditions, a state with wave number k has a small cubic volume $(2\pi/L)^3$ with one side of $2\pi/L$ in k-space. The probability of the existence of an electron is given by the square

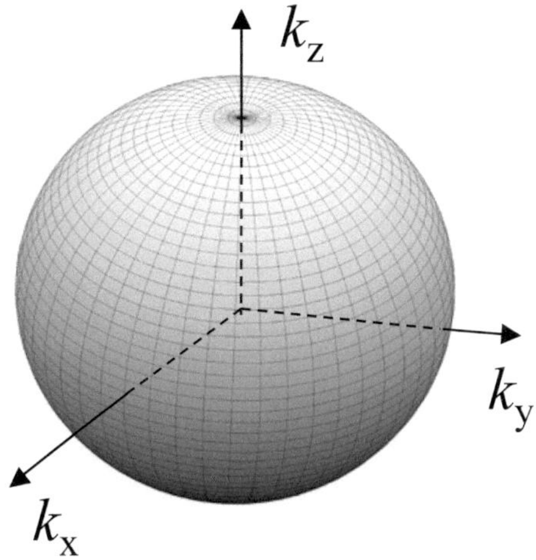

Fig. 1.1 Spherical Fermi surface formed by electronic states in vacuum

of its wave function, $|\varphi(\boldsymbol{r})|^2 = 1/L^3 = 1/V$. This probability is independent of $\boldsymbol{k}$ and $\boldsymbol{r}$. All the electrons are uniformly present in the entire volume V.

In the following, we consider this free electron state. If the number of states per unit volume in real space below the Fermi energy E_F is $n(E_F)$, then the total number of states in the Fermi surface is $L^3 n(E_F)$. Dividing the volume $(4\pi/3)k_F^3$ of the Fermi sphere by the volume $(2\pi/L)^3$ occupied by one electron, the total number of states $n(E_F)$ per unit volume ($L = 1$) can be calculated. Using $E_F = \hbar^2 k_F^2/2m$ we have

$$n(E_F) = \frac{2}{(2\pi)^3}\frac{4\pi}{3}k_F^3 = \frac{2}{(2\pi)^3}\frac{4\pi}{3}\left(\frac{2mE_F}{\hbar^2}\right)^{3/2} = \frac{1}{3\pi^2}\left(\frac{2mE_F}{\hbar^2}\right)^{3/2}. \tag{1.5}$$

Equation (1.5) has been doubled by including the spin degree of freedom. The density of the free electron states per unit energy, $D(E_F)$, is obtained by differentiating $n\ (E_F)$ by E_F,

$$D(E_F) = \frac{\mathrm{d}n(E_F)}{\mathrm{d}E_F} = \frac{1}{2\pi^2}\left(\frac{2m}{\hbar^2}\right)^{3/2}\sqrt{E_F}. \tag{1.6}$$

The density of states $D(E_F)$ depends on the energy. This is rewritten as

$$D(E_F) = \frac{mk_F}{\pi^2\hbar^2}. \tag{1.7}$$

The density of states is proportional to m and k_F. For a two-dimensional (2D) electron system, we obtain

$$n(E_F) = \frac{mE_F}{\pi\hbar^2} = \frac{k_F^2}{2\pi}, D(E_F) = \frac{m}{\pi\hbar^2}. \tag{1.8}$$

For a one-dimensional (1D) electron system, we have

$$n(E_F) = \frac{2}{\pi}\left(\frac{2mE_F}{\hbar^2}\right)^{1/2} = \frac{2k_F}{\pi}, D(E_F) = \frac{2m}{\pi\hbar^2 k_F}. \tag{1.9}$$

The spin degree of freedom is also included in both the 2D and 1D cases above.

1.1.2 Energy Band

What we need to know is the electronic state in a crystal, where the electrons move in the periodic lattice potentials formed by the crystal. Reflecting the periodicity, the wave functions describing the electronic states also have periodic functional forms. By introducing the periodic lattice potential $V(\boldsymbol{r})$, the Hamiltonian of the electron system can be written as

$$\mathcal{H} = -\frac{\hbar^2}{2m}\nabla^2 + V(\boldsymbol{r}). \tag{1.10}$$

Here, we define the reciprocal lattice vector $\boldsymbol{G}$ as follows:

$$\boldsymbol{R}\cdot\boldsymbol{G} = 2\pi m, m : \text{integer}. \tag{1.11}$$

The vector $\boldsymbol{R}$ shows the lattice point given by

$$\boldsymbol{R} = n_1\boldsymbol{a}_1 + n_2\boldsymbol{a}_2 + n_3\boldsymbol{a}_3, n_i : \text{integer}, \tag{1.12}$$

where $\boldsymbol{a}_\mathrm{i}(i = 1, 2, 3)$ are the unit vectors of the lattice. The reciprocal lattice vector $\boldsymbol{G}$ is defined as

$$\boldsymbol{G} = l_1\boldsymbol{g}_1 + l_2\boldsymbol{g}_2 + l_3\boldsymbol{g}_3, l_i : \text{integer}, \tag{1.13}$$

where $\boldsymbol{g}_\mathrm{i}(i = 1, 2, 3)$ are the unit vectors of the reciprocal lattice. The space spanned by the three unit vectors is called the reciprocal lattice space or k-space. When the three unit vectors $\boldsymbol{a}_\mathrm{i}$ are orthogonal to each other, then the unit vectors of the reciprocal lattice $\boldsymbol{g}_\mathrm{i}$ are also orthogonal to each other and their lengths are given by $g_i = 2\pi/a_i$ (i = 1, 2, 3). This expression satisfies Eq. (1.11) and thus $e^{i\boldsymbol{G}\cdot\boldsymbol{R}} = 1$. When the three unit vectors $\boldsymbol{a}_\mathrm{i}$ are not orthogonal, the unit vectors of the reciprocal lattice are generally given by

$$\boldsymbol{g}_1 = 2\pi\frac{\boldsymbol{a}_2\times\boldsymbol{a}_3}{\boldsymbol{a}_1\cdot(\boldsymbol{a}_2\times\boldsymbol{a}_3)}, \boldsymbol{g}_2 = 2\pi\frac{\boldsymbol{a}_3\times\boldsymbol{a}_1}{\boldsymbol{a}_1\cdot(\boldsymbol{a}_2\times\boldsymbol{a}_3)}, \boldsymbol{g}_3 = 2\pi\frac{\boldsymbol{a}_1\times\boldsymbol{a}_2}{\boldsymbol{a}_1\cdot(\boldsymbol{a}_2\times\boldsymbol{a}_3)}. \tag{1.14}$$

The denominator $\boldsymbol{a}_1 \cdot (\boldsymbol{a}_2 \times \boldsymbol{a}_3)$ shows the volume of the unit cell in real space. In general, it is possible to expand $V(\boldsymbol{r})$ by $\boldsymbol{G}$ as follows:

$$V(\boldsymbol{r}) = \sum_{\boldsymbol{G}} V_G e^{i\boldsymbol{G}\cdot\boldsymbol{r}}. \tag{1.15}$$

Note that $V(\boldsymbol{r})$ has the lattice periodicity, the translational symmetry $V(\boldsymbol{r}) = V(\boldsymbol{r}+\boldsymbol{R})$. This equation can be rewritten as

$$\sum_{\boldsymbol{G}} V_G e^{i\boldsymbol{G}\cdot\boldsymbol{r}} = V_0 + \sum_{\boldsymbol{G}<0} V_G e^{i\boldsymbol{G}\cdot\boldsymbol{r}} + \sum_{\boldsymbol{G}>0} V_G e^{i\boldsymbol{G}\cdot\boldsymbol{r}} = V_0 + 2\sum_{\boldsymbol{G}>0} V_G \cos(\boldsymbol{G}\cdot\boldsymbol{r}). \tag{1.16}$$

Here we set $V_{G=0} = V_0$. Since the potential is a real number, the relation $V_G = V_{-G}$ is used. The lattice has a positive charge, so the potential is negative for the electrons, $V(\boldsymbol{r}) < 0$, and thus $V_G < 0$. Because of the periodic lattice potential, the wave function $\varphi_{\boldsymbol{k}}(\boldsymbol{r})$ of the electron must satisfy a very general relation, the Bloch theorem (Appendix 1.1),

$$\varphi_{\boldsymbol{k}}(\boldsymbol{r}+\boldsymbol{R}) = e^{i\boldsymbol{k}\cdot\boldsymbol{R}}\varphi_{\boldsymbol{k}}(\boldsymbol{r}). \tag{1.17}$$

It is easy to verify that the free-electron wave function, Eq. (1.2) satisfies the Bloch theorem. By rewriting $\boldsymbol{k} = \boldsymbol{k}' + \boldsymbol{G}$, the Bloch theorem can be modified as follows:

$$\varphi_{\boldsymbol{k}}(\boldsymbol{r}+\boldsymbol{R}) = e^{i\boldsymbol{k}\cdot\boldsymbol{R}}\varphi_{\boldsymbol{k}}(\boldsymbol{r}) = e^{i(\boldsymbol{k}'+\boldsymbol{G})\cdot\boldsymbol{R}}\varphi_{\boldsymbol{k}'+\boldsymbol{G}}(\boldsymbol{r}) = e^{i\boldsymbol{k}'\cdot\boldsymbol{R}}\varphi_{\boldsymbol{k}'+\boldsymbol{G}}(\boldsymbol{r}).$$

The first term is written as $\varphi_{\boldsymbol{k}'+\boldsymbol{G}}(\boldsymbol{r}+\boldsymbol{R})$. By replacing $\boldsymbol{k}' \to \boldsymbol{k}$, we obtain

$$\varphi_{\boldsymbol{k}+\boldsymbol{G}}(\boldsymbol{r}+\boldsymbol{R}) = e^{i\boldsymbol{k}\cdot\boldsymbol{R}}\varphi_{\boldsymbol{k}+\boldsymbol{G}}(\boldsymbol{r}). \tag{1.18}$$

Even if the wave number $\boldsymbol{k}$ is changed by $\boldsymbol{G}$, the wave function obviously satisfies the Bloch theorem. This means that the choice of $\boldsymbol{k}$ is arbitrary: we can deduce the important fact that "the wave function and the energy band structure can be specified by $\boldsymbol{k}$ shifted by the reciprocal lattice vector $\boldsymbol{G}$." The simplest case is to specify $\boldsymbol{k}$ in the range $-g_i/2 < k_i \le g_i/2 (i = x, y, z)$. This reduced k-space range is called the first Brillouin zone.

For simplicity, we now consider a 1D lattice as shown in Fig. 1.2 and describe $\boldsymbol{r}, \boldsymbol{G}, \boldsymbol{g},$ and $\boldsymbol{k}$ as scalar quantities. For $G = g = 2\pi/a$, the first Brillouin zone is defined in the range $-\pi/a < k \le \pi/a$. By taking $l = 2$ in Eq. (1.13) and then $G = 2g = 4\pi/a$, we can define another range given by $-2\pi/a < k \le 2\pi/a$, which is twice the size of the first Brillouin zone. The range excluding the first Brillouin zone from $-2\pi/a < k \le 2\pi/a$ is defined as the second Brillouin zone. The third and fourth Brillouin zones can also be defined in a similar way. The Brillouin zones for a 2D lattice are illustrated in Appendix 1.2.

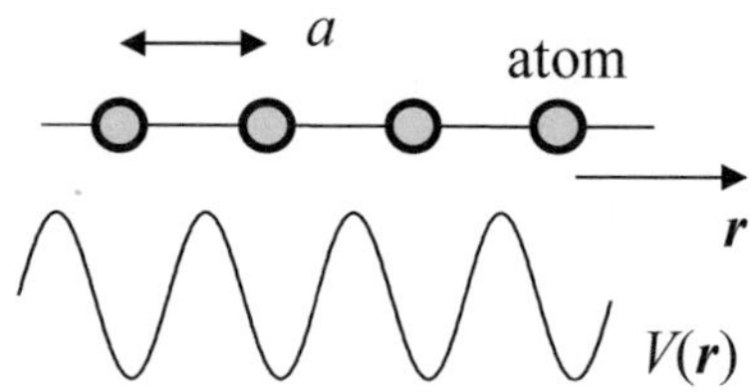

Fig. 1.2 1D lattice and schematic of its potential curve $V(r)$

Next, in the Hamiltonian of Eq. (1.10), let $\mathcal{H}_0 = -\hbar^2\nabla^2/2m$ be the non-perturbative term, assuming that $V(r)$ is sufficiently small to be treated as a perturbation. The eigenstate of the non-perturbative term with the periodic boundary condition in the range ($r = 0 \sim L$) and its eigenvalue (energy) are, respectively, given by

$$\varphi_k^{(0)}(r) = \frac{e^{ikr}}{\sqrt{L}}, \text{ and } E^{(0)}(k) = \frac{\hbar^2 k^2}{2m}.$$

The electron in the k state is scattered to the k' state by the perturbative term, the lattice potential $V(r)$. Its transition amplitude $S_{k'k}$ is written as

$$\begin{aligned} S_{k'k} &= \langle \varphi_{k'}^{(0)}(r)|V(r)|\varphi_k^{(0)}(r)\rangle = V_G \text{ for } k' = k + G, \\ S_{k'k} &= 0 \text{ for } k' \neq k + G. \end{aligned} \tag{1.19}$$

Here, $\langle X \rangle$ denotes the integration of X over the whole space. In other words, $V(r)$ causes a hybridization between the states with k and k' differing by G. The first-order perturbation energy is given by

$$E^{(1)}(k) = \langle \varphi_k^{(0)}(r)|V(r)|\varphi_k^{(0)}(r)\rangle = V_0. \tag{1.20}$$

This gives an offset independent of k for all the states. In the following, we take $V_0 = 0$ for simplicity. The second-order perturbation energy is given by

$$E^{(2)}(k) = \sum_G \frac{|V_G|^2}{E^{(0)}(k) - E^{(0)}(k+G)}. \tag{1.21}$$

The first-order correction term of the wave function is

$$\varphi_k^{(1)}(r) = \sum_{G\neq 0} \frac{V_G}{E^{(0)}(k) - E^{(0)}(k+G)} \varphi_{k+G}^{(0)}(r). \tag{1.22}$$

At the wave number $k = -G/2$, the eigenstates $\varphi_k^{(0)}(r)$ and $\varphi_{k+G}^{(0)}(r)$ are degenerate, $E^{(0)}(k) = E^{(0)}(k+G)$. The denominator in Eq. (1.21) becomes zero, showing that this calculation is not appropriate. Therefore, we need to obtain the eigenstate according to the perturbation theory for the degenerate case. The eigenstate is given by a linear combination of the two non-perturbative states,

$$\Psi_k(r) = u_k(0)\varphi_k^{(0)}(r) + u_k(G)\varphi_{k+G}^{(0)}(r), \tag{1.23}$$

where $u_k(0)$ and $u_k(G)$ are the coefficients satisfying the normalization condition, $|\Psi_k(r)|^2 = |u_k(0)|^2 + |u_k(G)|^2 = 1$. By multiplying the left-hand side of the Schrödinger equation $\mathcal{H}\Psi_k(r) = E(k)\Psi_k(r)$ by $\varphi_k^{(0)}(r)^*$ or $\varphi_{k+G}^{(0)}(r)^*$, and then integrating it, we obtain two eigenequations. The eigenvalues $E(k)$ are obtained by solving the following secular equation.

$$\begin{vmatrix} E^{(0)}(k) - E(k) & V_{-G} \\ V_G & E^{(0)}(k+G) - E(k) \end{vmatrix} = 0. \tag{1.24}$$

Note $V_G = V_{-G}$. The eigenvalues are

$$E_{\pm}(k) = \frac{E^{(0)}(k) + E^{(0)}(k+G)}{2} \pm \sqrt{\left(\frac{E^{(0)}(k) - E^{(0)}(k+G)}{2}\right)^2 + V_G^2}. \tag{1.25}$$

Since $V_G < 0$, the eigenvalues for $k = -G/2$ are simply given by

$$E_{\pm} = E^{(0)}(-G/2) \pm |V_G|. \tag{1.26}$$

The degeneracy of the two states $\varphi_k^{(0)}(r)$ and $\varphi_{k+G}^{(0)}(r)$ at $k = -G/2$ is lifted, and the energy gap $|2V_G|$ opens. The eigenvalues representing the energy bands $E(k)$ are depicted in three zone schemes in Fig. 1.3. In the extended zone scheme **a**, we see a parabolic band approximately given by $E^{(0)}(k) = \hbar^2k^2/2m$ with energy gaps $|2V_G|$. Since G can take many values, $G = 2\pi l/a (l : \text{integer})$, the band has periodic gaps at $k = -G/2 = \pi l/a$. The bands separated by the gap are called the first, second, third, bands from the lowest level. As seen in Eq. (1.19), the scattering of the electron from k to $k+G$ is called Bragg reflection. By shifting k by G, we can draw all bands in the first Brillouin zone ($-\pi/a < k \le \pi/a$) as shown in **b**; the right/left second band in **a** is shifted to the left/right by $G = g = 2\pi/a$, the right/left third band is shifted by $G = 2g = 4\pi/a$, and so on. The scheme shown in **b** is called the reduced zone scheme. As shown in **c**, the scheme in which the reduced zones are connected is called the repeated zone scheme. In the repeated zone scheme, note that the bands connect smoothly at the zone boundaries $k = \pm G/2$, since $\partial E(k)/\partial k = 0$. As we will see in Sect. 1.3, this result means that the velocity of the electron at the zone boundaries vanishes. From the above discussion, we obtain the two important rules:

1. The periodic lattice potential removes the band degeneracy at the boundaries of the Brillouin zone, resulting in the formation of energy bands.

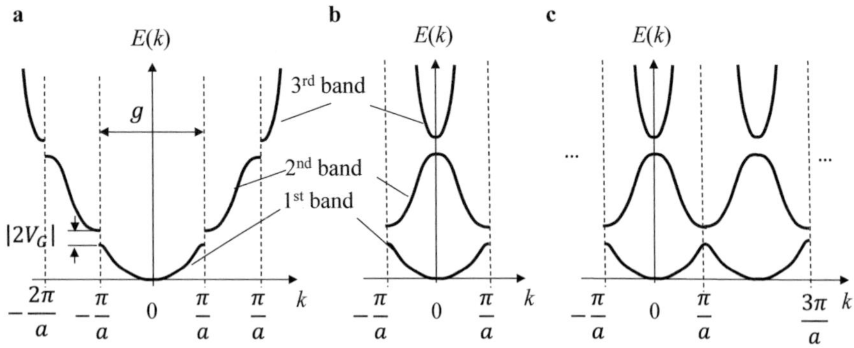

Fig. 1.3 Energy bands in free electron approximation for **a** extended, **b** reduced, and **c** repeated zone schemes

2. When the band degeneracy is removed, the energy band is flattened at the Brillouin zone boundaries; *i.e.*, the energy bands are smoothly connected at the boundaries in the repeated zone scheme.

Now consider the eigenstates with the eigenvalues given by Eq. (1.26). For $E_+ = E^{(0)} + |V_G|$, we obtain $u(0) = -u(G)$ and then the eigenstate is expressed as

$$\Psi_+(r) = \left(e^{-i\frac{\pi r}{a}} - e^{i\frac{\pi r}{a}}\right)/2 = -i\sin\left(\frac{\pi r}{a}\right). \tag{1.27}$$

For $E_- = E^{(0)} - |V_G|$, we obtain $u(0) = u(G)$, and then the eigenstate is

$$\Psi_-(r) = \left(e^{-i\frac{\pi r}{a}} + e^{i\frac{\pi r}{a}}\right)/2 = \cos\left(\frac{\pi r}{a}\right). \tag{1.28}$$

The electron densities are given by the squares of the amplitudes of the wave functions, $|\Psi_+(r)|^2 \propto \sin^2(\pi r/a)$ and $|\Psi_-(r)|^2 \propto \cos^2(\pi r/a)$ as depicted in Fig. 1.4. At the lattice points ($r = na$, n: integer), we obtain the potential $\mathrm{V}(na) = \sum_G V_G e^{iGna} = \sum_G V_G$, Since $V_G < 0$, the eigenstate $\Psi_-(r)$, which has a larger electron density at the lattice points, has the lower eigenvalue than $\Psi_+(r)$. The eigenstates $\Psi_+(r)$ and $\Psi_-(r)$ express standing waves formed by the $\pm k$ states; i.e., the electrons in these states are localized.

If N_l is the number of the lattice points with the periodic boundary condition, $\varphi_k(r) = \varphi_k(r + N_l a)$, we obtain the relation $e^{ikN_l a} = 1$, giving $k = 2\pi l/N_l a$ (l: integer). Therefore, one electron occupies the state with $\Delta k = 2\pi/N_l a$. In the first Brillouin zone $-\pi/a < k \le \pi/a$, the total number of the electronic states is

$$\left(\frac{2\pi}{a}\right)/\Delta k = N_l. \tag{1.29}$$

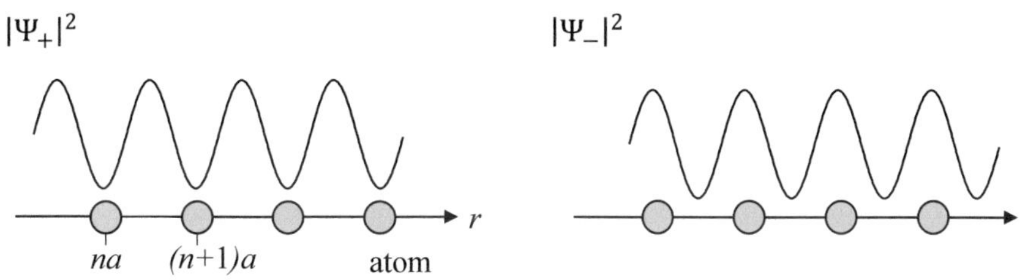

Fig. 1.4 Electron densities of wave functions, $|\Psi_+(r)|^2 \propto \sin^2(\pi r/a)$ and $|\Psi_-(r)|^2 \propto \cos^2(\pi r/a)$

Taking into account the spin degree of freedom, this number is doubled, $2N_l$. In Fig. 1.3, the first band is fully filled with $2N_l$ electrons. Similarly, the second band is also fully filled with another $2N_l$ electrons and so on. In an electric field, k of all electronic states is effectively shifted in the opposite direction of the electric field. For a partially filled band called the conduction band, the k shift results in charge transport; i.e., this band represents a metal. However, for a fully filled band called the valence band, k is not effectively shifted, indicating no charge transport. This state is an insulator, specifically called a band insulator. For charge transport in band insulators, the electrons in the fully filled (valence) bands must be thermally excited to the upper empty (conduction) bands.

1.1.3 Fermi Surface in Free Electron Approximation

How does the Fermi surface change if the number of electrons is hypothetically increased in the free electron approximation? The crystal must be neutral; therefore, to actually increase or decrease the number of electrons, the atoms in the crystal would have to be replaced by other atoms with different valences. For simplicity, let us look at an example of a 2D square lattice with the lattice constant a. The Hamiltonian Eq. (1.10) is rewritten as

$$\mathcal{H} = -\frac{\hbar^2}{2m}\left[\left(\frac{\partial}{\partial x}\right)^2 + \left(\frac{\partial}{\partial y}\right)^2\right] + V(\boldsymbol{r}). \tag{1.30}$$

Taking the first term as a non-perturbative term, we obtain the non-perturbative energy,

$$E^{(0)}(k) = \frac{\hbar^2\left(k_x^2 + k_y^2\right)}{2m}.$$

The energy bands along the k_x- and k_y-axes are parabolic, as shown in Fig. 1.3 **a**. The first Brillouin zone is indicated by a square with side $2\pi/a$ in Fig. 1.5. When the Fermi energy $E_F = \hbar^2 k_F^2/2m$ is within the first band (sufficiently small), the

Fermi surface (actually Fermi line) is a circle centered at the origin of the k-space Γ (Fig. 1.5a). We set periodic boundary conditions, assuming N_l lattice points along the x- and y-directions. Taking into account of the spin degree of freedom, we consider the case for $2N_l^2$ electrons. The area of the first Brillouin zone is $S_{BZ} = (2\pi/a)^2$, and there is one electronic state in the area $\delta S = \Delta k^2/2 = (2\pi/aN_l)^2/2$ from Eq. (1.29). Therefore, $2N_l^2$ electrons occupy the area equal to the first Brillouin zone, $\delta S \times 2N_l^2 = S_{BZ}$. If the energy band is gapless at the zone boundaries, we obtain a circular Fermi surface in the extended zone scheme as shown in Fig. 1.5b, whose area is equal to S_{BZ}. As discussed in Sect. 1.1, the band is gapped at the boundaries, as illustrated in Fig. 1.5c. By shifting the Fermi surfaces by the reciprocal lattice vectors $\boldsymbol{g}_x$ or $\boldsymbol{g}_y$, we obtain the Fermi surfaces in the reduced zone scheme as shown in Fig. 1.5d. To obtain the closed Fermi surfaces, we draw the Fermi surfaces in the repeated zone scheme in Fig. 1.5e. In this scheme, we can see that two types of closed Fermi surfaces are formed: elliptical closed Fermi surfaces α and square closed Fermi surfaces β. Since the circular Fermi surface in Fig. 1.5b is equal to S_{BZ}, the closed area of β is twice that of α.

For a 3D system, the volume of the k-space occupied by $2N_l^3$ electrons is equal to that of the first Brillouin zone. The Brillouin zone has different shapes depending on the crystal symmetry of the material, and thus the Fermi surfaces have different shapes depending on the symmetry. The k-points with high symmetry are

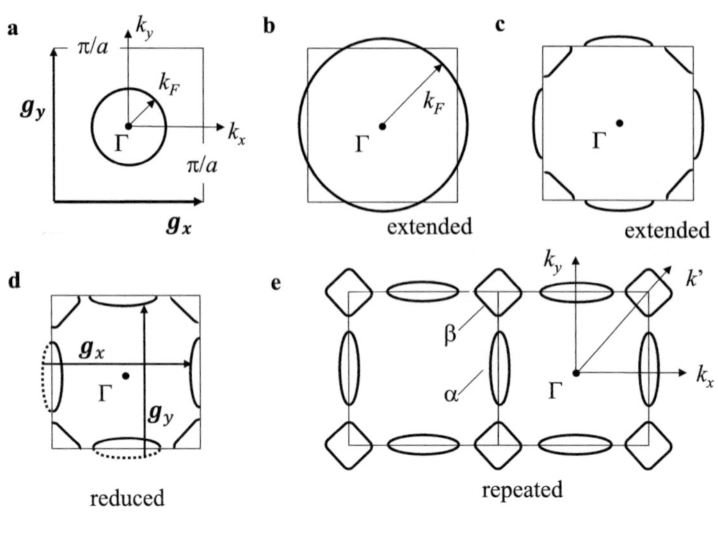

Fig. 1.5 Fermi surface of 2D square lattice with lattice constant a in free electron approximation. The first Brillouin zone is given by the square with side $2\pi/a$. **a** Fermi surface for low carrier concentration (low Fermi energy), **b** Circular Fermi surface in the extended zone scheme, whose area is equal to the first Brillouin zone S_{BZ}, **c** Fermi surface with gaps at the zone boundaries in the extended zone scheme, **d** Fermi surface in the reduced zone scheme, **e** Fermi surface in the repeated zone scheme—two closed Fermi surfaces α and β are formed

designated by letters according to a rule. The origin of the Brillouin zone is called the Γ point. Points inside the Brillouin zone are designated by Greek letters such as Γ, Σ, Δ, ... and points at the boundaries of the Brillouin zone are designated by the alphabet *K*, *M*, *U*,...[1]. In Appendix 1.3, we see the energy bands of the Fermi surface shown in Fig. 1.5e. In Appendix 1.4, the Fermi surface for a 2D system with anisotropic mass is depicted.

1.2 Tight Binding Approximation

1.2.1 Wave Function and Energy Band

The *d*- and *f*-orbitals in metals and the molecular orbitals in organic conductors cannot be described by plane waves like free electrons. This is because the electrons in these orbitals are strongly bound to the atoms or molecules. Therefore, in such materials, the wave function with a large amplitude at each lattice point as shown in Fig. 1.6 can describe the electronic state more accurately than a plane wave. The approximation using such a wave function is called the tight binding approximation. Unlike the free electron approximation, the tight binding approximation explicitly includes the effect of the lattice potential and therefore the band dispersion (the wave number dependence of the band) is generally different from a quadratic function.

In the band calculations, we make the following assumptions:

(1) The Hamiltonian $\mathcal{H}$ of the whole crystal can be approximated by the local Hamiltonian $\mathcal{H}_a$ at the lattice point *a*.
(2) The eigenstate at the lattice point *a* is described by a single orbit.
(3) The wave function is localized around each lattice point *a*, and the overlap of the wave functions between the neighboring lattice points is small enough.

In this case the wave function of the entire crystal should be represented by a linear combination of the wave functions $\varphi_a(\boldsymbol{r}-\boldsymbol{R})$ at each lattice point *a*. Summing over the position $\boldsymbol{R}$ of each lattice point, we express the wave function $\Psi_{\boldsymbol{k}}(\boldsymbol{r})$ of

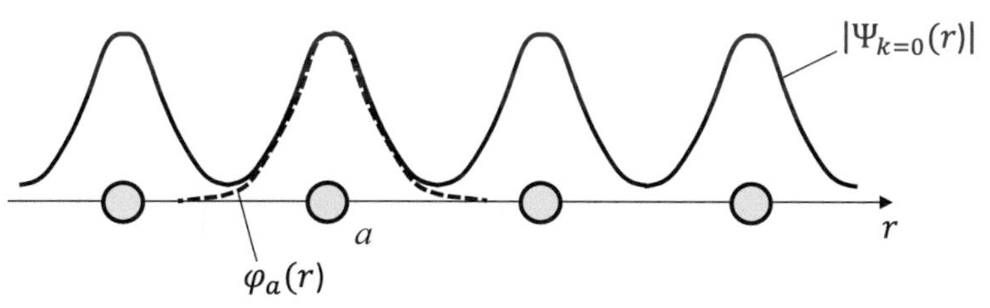

Fig. 1.6 1D lattice and schematic of the wave function with large amplitude at each lattice point. Here an *s*-like wave function is assumed at each site

the entire crystal as

$$\Psi_k(\boldsymbol{r}) = C\sum_{\boldsymbol{R}} e^{i\boldsymbol{k}\cdot\boldsymbol{R}}\varphi_a(\boldsymbol{r}-\boldsymbol{R}), \tag{1.31}$$

where C is the normalization constant. It is easy to show that the wave function $\Psi_k(\boldsymbol{r})$ satisfies the Bloch theorem, Eq. (1.17). The wave function $\varphi_a(\boldsymbol{r})$ of an electron localized at the lattice point satisfies the Schrödinger equation,

$$\mathcal{H}_a\varphi_a(\boldsymbol{r}) = \varepsilon\varphi_a(\boldsymbol{r}), \tag{1.32}$$

where ε is the eigenvalue. This Hamiltonian is given by $\mathcal{H}_a = \mathcal{H}_0 + V_a(\boldsymbol{r})$ where $\mathcal{H}_0 = -\hbar^2\nabla^2/2m$ and $V_a(\boldsymbol{r})$ is the lattice potential at point a. What we need is the expectation value of the total Hamiltonian $\mathcal{H} = \mathcal{H}_0 + V(\boldsymbol{r})$, where $V(\boldsymbol{r})$ is the potential of the whole crystal including $V_a(\boldsymbol{r})$,

$$\langle\Psi_k(\boldsymbol{r})|\mathcal{H}|\Psi_k(\boldsymbol{r})\rangle = |C|^2\sum_{\boldsymbol{R}\boldsymbol{R}'}\int e^{-i\boldsymbol{k}\cdot\boldsymbol{R}'}\varphi_a^*(\boldsymbol{r}-\boldsymbol{R}')\mathcal{H}e^{i\boldsymbol{k}\cdot\boldsymbol{R}}\varphi_a(\boldsymbol{r}-\boldsymbol{R})\mathrm{d}\boldsymbol{r}. \tag{1.33}$$

By taking $\boldsymbol{r}' = \boldsymbol{r}-\boldsymbol{R}$ and $\boldsymbol{R}'' = \boldsymbol{R}-\boldsymbol{R}'$, the right side of the above equation is written as

$$|C|^2\sum_{\boldsymbol{R}\boldsymbol{R}''}\int\varphi_a^*(\boldsymbol{r}'+\boldsymbol{R}'')\mathcal{H}e^{i\boldsymbol{k}\cdot\boldsymbol{R}''}\varphi_a(\boldsymbol{r}')\mathrm{d}\boldsymbol{r}' = |C|^2N_l\sum_{\boldsymbol{R}}\int\varphi_a^*(\boldsymbol{r}+\boldsymbol{R})\mathcal{H}e^{i\boldsymbol{k}\cdot\boldsymbol{R}}\varphi_a(\boldsymbol{r})\mathrm{d}\boldsymbol{r}.$$

Note that the integral depends only on the relative distance between $\varphi_a^*(\boldsymbol{r}'+\boldsymbol{R}'')$ and $\varphi_a(\boldsymbol{r}')$. On the right side, the equation is summed over $\boldsymbol{R}$, where N_l is the total number of the lattice points. The above equation can be divided into two parts,

$$\begin{aligned}\sum_{\boldsymbol{R}}\int\varphi_a^*(\boldsymbol{r}+\boldsymbol{R})\mathcal{H}e^{i\boldsymbol{k}\cdot\boldsymbol{R}}\varphi_a(\boldsymbol{r})\mathrm{d}\boldsymbol{r} &= \int\varphi_a^*(\boldsymbol{r})\mathcal{H}\varphi_a(\boldsymbol{r})\mathrm{d}\boldsymbol{r}\\ &\quad+\sum_{\boldsymbol{R}\neq 0}\int\varphi_a^*(\boldsymbol{r}+\boldsymbol{R})\mathcal{H}e^{i\boldsymbol{k}\cdot\boldsymbol{R}}\varphi_a(\boldsymbol{r})\mathrm{d}\boldsymbol{r}.\end{aligned} \tag{1.34}$$

Since $\mathcal{H} = \mathcal{H}_0 + V(\boldsymbol{r}) = \mathcal{H}_a + V(\boldsymbol{r}) - V_a(\boldsymbol{r})$, and thus $\mathcal{H} \approx \mathcal{H}_a$ for $\varphi_a(\boldsymbol{r})$, the first term of Eq. (1.34) is approximated as

$$\int\varphi_a^*(\boldsymbol{r})\mathcal{H}\varphi_a(\boldsymbol{r})\mathrm{d}\boldsymbol{r} \approx \int\varphi_a^*(\boldsymbol{r})\mathcal{H}_a\varphi_a(\boldsymbol{r})\mathrm{d}\boldsymbol{r} = \varepsilon.$$

From Eq. (1.32), the second term of Eq. (1.34) is rewritten as

$$\sum_{\boldsymbol{R}\neq 0}\int\varphi_a^*(\boldsymbol{r}+\boldsymbol{R})\mathcal{H}e^{i\boldsymbol{k}\cdot\boldsymbol{R}}\varphi_a(\boldsymbol{r})\mathrm{d}\boldsymbol{r} = \sum_{\boldsymbol{R}\neq 0}\int\varphi_a^*(\boldsymbol{r}+\boldsymbol{R})[\varepsilon + V(\boldsymbol{r}) - V_a(\boldsymbol{r})]e^{i\boldsymbol{k}\cdot\boldsymbol{R}}\varphi_a(\boldsymbol{r})\mathrm{d}\boldsymbol{r}.$$

By using the above relations, we can express Eq. (1.33) as

$$\langle \Psi_k(\boldsymbol{r})|\mathcal{H}|\Psi_k(\boldsymbol{r})\rangle = |C|^2 N_l \left[\varepsilon \left(1 + \sum_{\boldsymbol{R}\neq 0} s(\boldsymbol{R}) e^{\mathrm{i}\mathrm{k}\cdot\mathrm{R}} \right) - \sum_{\boldsymbol{R}\neq 0} t(\boldsymbol{R}) e^{\mathrm{i}\mathrm{k}\cdot\mathrm{R}} \right], \tag{1.35}$$

where

$$s(\boldsymbol{R}) = \int \varphi_a^*(\boldsymbol{r}+\boldsymbol{R})\varphi_a(\boldsymbol{r})\mathrm{d}\boldsymbol{r}, \tag{1.36}$$

$$t(\boldsymbol{R}) = -\int \varphi_a^*(\boldsymbol{r}+\boldsymbol{R})[V(\boldsymbol{r}) - V_a(\boldsymbol{r})]\varphi_a(\boldsymbol{r})\mathrm{d}\boldsymbol{r}. \tag{1.37}$$

The dimensionless quantity $s(\boldsymbol{R})$, called the overlap integral, represents the overlap of the wave functions between the neighboring lattice points. The quantity $t(\boldsymbol{R})$ is called the resonance integral or transfer integral and has a unit of energy. It also corresponds to the scattering amplitude of an electron from position $\boldsymbol{r}$ to $\boldsymbol{r}+\boldsymbol{R}$ by the potential $V(\boldsymbol{r}) - V_a(\boldsymbol{r})$. If $\varphi_a(\boldsymbol{r})$ has s-orbital symmetry as shown in Fig. 1.6, we have a positive value of $t(\boldsymbol{R}) > 0$ since $V(\boldsymbol{r}) - V_a(\boldsymbol{r}) < 0$. For lower symmetry cases, $t(\boldsymbol{R})$ can be positive or negative, depending on the phase of the wave function.

The normalization condition of the wave function is written as

$$\langle \Psi_k(\boldsymbol{r})|\Psi_k(\boldsymbol{r})\rangle = |C|^2 N_l \left(1 + \sum_{\boldsymbol{R}\neq 0} s(\boldsymbol{R}) e^{i\boldsymbol{k}\cdot\boldsymbol{R}} \right) = 1.$$

If the electron is localized at the lattice point, $s(\boldsymbol{R})$ is sufficiently less than unity, and then the above condition is approximated as $|C|^2 N_l = 1$. If the crystal has inversion symmetry, $V(\boldsymbol{r}) = V(-\boldsymbol{r})$, the eigenvalue for the total Hamiltonian $\mathcal{H}$ is simplified as

$$E(\boldsymbol{k}) = \frac{\langle \Psi_k(\boldsymbol{r})|\mathcal{H}|\Psi_k(\boldsymbol{r})\rangle}{\langle \Psi_k(\boldsymbol{r})|\Psi_k(\boldsymbol{r})\rangle} = \varepsilon - 2\sum_{\boldsymbol{R}>0} t(\boldsymbol{R})\cos(\boldsymbol{k}\cdot\boldsymbol{R}). \tag{1.38}$$

By summing the transfer integral $t(\boldsymbol{R})$ over the positions $\boldsymbol{R}$ of the neighboring lattice points, the energy bands $E(\boldsymbol{k})$ can be calculated. The band $E(\boldsymbol{k})$ reflects the symmetry of the potential, i.e., the crystal structure. From the relation $\cos[\boldsymbol{k}\cdot\boldsymbol{R}] = \cos[(\boldsymbol{k}+\boldsymbol{G})\cdot\boldsymbol{R}]$, the band given by Eq. (1.38) is obviously periodic with the reciprocal lattice vector $\boldsymbol{G}$. If the crystal has an orthorhombic structure with the lattice constants (a, b, c) as shown in Fig. 1.7 and $t(\boldsymbol{R})$ is summed over the nearest lattice points along each axis, we obtain the energy band,

$$E(\boldsymbol{k}) = \varepsilon - 2t_a\cos(k_x a) - 2t_b\cos(k_y b) - 2t_c\cos(k_z c), \tag{1.39}$$

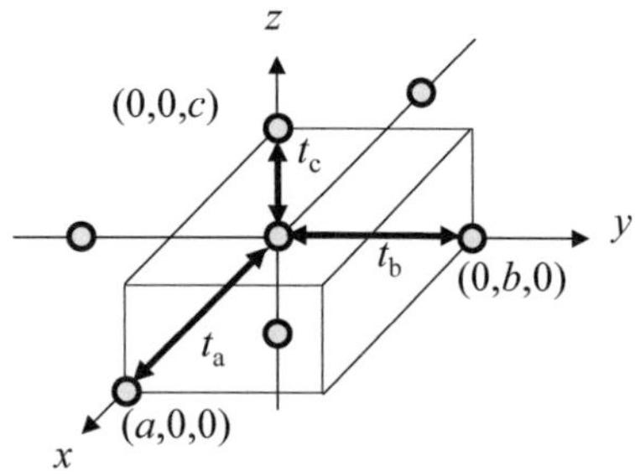

Fig. 1.7 Orthorhombic crystal structure with lattice constants (a, b, c) and transfer integrals between neighboring sites (t_a, t_b, t_c)

where t_i is the transfer integral between the neighboring lattice points along the i-axis. Although the energy ε originally arises from Eq. (1.32), we can set an arbitrary offset in the energy band. Taking $\varepsilon = 2(t_a + t_b + t_c)$, we obtain $E(\boldsymbol{k} = 0) = 0$ at the Γ point in the Brillouin zone. If the Fermi energy is sufficiently small, we have a parabolic band, $E(\boldsymbol{k}) \approx t_a(k_x a)^2 + t_b(k_y b)^2 + t_c(k_z c)^2$, similar to the result for the free electron approximation.

1.2.2 Fermi Surface in Tight Binding Approximation

Let us see how Fermi surfaces are formed in the tight binding approximation. We assume a 2D square lattice with lattice constant a as shown in Fig. 1.8a. The energy band is obtained by summing $t(\boldsymbol{R})$ over the nearest neighboring lattice points in Eq. (1.38). Taking $\varepsilon = 4t_a$, we have

$$E(\boldsymbol{k}) = 4t_a - 2t_a\cos(k_x a) - 2t_a\cos(k_y a). \tag{1.40}$$

Here, we define k' as the wave number in the diagonal direction; the $\boldsymbol{k}'$ vector is along $(\pi/a, \pi/a)$ in the Brillouin zone as shown in Fig. 1.8c. The energy bands along k_x and k' are depicted in Fig. 1.8b. When the Fermi level E_F is at $E_1 < 4t_a$

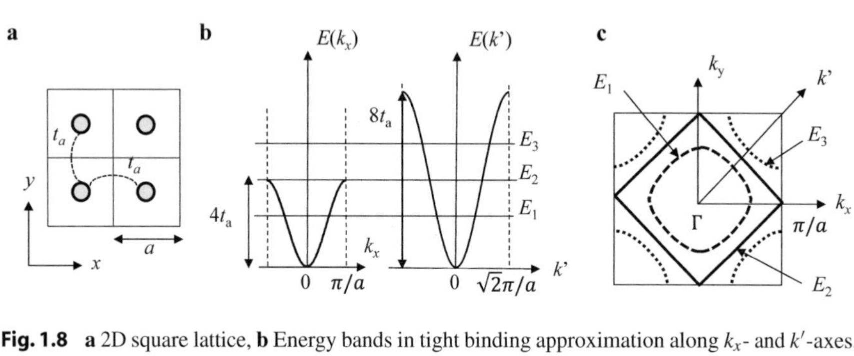

Fig. 1.8 **a** 2D square lattice, **b** Energy bands in tight binding approximation along k_x- and k'-axes, **c** Fermi surfaces for different Fermi levels (E_1, E_2, E_3)

the band crosses E_{F} in the directions of k_x and k'. This results in a slightly distorted circular Fermi surface as shown by the dashed curve in Fig. 1.8c. This shape is different from the circular Fermi surface in the free electron approximation. For $E_F = E_2 = 4t_a$, the Fermi wave number is given by the solution of $\cos(k_x a) + \cos(k_y a) = 0$. This Fermi surface extends to the zone boundary $k_x, k_y = \pm\pi/a$, resulting in a square Fermi surface as shown by the solid line in Fig. 1.8c. This area is exactly half the area of the Brillouin zone. This half-filled band corresponds to the density of one electron per site when the spin degree of freedom is included.

This situation, a half-filled band, is realized in the parent material of high-temperature superconductors such as La_2CuO_4. In La_2CuO_4, the Cu atoms form a square lattice as shown in Fig. 1.8a and the energy band is formed by a $3d$ orbital of Cu. Since this band is half-filled, the presence of a square Fermi surface, showing a metallic state, is expected as shown by the solid line in Fig. 1.8c in the band calculation. However La_2CuO_4 is an insulator. This insulating state, called a Mott insulator, is caused by the Coulomb repulsion between the electrons which is sufficiently larger than the bandwidth ($4t_a$), corresponding to the kinetic energy of the electrons; the strong Coulomb repulsion localizes one electron at each Cu atom. In La_2CuO_4, it is very interesting that the Mott insulating state breaks down and superconductivity appears when La is partially substituted with Sr; the hole doping in the band. The mechanism of this superconductivity is still not fully understood and remains one of the major research issues.

When the Fermi level rises to $E_3 > 4t_a$, there is no Fermi surface on the k_x-axis because the band does not cross E_3 on the k_x-axis, but there is a Fermi surface on the k'-axis. As a result, a Fermi surface is formed at the corners of the Brillouin zone as indicated by the dotted curve in Fig. 1.8c. From $\cos(k_x a) + \cos(k_y a) =$ const, we obtain

$$\frac{\partial k_y}{\partial k_x} = -\frac{\sin(k_x a)}{\sin(k_y a)}. \tag{1.41}$$

For $E_F > 4t_a$, this relation gives $\partial k_y/\partial k_x = 0$ at the zone boundaries $k_x = \pm\pi/a$ and $\partial k_y/\partial k_x = \infty$ at $k_y = \pm\pi/a$. The results mean that the Fermi surface intersects perpendicular to the zone boundaries, and a smooth closed orbit is formed in the repeated zone scheme. Another example of a 2D Fermi surface is given in Appendix 1.5.

1.2.3 Energy Band for Two Atoms in Unit Cell

So far we have assumed that there is one atom in the unit cell. Now let us consider the case where there are two atoms, A and B. In this case the wave function $\varphi(\mathrm{r})$ of the unit cell is given by a linear combination of the wave functions $\varphi_A(\mathrm{r})$ and $\varphi_B(\mathrm{r})$ of the respective atoms,

$$\varphi(\boldsymbol{r}) = C_A\varphi_A(\boldsymbol{r}) + C_B\varphi_B(\boldsymbol{r}). \tag{1.42}$$

The wave function $\varphi(\boldsymbol{r})$ satisfies the Schrödinger equation $\mathcal{H}\varphi(\boldsymbol{r}) = \varepsilon\varphi(\boldsymbol{r})$. The coefficients C_A and C_B are the normalized constants of the wave function. The total wave function of the crystal $\Psi_{\boldsymbol{k}}(\boldsymbol{r})$ is given by $\Psi_{\boldsymbol{k}}(\boldsymbol{r}) = C\sum_{\boldsymbol{R}} e^{i\boldsymbol{k}\cdot\boldsymbol{R}}\varphi(\boldsymbol{r}-\boldsymbol{R})$. By multiplying the left-hand side of the Schrödinger equation $\mathcal{H}\Psi_{\boldsymbol{k}}(\boldsymbol{r}) = E(\boldsymbol{k})\Psi_{\boldsymbol{k}}(\boldsymbol{r})$ by $\varphi_A^*(\boldsymbol{r})$, and integrating, we obtain

$$\begin{aligned}\langle\varphi_A(\boldsymbol{r})|\mathcal{H}|\Psi_{\boldsymbol{k}}(\boldsymbol{r})\rangle/C = {} & C_A\langle\varphi_A(\boldsymbol{r})|\mathcal{H}|\varphi_A(\boldsymbol{r})\rangle \\ & + C_A\sum_{\boldsymbol{R}\neq 0}\langle\varphi_A(\boldsymbol{r})|\mathcal{H}|\varphi_A(\boldsymbol{r}-\boldsymbol{R})\rangle e^{i\boldsymbol{k}\cdot\boldsymbol{R}} \\ & + C_B\sum_{\boldsymbol{R}}\langle\varphi_A(\boldsymbol{r})|\mathcal{H}|\varphi_B(\boldsymbol{r}-\boldsymbol{R})\rangle e^{i\boldsymbol{k}\cdot\boldsymbol{R}} = C_A E(\boldsymbol{k}).\end{aligned}$$

Similarly, by multiplying the Schrödinger equation by $\varphi_B^*(\boldsymbol{r})$ and integrating, we obtain

$$\begin{aligned}\langle\varphi_B(\boldsymbol{r})|\mathcal{H}|\Psi_{\boldsymbol{k}}(\boldsymbol{r})\rangle/C = {} & C_B\langle\varphi_B(\boldsymbol{r})|\mathcal{H}|\varphi_B(\boldsymbol{r})\rangle \\ & + C_B\sum_{\boldsymbol{R}\neq 0}\langle\varphi_B(\boldsymbol{r})|\mathcal{H}|\varphi_B(\boldsymbol{r}-\boldsymbol{R})\rangle e^{i\boldsymbol{k}\cdot\boldsymbol{R}} \\ & + C_A\sum_{\boldsymbol{R}}\langle\varphi_B(\boldsymbol{r})|\mathcal{H}|\varphi_A(\boldsymbol{r}-\boldsymbol{R})\rangle e^{i\boldsymbol{k}\cdot\boldsymbol{R}} = C_B E(\boldsymbol{k}).\end{aligned}$$

By using the relations,

$$\langle\varphi_A(\boldsymbol{r})|\mathcal{H}|\varphi_A(\boldsymbol{r})\rangle = \langle\varphi_B(\boldsymbol{r})|\mathcal{H}|\varphi_B(\boldsymbol{r})\rangle = \varepsilon,$$

$$\langle\varphi_A(\boldsymbol{r})|\mathcal{H}|\varphi_A(\boldsymbol{r}-\boldsymbol{R})\rangle = t_{AA}(\boldsymbol{R}),$$

$$\langle\varphi_A(\boldsymbol{r})|\mathcal{H}|\varphi_B(\boldsymbol{r}-\boldsymbol{R})\rangle = t_{AB}(\boldsymbol{R}),$$

we have the Schrödinger equation in the matrix representation,

$$\begin{pmatrix} \varepsilon + \sum_{\boldsymbol{R}\neq 0} t_{AA}(\boldsymbol{R})e^{i\boldsymbol{k}\cdot\boldsymbol{R}} & \sum_{\boldsymbol{R}} t_{AB}(\boldsymbol{R})e^{i\boldsymbol{k}\cdot\boldsymbol{R}} \\ \sum_{\boldsymbol{R}} t_{BA}(\boldsymbol{R})e^{i\boldsymbol{k}\cdot\boldsymbol{R}} & \varepsilon + \sum_{\boldsymbol{R}\neq 0} t_{BB}(\boldsymbol{R})e^{i\boldsymbol{k}\cdot\boldsymbol{R}} \end{pmatrix}\begin{pmatrix} C_A \\ C_B \end{pmatrix} = E(\boldsymbol{k})\begin{pmatrix} C_A \\ C_B \end{pmatrix}. \tag{1.43}$$

Here, the wave function in Eq. (1.42) is expressed as

$$\varphi(\boldsymbol{r}) = C_A\varphi_A(\boldsymbol{r}) + C_B\varphi_B(\boldsymbol{r}) = C_A\begin{pmatrix} 1 \\ 0 \end{pmatrix} + C_B\begin{pmatrix} 0 \\ 1 \end{pmatrix} = \begin{pmatrix} C_A \\ C_B \end{pmatrix}$$

.

The eigenvalues $E(\boldsymbol{k})$ are calculated from the following secular equation,

$$\begin{vmatrix} \varepsilon + \sum\limits_{\boldsymbol{R}\neq 0} t_{AA}(\boldsymbol{R})e^{i\boldsymbol{k}\cdot\boldsymbol{R}} - E(\boldsymbol{k}) & \sum\limits_{\boldsymbol{R}} t_{AB}(\boldsymbol{R})e^{i\boldsymbol{k}\cdot\boldsymbol{R}} \\ \sum\limits_{\boldsymbol{R}} t_{BA}(\boldsymbol{R})e^{i\boldsymbol{k}\cdot\boldsymbol{R}} & \varepsilon + \sum\limits_{\boldsymbol{R}\neq 0} t_{BB}(\boldsymbol{R})e^{i\boldsymbol{k}\cdot\boldsymbol{R}} - E(\boldsymbol{k}) \end{vmatrix} = 0. \tag{1.44}$$

Examples of energy bands for two atoms in the unit cell are given in Appendices 1.6 and 1.7.

1.3 Band Mass and Classification of Carrier

1.3.1 Band Mass

We are not interested in electrons in vacuum, but in electrons in a solid with a periodic crystal potential. The mass of electrons in a solid is different from that of free electrons. The velocity of the electron in a solid is given by the Hamilton equation with respect to the conjugate variables, r and $\boldsymbol{p}(=\hbar\mathrm{k})$,

$$v_i = \frac{\mathrm{d}r_i}{\mathrm{d}t} = \frac{\partial\mathcal{H}}{\partial p_i} = \frac{1}{\hbar}\frac{\partial E(k)}{\partial k_i}(i = x, y, z) \text{ or } \boldsymbol{v} = \frac{1}{\hbar}\nabla_{\boldsymbol{k}}E(\boldsymbol{k}). \tag{1.45}$$

In general, the band $E(\boldsymbol{k})$ has a complex functional form depending on the material, but Eq. (1.45) is valid no matter what functional form $E(\boldsymbol{k})$ has and does not depend on the band approximation method. It is easy to prove that the velocity vector $\boldsymbol{v}$ is perpendicular to the Fermi surface. Since the Fermi surface is isoenergetic surface, we have the relation on the surface, $E(\boldsymbol{k}+\Delta\boldsymbol{k}) - E(\boldsymbol{k}) = \nabla_{\boldsymbol{k}}E(\boldsymbol{k})\cdot\Delta\boldsymbol{k} = 0$, where $\Delta\boldsymbol{k}$ is an infinitesimally small vector from point $\boldsymbol{k}$ on the Fermi surface, parallel to the Fermi surface. The relation $\nabla_{\boldsymbol{k}}E(\boldsymbol{k})\cdot\Delta\boldsymbol{k} = 0$ shows that the vector $\nabla_{\boldsymbol{k}}E(\boldsymbol{k})$ is perpendicular to all $\Delta\boldsymbol{k}$ vectors and thus the velocity of the electron $\boldsymbol{v}$ in real space is perpendicular to the Fermi surface. Electronic states with wave vectors $\boldsymbol{k}$ can be spread over the entire crystal. However, when we actually describe the motion of electrons as particles, we deal with a wave packet, which is a superposition of states in a narrow region centered on a particular wave vector $\boldsymbol{k}$. In other words, the electron velocity $\boldsymbol{v}$ is the velocity of the wave packet, the group velocity defined by Eq. (1.45).

The velocity is rewritten as

$$\begin{aligned} \frac{\partial v_i}{\partial t} &= \frac{\partial}{\partial t}\left\{\frac{1}{\hbar}\frac{\partial E(k)}{\partial k_i}\right\} = \frac{1}{\hbar}\sum_j\left\{\frac{\partial^2 E(k)}{\partial k_j\partial k_i}\right\}\frac{\partial k_j}{\partial t} \\ &= \frac{1}{\hbar^2}\sum_j\left\{\frac{\partial^2 E(k)}{\partial k_j\partial k_i}\right\}F_j, \end{aligned}$$

where the relation $\hbar \partial k_j/\partial t = F_j$ is used. By comparing this with $\boldsymbol{F}/m = \partial \boldsymbol{v}/\partial t$, we can define the mass as

$$\frac{1}{m_{ij}} = \frac{1}{\hbar^2}\left\{\frac{\partial^2 E(k)}{\partial k_j \partial k_i}\right\}. \tag{1.46}$$

Thus, the mass of the electron is a tensor, which can be expressed as the second derivative with respect to k of the energy band. The equation of motion in an electric field $\boldsymbol{E}$ is given by $F_j = -eE_j = m_{ij}(\partial v_i/\partial t)$. If the electron is accelerated in the i-axis by $\boldsymbol{E}$ in the j-axis, the proportionality factor gives m_{ij}. For the free electron energy band $E(\boldsymbol{k}) = \hbar^2\boldsymbol{k}^2/2m_0$ we obtain the isotropic mass $m_{ii}(i = x, y, z) = m_0$. Since the k-dependence of the energy band $E(\boldsymbol{k})$ determines the mass of the electron, it is called the band mass. In general, the band mass can be lighter or heavier than the free electron, depending on the functional form of $E(\boldsymbol{k})$. As seen in Sect. 1.1.2, the electron velocity at the zone boundaries vanishes, $v_i \propto \partial E(k)/\partial k_i = 0$. In other words, the mass becomes infinite, $1/m_{ii} \propto \partial^2 E(k)/\partial k_i \partial k_i = 0$; the electron is localized in the k_i-axis.

1.3.2 Electron and Hole

Electrons carry the negative charge ($-e$) in a crystal, but it is often more convenient to think of particles with the positive charge ($+e$) in states unoccupied by electrons. Consider a simple energy band shown in Fig. 1.9a. From Eq. (1.45), the electron at the Fermi level has a velocity vector in the positive direction of the k_x-axis for $k_x > 0$, $v_x \propto \partial E(k)/\partial k_x > 0$ and a positive mass $m_{xx} \propto \left(\partial^2 E/\partial k_x^2\right)^{-1} > 0$. However, for the band in Fig. 1.9b, we have a negative velocity $v_x \propto \partial E(k)/\partial k_x < 0$ and a negative mass $m_{xx} \propto \left(\partial^2 E/\partial k_x^2\right)^{-1} < 0$. Thus the signs of the electron velocity and band mass are reversed depending on the slope and curvature of the band. Since it is troublesome to handle electrons with such negative mass in the same material, a "hole" is introduced by taking the positive mass and charge ($+e$) and by reversing the velocity. The equation of motion $\boldsymbol{F} = m\dot{\boldsymbol{v}} = -e\boldsymbol{E}$ in an electric field ($\boldsymbol{E}$) for the electron remains the same as that for the hole with the positive mass, $-m\dot{\boldsymbol{v}} = e\boldsymbol{E}$ ($-e \rightarrow e$ and $\boldsymbol{v} \rightarrow -\boldsymbol{v}$). By measuring the energy downward for the holes $E(\boldsymbol{k}) \rightarrow -E(\boldsymbol{k})$, the thermal excitation of the holes can be discussed by Fermi statistics as for the electrons.

Since the Fermi surface is formed by an energy band, we can define the major carrier of the Fermi surface from the character of the band. It is easily determined by virtually filling the band with electrons. For the band in Fig. 1.9a, the Fermi surface is the electron surface because it expands by filling the band with electrons. For the band in Fig. 1.9b, the Fermi surface is the hole surface because it shrinks. By drawing the band in Fig. 1.9b upside down, $E(\boldsymbol{k}) \rightarrow -E(\boldsymbol{k})$, we can see that the holes occupy the band from the bottom up to E_F.

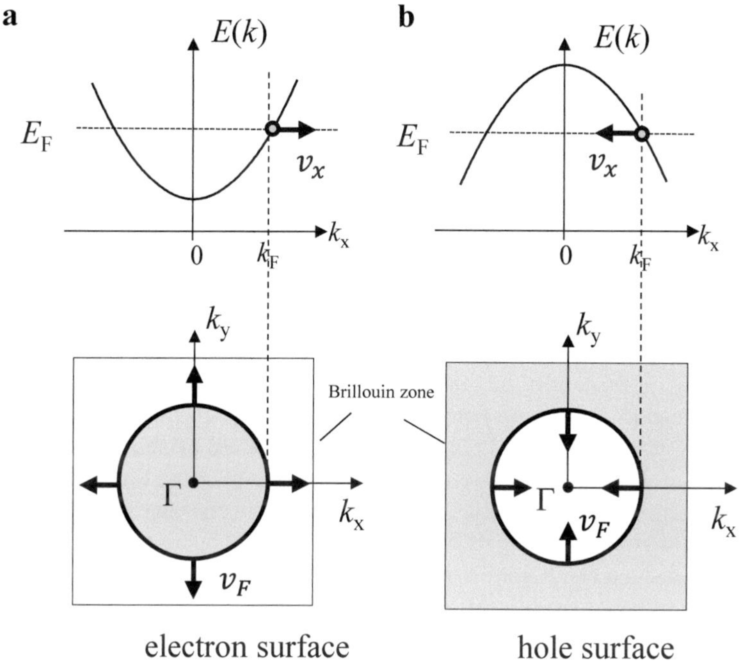

Fig. 1.9 Energy bands and Fermi surfaces for **a** electron and **b** hole carriers. Thick arrows indicate the velocity vectors. Gray areas indicate the states occupied by electrons

1.4 Reconstruction of Fermi Surface

In real crystals, a structural phase transition often occurs with decreasing temperature, resulting in a symmetry breaking of the electronic state. For example, if a superlattice potential appears below the transition temperature T_c, the lattice constants change. As the Brillouin zone is redefined below T_c, the energy bands forming the Fermi surface change. Such a change in the Fermi surface is called Fermi surface reconstruction. If the Fermi surface structure above T_c is known, it is relatively easy to predict the reconstructed Fermi surface below T_c using the following procedure. The procedure will be helpful in understanding the changes of the physical property below T_c.

Consider a Fermi surface of a 2D rectangular lattice with the lattice constants, $a\|x$ and $b\|y$. Taking $t_c = 0$ and $\varepsilon = (2t_a + 2t_b)$ in the tight binding approximation, Eq. (1.39), we write the band as

$$E(\boldsymbol{k}) = (2t_a + 2t_b) - 2t_a\cos(k_x a) - 2t_b\cos\left(k_y b\right). \tag{1.47}$$

For $t_a \gg t_b$, a pair of the open Fermi surfaces (quasi-1D Fermi surfaces) is formed for the Fermi energy $E_{\mathrm{F}} = (2t_a + 2t_b)$ as shown in Fig. 1.10a. The first Brillouin zone is defined in the range $-\pi/a < k_x \leq \pi/a$ and $-\pi/b < k_y \leq \pi/b$. Taking into account the spin degree of freedom, the Fermi energy $E_F = (2t_a + 2t_b)$ corresponds to the electron number equal to the lattice points. The energy band is half-filled; the half of the Brillouin zone (gray area) is occupied by the electrons. The electrons on the right Fermi surface ($k_x > 0$) have the positive x-axis velocity $v_x = (1/\hbar)\partial E(\boldsymbol{k})/\partial k_x = 2at_a \sin(k_F a)/\hbar$, which is much larger than the y-axis velocity $v_y = 2bt_b \sin(k_F b)/\hbar$ as shown in Fig. 1.10b. Similarly, the electrons on the left Fermi surface have the negative velocity $v_x = -2at_a \sin(k_F a)/\hbar$. In this electronic system, all the electron motion is almost limited in the $k_x \| x$ direction; thus it is a quasi-1D system. Here we suppose that a superlattice potential with a double lattice period in the x-axis is superimposed on the lattice. Since the lattice constant is doubled ($a \to 2a$), the wave number of the superlattice potential is written as $q_x = \pi/a$, and thus the first Brillouin zone is halved, $-\pi/2a < k_x \leq \pi/2a$. The energy bands and Fermi surfaces must have the new periodicity of q_x. The reconstructed bands and Fermi surfaces are obtained by shifting them by $\pm q_x$, as depicted in Fig. 1.10c. The left 1D Fermi surface (FS_2) in Fig. 1.10a is shifted by q_x, and then the pair of the 1D Fermi surfaces (FS_1 and FS_2) is formed at the new right zone boundary $k_x = \pi/2a$. The same happens at the left boundary $k_x = -\pi/2a$. As a result, the Fermi surfaces mirrored at the new boundaries $k_x = \pm\pi/2a$ are added to the original Fermi surfaces. This reconstruction procedure is often referred to as "folding back" of the band or Fermi surface. Note that the reconstructed Fermi surface in Fig. 1.10c has the periodicity of q_x.

In Fig. 1.10c, we see that the two pairs of the 1D Fermi surfaces cross the zone boundaries. In reality, this degeneracy is removed and then gaps open as depicted in Fig. 1.10d. We can understand how the gaps open at the boundaries by following the electron motion. The electrons on FS_1 and FS_2 have the velocities in the opposite directions. Therefore, the electrons can be smoothly transferred from FS_1 to FS_2 at the boundaries by opening the gaps as shown in Fig. 1.10d. This reconstruction results in two closed Fermi surfaces. By filling the energy band with electrons, one closed area expands and the other shrinks; the two closed Fermi surfaces are the electron and hole surfaces. Both areas are the same; the reconstructed Fermi surfaces represent a compensated metal. In this way, the superlattice potential drastically reconstructs the Fermi surfaces, accompanied by the change of the dimensionality from 1 to 2D.

Next, we suppose a superlattice potential with a double period in the y-axis ($b \to 2b$). The wave number of the superlattice potential is given by $q_y = \pi/b$ and then the first Brillouin zone is redefined in the range, $-\pi/2b < k_y \leq \pi/2b$. By shifting FS_1 and FS_2 by $\pm q_y$, two pairs of the 1D Fermi surfaces are formed as shown in Fig. 1.10e. Both FS_1 and FS_2 cross the zone boundaries $k_y = \pm\pi/2b$. The degeneracy is removed, and then gaps open as depicted in Fig. 1.10f. Since the electrons on the original and shifted FS_1 have the velocities in the same direction, two pairs of the open Fermi surfaces are resultantly formed. Note that the total area of the Fermi surfaces is unchanged since the Brillouin zone is halved.

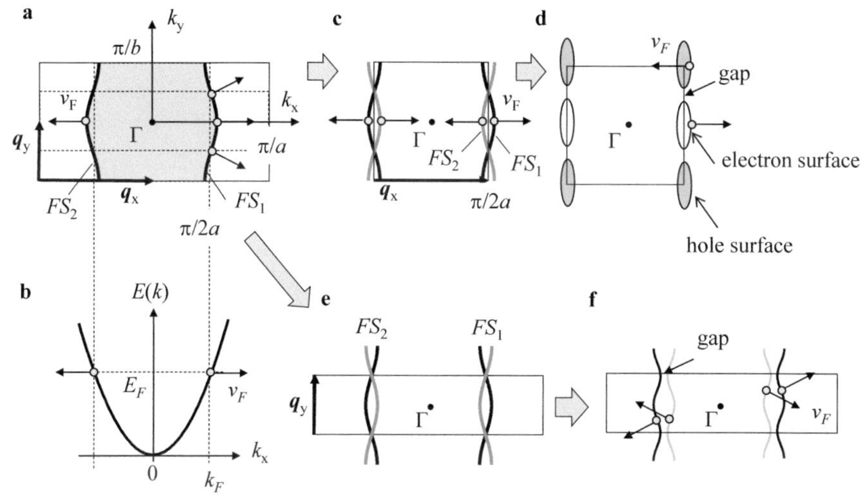

Fig. 1.10 **a** Pair of open Fermi surfaces for a 2D rectangular lattice with the lattice constants, $a\|x$ and $b\|y$. The electron number is equal to the lattice points, N_l^2. Gray area shows the states occupied by the electrons. Fermi velocity vectors v_F are indicated by thin arrows. **b** Energy band forming Fermi surface in **a**. **c** Two pairs of 1D Fermi surfaces (FS_1 and FS_2) at the zone boundaries $k_x = \pm\pi/2a$ formed by the superlattice potential with $q_x = \pi/a$. **d** Fermi surfaces with gaps at the zone boundaries, where two closed Fermi surfaces are formed. **e** Two pairs of 1D Fermi surfaces near $k_x = \pm\pi/2a$ for $q_y = \pi/a$. **f** Fermi surfaces with gaps at the zone boundaries $k_y = \pm\pi/2b$

Such Fermi surface reconstruction has been observed in the quasi-1D organic conductor $(TMTSF)_2X$. This material is composed of the TMTSF (*tetramethylselenafulvalene*) planar organic molecule and anion X, whose schematic crystal structure is presented in Fig. 1.11 [2]. The molecular orbital of TMTSF overlaps predominantly with its neighbors along the *a*-axis. The overlap along the *b*-axis is smaller than that along the *a*-axis and the overlap along the *c*-axis is extremely small. Due to the anisotropic overlap, quasi-1D Fermi surfaces are formed in $(TMTSF)_2X$, as depicted in Fig. 1.10a with $a\|x$ and $b\|y$. The anion X = ClO_4 has a tetragonal structure with two degrees of freedom (up and down). At 24 K, the ClO_4 anions are alternately aligned along the *b*-axis, forming the superlattice potential with $q_y = \pi/b$, and thus the Fermi surface is reconstructed as shown in Fig. 1.10f. In $(TMTSF)_2ClO_4$, when a magnetic field is applied along the *c*-axis, a spin-density wave is induced and then the quantum Hall effect is observed [3, 4]. In a higher magnetic field, the spin-density wave is broken and a new type of quantum oscillation is observed in the Hall resistance [5]. In spite of such a simple Fermi surface, it is surprising that various fascinating phase transitions are observed.

For $(TMTSF)_2X$ ($X = ReO_4$), the ReO_4 ions are alternately aligned along the *a*- and *b*-axes at 160 K, whose superlattice potential is given by a wave vector $\boldsymbol{q}$ in Fig. 1.12 [6]. The Fermi surface is reconstructed by $\boldsymbol{q}$ and then the pair of the

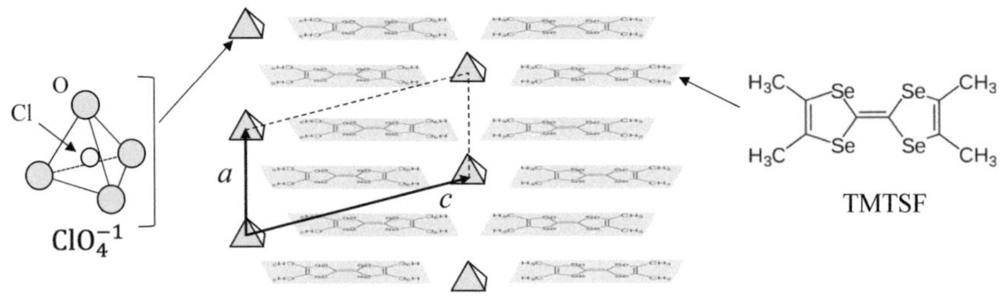

Fig. 1.11 Schematic structure of quasi-1D organic conductor $(TMTSF)_2ClO_4$, composed of ClO_4 anion and TMTSF molecule [2]. The dashed parallelogram indicates the unit cell

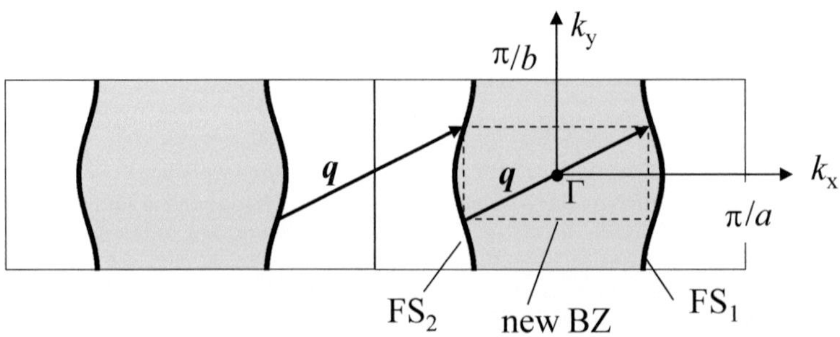

Fig. 1.12 Pair of 1D Fermi surfaces and nesting vector $\boldsymbol{q}$ in the repeated zone scheme for quasi-1D organic conductor $(TMTSF)_2ReO_4$ [6]. The perfect nesting by $\boldsymbol{q}$ occurs below ~160 K, leading to a metal–insulator transition. The dashed rectangle indicates the new Brillouin zone (BZ) formed by the nesting

Fermi surfaces (FS_1 and FS_2) completely disappears. This is a phase transition from a metal to an insulator. Such overlapping and resulting disappearance of the Fermi surfaces is called "nesting" of Fermi surfaces and the wave number vector $\boldsymbol{q}$ characterizing the periodic potential is called "nesting vector." In particular, the case where the Fermi surface completely disappears is called "perfect nesting." Another example of Fermi surface reconstruction is presented in Appendix 1.8.

1.5 Summary

In this chapter, we have seen how energy bands and Fermi surfaces are formed in the free electron and tight binding approximations. In particular, it is important to know how the Fermi surface is reconstructed and gapped at the Brillouin zone boundaries by a superlattice potential. Although there are many other sophisticated methods for determining energy bands, the above approximations are useful for understanding Fermi surfaces in many materials and are easy to understand. The

Table 1.1 Typical Fermi surface structures and their energy bands in free electron and/or tight binding approximations

Fermi surface structure	Energy band
3D case 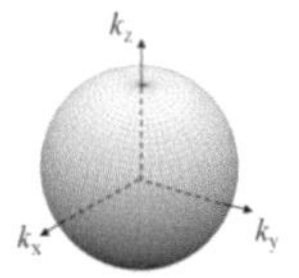	$E(\boldsymbol{k}) = \hbar^2\boldsymbol{k}^2/2m = \hbar^2\left(k_x^2 + k_y^2 + k_z^2\right)/2m$ or $E(\boldsymbol{k}) = \varepsilon - 2t_a\cos(k_x a) - 2t_a\cos(k_y a) - 2t_a\cos(k_z a)$
Quasi 2D case 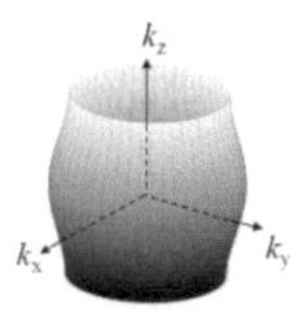	$E(\boldsymbol{k}) = \frac{\hbar^2}{2m}\left(k_x^2 + k_y^2\right) - 2t_c\cos(k_z c)$ for $\frac{\hbar^2}{2m}\left(k_x^2 + k_y^2\right) \gg 2t_c > 0$ or $E(\boldsymbol{k}) = \varepsilon - 2t_a\cos(k_x a) - 2t_b\cos(k_y a) - 2t_c\cos(k_z c)$ for $t_a \approx t_b \gg t_c > 0$
Quasi 1D case 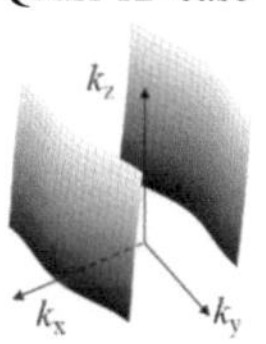	$E(\boldsymbol{k}) = \hbar v_F(\lvert k_x\rvert - k_F) - 2t_b\cos(k_y b) - 2t_c\cos(k_z c)$ for $\hbar v_F k_F > 2t_b \gg 2t_c > 0$ or $E(\boldsymbol{k}) = \varepsilon - 2t_a\cos(k_x a) - 2t_b\cos(k_y b) - 2t_c\cos(k_z c)$ for $t_a > t_b \gg t_c > 0$

size, shape, and symmetry of the Fermi surface are important factors in determining the physical properties. Table 1.1 lists examples of Fermi surface structures and their energy band models that will be used as examples in the following chapters.

Appendix 1.1

The Bloch theorem is a general theory in which only periodic boundary conditions are imposed on the wave function and does not depend on any approximation method of the wave function.

We define an operator $T_{\boldsymbol{a}}$ that moves the wave function $\varphi_{\boldsymbol{k}}(\boldsymbol{r})$ to the neighboring lattice point with distance $\boldsymbol{a}$, $T_{\boldsymbol{a}}\varphi_{\boldsymbol{k}}(\boldsymbol{r}) = \varphi_k(\boldsymbol{r}+\boldsymbol{a})$. Since the Hamiltonian $\mathcal{H} = -\hbar^2\nabla^2/2m + V(\boldsymbol{r})$ should be invariant by this operation, we have $V(\boldsymbol{r}) = V(\boldsymbol{r}+\boldsymbol{a})$ and then $T_{\boldsymbol{a}}\mathcal{H}\varphi_{\boldsymbol{k}}(\boldsymbol{r}) = \mathcal{H}T_{\boldsymbol{a}}\varphi_{\boldsymbol{k}}(\boldsymbol{r})$. By multiplying the left-hand side of $\mathcal{H}\varphi_{\boldsymbol{k}}(\boldsymbol{r}) = E\varphi_{\boldsymbol{k}}(\boldsymbol{r})$ by $T_{\boldsymbol{a}}$, we get $T_{\boldsymbol{a}}\mathcal{H}\varphi_{\boldsymbol{k}}(\boldsymbol{r}) = T_{\boldsymbol{a}}E\varphi_{\boldsymbol{k}}(\boldsymbol{r}) = ET_{\boldsymbol{a}}\varphi_{\boldsymbol{k}}(\boldsymbol{r})$ and thus $\mathcal{H}T_{\boldsymbol{a}}\varphi_{\boldsymbol{k}}(\boldsymbol{r}) = ET_{\boldsymbol{a}}\varphi_{\boldsymbol{k}}(\boldsymbol{r})$. This result shows that $T_{\boldsymbol{a}}\varphi_{\boldsymbol{k}}(\boldsymbol{r})$ is also the eigenstate, and $\varphi_{\boldsymbol{k}}(\boldsymbol{r})$ and $T_{\boldsymbol{a}}\varphi_{\boldsymbol{k}}(\boldsymbol{r})$ are degenerate. The difference between the two states is only the phase, $T_{\boldsymbol{a}}\varphi_{\boldsymbol{k}}(\boldsymbol{r}) = e^{i\theta}\varphi_{\boldsymbol{k}}(\boldsymbol{r})$. The periodic boundary condition, $\varphi_{\boldsymbol{k}}(\boldsymbol{r}) = \varphi_{\boldsymbol{k}}(\boldsymbol{r}+N_l\boldsymbol{a})$ (N_l : integer), gives the relation,

$$\varphi_{\boldsymbol{k}}(\boldsymbol{r}+N_l\boldsymbol{a}) = T_{N_l\boldsymbol{a}}\varphi_{\boldsymbol{k}}(\boldsymbol{r}) = T_{\boldsymbol{a}}^{N_l}\varphi_{\boldsymbol{k}}(\boldsymbol{r}) = e^{iN_l\theta}\varphi_{\boldsymbol{k}}(\boldsymbol{r}) = \varphi_{\boldsymbol{k}}(\boldsymbol{r}),$$

which requires $N_l\theta = 2\pi l(l : \text{integer})$. From the periodic boundary condition of length $N_l a$, an integer number of the wave with the wavelength λ are included in the length $N_l a$, $\lambda l = N_l a$. Its wave number is given by $k = 2\pi/\lambda = 2\pi l/N_l a$, and thus $\theta = 2\pi l/N_l = ka$. Therefore, the operator T_a is written as

$$T_a\varphi_k(\boldsymbol{r}) = e^{i\theta}\varphi_k(\boldsymbol{r}) = e^{i\boldsymbol{k}\cdot\boldsymbol{a}}\varphi_k(\boldsymbol{r}).$$

Since any lattice point is expressed as $\boldsymbol{R} = n\boldsymbol{a}(n : \text{integer})$, we obtain the Bloch theorem

$$\varphi_k(\boldsymbol{r}+\boldsymbol{R}) = \varphi_k(\boldsymbol{r}+n\boldsymbol{a}) = T_a^n\varphi_k(\boldsymbol{r}) = e^{i\boldsymbol{k}\cdot n\boldsymbol{a}}\varphi_k(\boldsymbol{r}) = e^{i\boldsymbol{k}\cdot\boldsymbol{R}}\varphi_k(\boldsymbol{r}).$$

Appendix 1.2

Let us see how Brillouin zones are formed in a 2D system. As the simplest case, consider a 2D square lattice with lattice constant a as shown in Fig. 1.13a. In this case, the reciprocal lattice points are also square as shown in Fig. 1.13b.

First Brillouin zone: Lines are drawn from one lattice point to other points around it, and the perpendicular bisectors are drawn. The innermost area enclosed by the perpendicular bisectors is the first Brillouin zone.

Second Brillouin zone: The next inner area enclosed by the perpendicular bisectors, adjacent to the first Brillouin zone. Note that the first and second Brillouin zones have the same area.

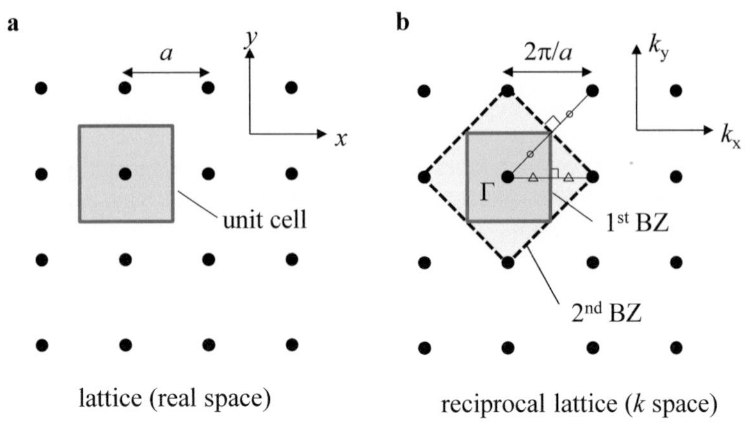

Fig. 1.13 **a** 2D square lattice with lattice constant a, **b** Reciprocal lattice of the 2D square lattice and its Brillouin zones (BZs)

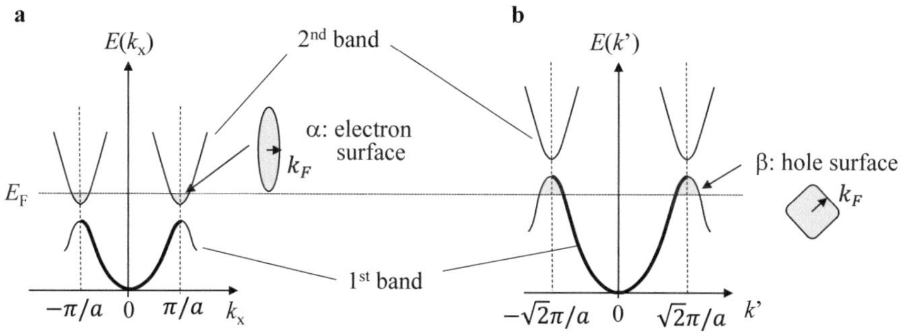

Fig. 1.14 Energy bands of a 2D square lattice in free electron approximation along **a** k_x-axis and **b** k'-axis in Fig. 1.5e. The first and second bands form the hole (β) and electron (α) surfaces shown in Fig. 1.5e, respectively. The vertical dashed lines show the zone boundaries

Appendix 1.3

Here we depict the energy bands of the Fermi surface shown in Fig. 1.5e. Along the k_x-axis, the second band crosses the Fermi level E_F as shown in Fig. 1.14a, forming an elliptical Fermi surface α near the zone boundary. As this band is filled with electrons, raising the Fermi level, the Fermi surface expands. Therefore, this Fermi surface is assigned to an electron surface. Along the k'-axis, the first band crosses E_F near the zone corners. Since this Fermi surface β shrinks with increasing electrons, it is assigned to a hole surface. The types of the carriers will be discussed in Sect. 1.4. As shown in Fig. 1.14, the Fermi wave numbers k_F are defined as those from the bottom (top) of the electron (hole) band.

Appendix 1.4

Here we consider a 2D electron system with a rectangular unit cell. Anisotropic mass $(m_x > m_y)$ is phenomenologically introduced into the Hamiltonian,

$$\mathcal{H} = -\frac{\hbar^2}{2m_x}\left(\frac{\partial}{\partial x}\right)^2 - \frac{\hbar^2}{2m_y}\left(\frac{\partial}{\partial y}\right)^2 + V(\boldsymbol{r}).$$

The energy of the non-perturbative terms (first two terms)

$$E^{(0)} = \frac{\hbar^2 k_x^2}{2m_x} + \frac{\hbar^2 k_y^2}{2m_y}$$

exhibits parabolic energy bands with different curvatures in the k_x- and k_y-axes; the energy band in the k_x-axis rises more gently from the Γ point than that in the k_y-axis. As shown in Fig. 1.15a, this anisotropic mass gives an elliptical Fermi surface

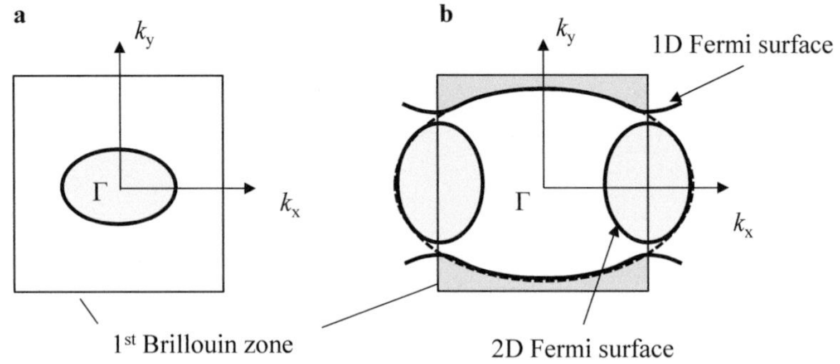

Fig. 1.15 Fermi surfaces of rectangular unit cell with anisotropic mass for **a** a low electron number and **b** the electron number equal to $2N_l^2$. The area of the dashed ellipsoid is equal to the first Brillouin zone

with the major axis along k_x around Γ if the Fermi energy is sufficiently small. Taking into account the spin degree of freedom, when the number of electrons is twice the number of the lattice points $2N_l^2$, the area of the Fermi surface (dashed ellipsoid) is equal to the Brillouin zone as shown in Fig. 1.15b. The bands crosses the zone boundaries and the gaps open. In this way, a pair of 1D (open) and 2D (closed) Fermi surfaces are formed.

Appendix 1.5

Let us consider another example of a Fermi surface in the tight binding approximation. Consider a 2D electron system with a rectangular unit lattice ($a > b$) as shown in Fig. 1.16a. This band has a from,

$$E(\boldsymbol{k}) = \varepsilon - 2t_a\cos(k_x a) - 2t_b\cos\left(k_y b\right).$$

Taking $\varepsilon = 2(t_a + t_b)$, we obtain $E(\boldsymbol{k} = 0) = 0$.

Assuming a wave function with s-orbital symmetry, the transfer integral should be smaller in the a-axis, $t_a < t_b$. The energy bands in the k_x- and k_y-axes will be given in Fig. 1.16b. For the Fermi level $E_F = E_1 < 4t_a$, the band crosses E_F in both the k_x- and k_y-axes, forming a closed Fermi surface around Γ as indicated by the dashed curve in Fig. 1.16c. For $E_F = E_2$ ($4t_a < E_2 < 4t_b$), the band does not cross E_F in k_x, and we have an open Fermi surface in k_x, as shown by the solid curve in Fig. 1.16c. For $E_F = E_3(4t_b < E_3 < 4t_a + 4t_b)$, the band does not cross E_F in either k_x or k_y but crosses around $(\pi/a, \pi/b)$ where a closed Fermi surface is formed as shown by the dotted curve in Fig. 1.16c. If E_F increases up to $4t_a + 4t_b$, no Fermi surface is formed, this electronic system becomes insulating. As the electron number increases (the band is filled with more electrons), the Fermi surface for $E_F = E_1$ (dashed curve) expands, showing the electron surface

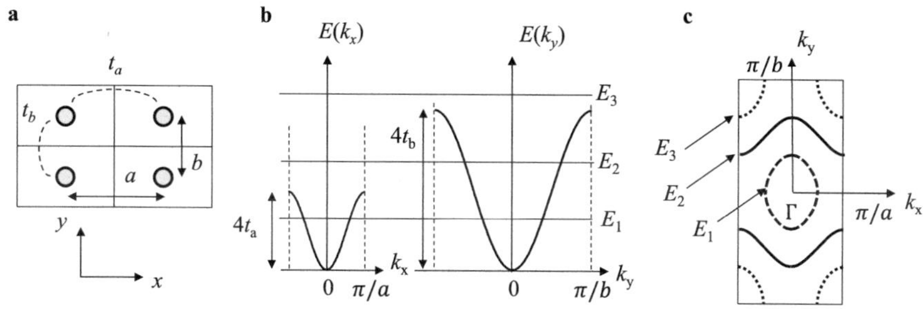

Fig. 1.16 **a** 2D rectangular lattice, **b** Energy bands in tight binding approximation along k_x- and k_y-axes, **c** Fermi surfaces for different Fermi levels (E_1, E_2, E_3)

(carriers are electrons). The Fermi surface for $E_F = E_3$ shrinks with increasing electrons, showing the hole surface (carriers are holes).

As we will see in Sect. 1.3, the electron velocity is given by $\boldsymbol{v} = \nabla_{\boldsymbol{k}} E(\boldsymbol{k})/\hbar$ where $\nabla_{\boldsymbol{k}}$ is the derivative with respect to $\boldsymbol{k}$. Since $\boldsymbol{v}$ is a vector perpendicular to the Fermi surface, the electron velocity on the open Fermi surface for $E_F = E_2$ is mostly in the k_y-axis, indicating that this is a 1D electron system in which the electrons tend to move in this direction. From the relation $t_a < t_b$, this makes sense. If the open Fermi surface is completely flat (a perfect 1D system, $t_a = 0$), the electron motion is limited only in the k_y-axis. In this way, as the number of electrons changes, the dimensionality of the Fermi surface can change between 2D and 1D.

Appendix 1.6

Let us actually calculate the energy bands for the case of two atoms in the unit cell. Suppose that two atoms A and B in the unit cell form a 1D chain, as shown in Fig. 1.17. We define the lattice constant a and the atomic spacings a_1 and a_2, $a = a_1 + a_2$. Assuming that the transfer integral between the same atoms is negligible, $t_{AA}(a) = t_{BB}(a) = 0$, we obtain

$$\sum_R t_{AB}(R)e^{ikR} = t_{AB}(a_1)e^{ika_1} + t_{AB}(-a_2)e^{-ik\cdot a_2},$$
$$\sum_R t_{BA}(R)e^{ikR} = t_{BA}(-a_1)e^{-ika_1} + t_{BA}(a_2)e^{ik\cdot a_2}. \tag{1.48}$$

Substituting these into Eq. (1.44), we have

$$\begin{vmatrix} \varepsilon - E(k) & t_{AB}(a_1)e^{ik\cdot a_1} + t_{AB}(-a_2)e^{-ika_2} \\ t_{BA}(-a_1)e^{-ik\cdot a_1} + t_{BA}(a_2)e^{ika_2} & \varepsilon - E(k) \end{vmatrix} = 0. \tag{1.49}$$

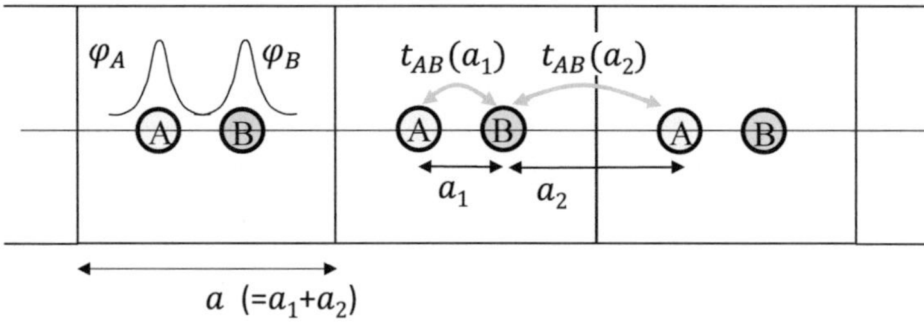

Fig. 1.17 1D lattice with two atoms in the unit cell and transfer integrals between neighboring sites A and B, $t_{AB}(a_1)$ and $t_{AB}(a_2)$

Since $t_{AB}(x) = t_{AB}(-x) = t_{BA}(-x)$, the energy band is calculated as

$$E(k) = \varepsilon \pm \sqrt{t_{AB}^2(a_1) + t_{AB}^2(a_2) + 2t_{AB}(a_1)t_{AB}(a_2)\cos(ka)}. \tag{1.50}$$

Taking $\varepsilon = t_{AB}(a_1) + t_{AB}(a_2)$, we obtain the bands as shown in Fig. 1.18a. Note that an energy gap E_g opens at the zone boundaries $k = \pm\pi/a$,

$$E_g = 2|t_{AB}(a_1) - t_{AB}(a_2)|.$$

Taking into account the spin degree of freedom, if each atom is occupied by one electron, the lower band is fully filled with electrons and we obtain a band insulator with the gap E_g. If the atoms A and B are equivalent, $a_1 = a_2 = a/2$, we have a simple form,

$$E(k) = \varepsilon \pm 2t\cos\left(\frac{ka}{2}\right),$$

where $t_{AB}(a_1) = t_{AB}(a_2) = t$, as depicted in Fig. 1.18b. No gap $\left(E_g = 0\right)$ at the zone boundaries means that the electronic state is metallic. Since the unit cell is halved ($a \to a/2$), we obtain $E(k) = \varepsilon - 2t\cos(ka/2)$ for the new first Brillouin zone $-2\pi/a < k < 2\pi/a$, as shown in Fig. 1.18c. Note that this band is equivalent to the 1D energy band with the lattice constant of $a/2$. The above discussion shows that if a superlattice potential (periodic lattice distortion) doubling the unit cell is added to a 1D metallic state with a half-filled band, a gap opens at E_F; a metal–insulator transition occurs.

Appendix 1.7

Let us calculate the energy bands of a layer material called graphene. Graphene forms a honeycomb lattice composed only of carbon (C) atoms, as shown in Fig. 1.19a. Taking $\boldsymbol{a}_1$ and $\boldsymbol{a}_2(|\boldsymbol{a}_1| = |\boldsymbol{a}_2| = a)$ as unit vectors, we can define a rhombus-shaped unit cell containing two C atoms (sites A and B). The mixing

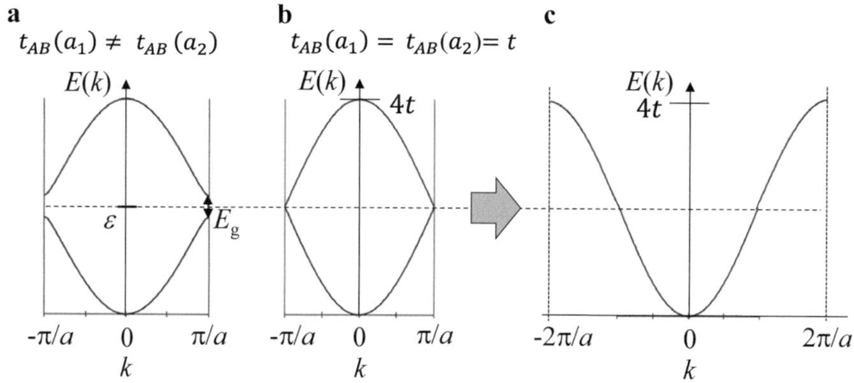

Fig. 1.18 Energy bands of 1D lattice with two atoms in unit cell for **a** $t_{AB}(a_1) \neq t_{AB}(a_2)$ and **b** $t_{AB}(a_1) = t_{AB}(a_2) = t$. Gaps open at the zone boundaries $k = \pm\pi/a$ for **a** but are closed for **b**. **c** Energy band of 1D lattice with the lattice constant of $a/2$

between the $2s$, $2p_x$ and $2p_y$ orbitals forms the sp^2 hybrid orbitals, which extend in the honeycomb plane and make an angle of 120° with one another. This orbital, called the σ orbital, bonds with those of the neighboring C atoms, thereby stabilizing the honeycomb structure. The conduction band is formed by the $2p_z$ orbitals. The orbitals extend perpendicular to the lattice plane and are called π orbitals. Each π orbital contains one electron, forming a half-filled band. We write the orbitals at sites A and B as $\varphi_A(\boldsymbol{r})$ and $\varphi_B(\boldsymbol{r})$, respectively. The σ orbital fully filled with the electrons is at a much lower energy level than the π orbital and thus does not contribute to the electrical conduction.

From the unit vectors $\boldsymbol{a}_1$ and $\boldsymbol{a}_2$, the reciprocal lattice with the unit vectors, $\boldsymbol{g}_1$ and $\boldsymbol{g}_2\left(|\boldsymbol{g}_1| = |\boldsymbol{g}_2| = 4\pi/\sqrt{3}a\right)$ is defined in Fig. 1.19b, forming a hexagonal Brillouin zone. By drawing the perpendicular bisectors of the lines connecting the neighboring reciprocal lattice points, the hexagonal first Brillouin zone is obtained. The high symmetry points at the zone boundaries are marked as K, K' and M.

We assume that the transfer integral between the same sites is negligible, $t_{AA}(\boldsymbol{R}) = t_{BB}(\boldsymbol{R}) = 0$. There are three transfer integrals from site A to B, whose directions $\boldsymbol{R}_i (i = 1, 2, 3)$ are indicated in Fig. 1.19a,

$$\boldsymbol{R}_1\left(\frac{a}{\sqrt{3}}, 0\right), \boldsymbol{R}_2\left(-\frac{a}{2\sqrt{3}}, \frac{a}{2}\right), \boldsymbol{R}_3\left(-\frac{a}{2\sqrt{3}}, -\frac{a}{2}\right).$$

By summing the three terms, we obtain

$$\sum_{\boldsymbol{R}} t_{AB}(\boldsymbol{R}) e^{i\boldsymbol{k}\cdot\boldsymbol{R}} = t_{AB}(\boldsymbol{R}_1)\exp\left[i\left(\frac{k_x}{\sqrt{3}}\right)a\right] + t_{AB}(\boldsymbol{R}_2)\exp\left[i\left(-\frac{k_x}{2\sqrt{3}} + \frac{k_y}{2}\right)a\right] + t_{AB}(\boldsymbol{R}_3)\exp\left[i\left(-\frac{k_x}{2\sqrt{3}} - \frac{k_y}{2}\right)a\right]. \quad (1.51)$$

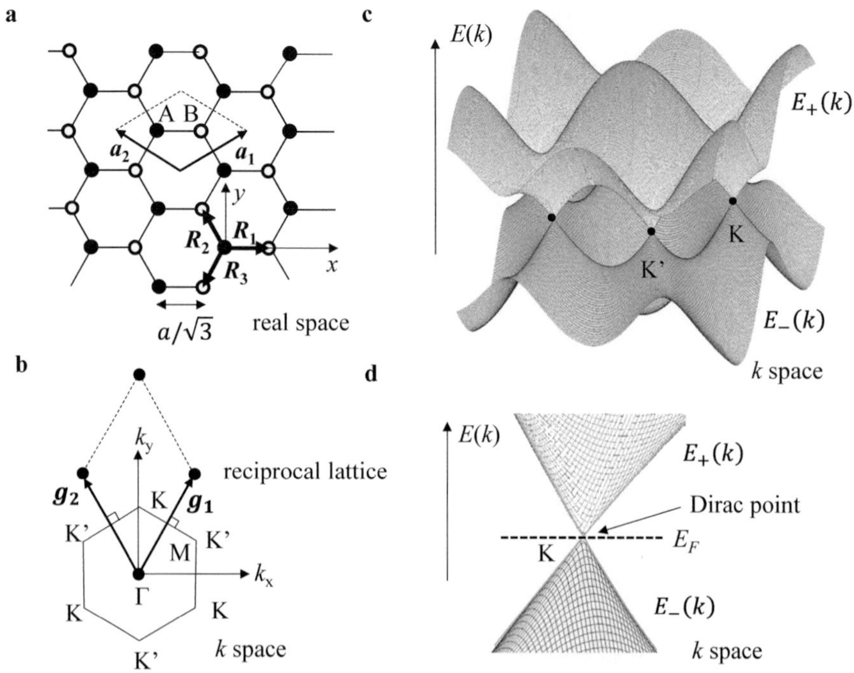

Fig. 1.19 **a** Honeycomb lattice of graphene and its rhombus-shaped unit cell containing two C atoms (A, B). The unit vectors are given by $\boldsymbol{a}_1$ and $\boldsymbol{a}_2$. **b** Reciprocal lattice and its hexagonal first Brillouin zone. The unit vectors of the reciprocal lattice are given by $\boldsymbol{g}_1$ and $g_2\left(|g_1| = |g_2| = 4\pi/\sqrt{3}a\right)$. **c** Energy bands of graphene $E_\pm(\boldsymbol{k})$ given by Eq. (1.52). **d** Dirac cone energy dispersion near K

Similarly, we can obtain $\sum_{\boldsymbol{R}} t_{BA}(\boldsymbol{R})e^{i\boldsymbol{k}\cdot\boldsymbol{R}}$ for $\boldsymbol{R} = -\boldsymbol{R}_1, -\boldsymbol{R}_2, -\boldsymbol{R}_3$. By substituting $\sum_{\boldsymbol{R}} t_{AB}(\boldsymbol{R})e^{i\boldsymbol{k}\cdot\boldsymbol{R}}$ and $\sum_{\boldsymbol{R}} t_{BA}(\boldsymbol{R})e^{i\boldsymbol{k}\cdot\boldsymbol{R}}$ into Eq. (1.44) and taking $\varepsilon = 0$, two eigenvalues are calculated as follows:

$$E_\pm(\boldsymbol{k}) = \pm t_{AB}\sqrt{3 + 2\cos\left[\left(\frac{\sqrt{3}k_x}{2} - \frac{k_y}{2}\right)a\right] + 2\cos\left[k_y a\right] + 2\cos\left[\left(\frac{\sqrt{3}k_x}{2} + \frac{k_y}{2}\right)a\right]}. \tag{1.52}$$

Both the energy bands are depicted in Fig. 1.19c. The lower band $E_-(\boldsymbol{k})$ is fully filled, but the upper band $E_+(\boldsymbol{k})$ is empty. At the following points,

$$\mathbf{K} = \left(0, \frac{4\pi}{3a}\right), \left(\frac{2\pi}{\sqrt{3}a}, -\frac{2\pi}{3a}\right), \left(-\frac{2\pi}{\sqrt{3}a}, -\frac{2\pi}{3a}\right),$$
$$\mathbf{K}' = \left(\frac{2\pi}{\sqrt{3}a}, \frac{2\pi}{3a}\right), \left(0, -\frac{4\pi}{3a}\right), \left(-\frac{2\pi}{\sqrt{3}a}, \frac{2\pi}{3a}\right),$$

we have $E_+(\boldsymbol{k}) = E_-(\boldsymbol{k}) = 0$, showing that the two energy bands $E_\pm(\boldsymbol{k})$ touch with each other (gapless). Taking the coordinate near the first K point as $(\delta k_x, 4\pi/3a + \delta k_y)$, we obtain the energy bands with the linear dispersion,

$$E_\pm(\boldsymbol{k}) = \pm\frac{\sqrt{3}}{2}a(\delta k)t_{AB}, \delta k^2 = \delta k_x^2 + \delta k_y^2. \tag{1.53}$$

A similar linear dispersion is obtained near the K' points. The linear dispersion has a conical structure as depicted in Fig. 1.19d and is identical to that of light, $E = \hbar c_l k$, where c_l is the speed of light. This band structure is called a Dirac cone, and the apex is called a Dirac point. Graphene has been extensively studied due to its very unique physical properties resulting from the linear dispersion.

Appendix 1.8

Let us look at another example of Fermi surface reconstruction in a real material. We assume the half-filled energy band given by Eq. (1.47) with $t_a < t_b$. This energy band forms a closed Fermi surface, whose area is half that of the first Brillouin zone, as shown in Fig. 1.20a. If a superlattice potential with a double period in the x-axis ($q_x = \pi/a$) is added, the Brillouin zone is halved as shown by the dashed line. The original Fermi surface is shifted by $\pm q_x$, or folded back at the new zone boundaries ($k_x = \pi/2a$), as depicted in Fig. 1.20b. The degeneracy at the boundaries is removed, and gaps open. The reconstruction results in a closed 2D and a pair of 1D Fermi surfaces as shown in Fig. 1.20c. Note the smooth curvature at the boundaries. A similar reconstruction of the Fermi surface actually occurs in some quasi-2D organic conductors.

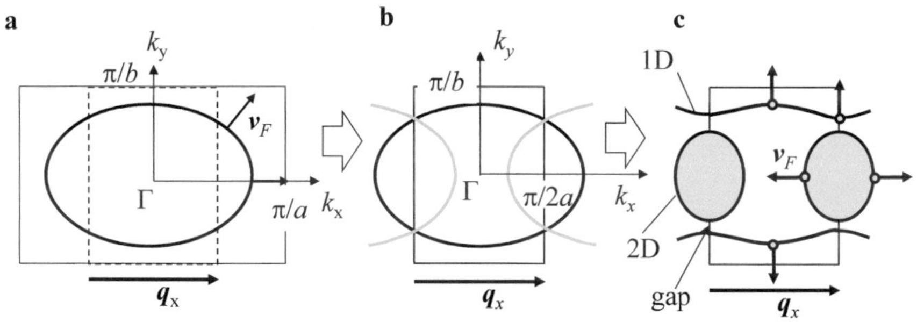

Fig. 1.20 **a** Closed Fermi surface in a rectangular Brillouin zone for a half-filled band, **b** Fermi surface folded back at new zone boundaries due to superlattice potential with $\boldsymbol{q}_x$, **c** Reconstructed Fermi surface with gaps at new zone boundaries. The arrows indicate the Fermi velocity vectors

References

1. Setyawan, W., Curtarolo, S.: High-throughput electronic band structure calculations: Challenges and tools. Comput. Mater. Sci. Mater. Sci. **49**, 299–312 (2010). https://doi.org/10.1016/j.commatsci.2010.05.010
2. For organic conductors, please refer to textbooks, Wosnitza, J: Fermi Surfaces of Low Dimensional Organic Metals and Superconductors. Springer-Verlag, Berlin (1996). Ishiguro, T.K., Yamaji, K.: Organic Superconductors, 2nd edn. Springer, Berlin (1997)
3. Ribault, M., Cooper, J., Jérome, D., Mailly, D., Moradpour, A., Bechgaard, K.: Quantum hall effect and fermi surface instabilities in (TMTSF)2ClO4. J. Phys. Lett. **45**, 935–941 (1984). https://doi.org/10.1051/jphyslet:019840045019093500
4. Kang, W., Jérome, D.: Quantized Hall effect in the organic superconductor $(TMTSF)_2ClO_4$ revisited. J. Phys. I **1**, 449 (1991). https://doi.org/10.1051/jp1:1991145
5. Uji, S., Yasuzuka, S., Konoike, T., Enomoto, K., Yamada, J., Choi, E.S., Graf, D., Brooks, J.S.: Quantum Oscillation of Hall Resistance in the Extreme Quantum Limit of an Organic Conductor $(TMTSF)_2ClO_4$. Phys. Rev. Lett. **94**, 077206 (2005). https://doi.org/10.1103/PhysRevLett.94.077206
6. Moret, R., Pouget, J.P., Comes, R., Bechgaard, K.: X-Ray Scattering Evidence for Anion Ordering and Structural Distortions in the Low-Temperature Phase of Di(Tetramethyltetraselanafulvalenium) Perrhenate $[(TMTSF)_2ReO_4]$. Phys. Rev. Lett. **49**, 1008 (1982). https://doi.org/10.1103/PhysRevLett.49.1008

Landau Quantization and Quantum Oscillation

2

Abstract

The electrons on the Fermi surface predominantly determine the physical properties of conductors. Therefore, the information on the Fermi surface is extremely important for understanding the origin of the properties. The conduction electrons undergo cyclotron motions in magnetic fields. The electronic states are then Landau-quantized, which leads to quantum oscillations in the free energy or thermodynamic grand potential as a function of the magnetic field. The quantum oscillations contain valuable information on the Fermi surface. In this chapter, the Landau quantization of electronic states in magnetic fields and its fundamentals are discussed, which are essential for understanding quantum oscillations. The observation conditions of quantum oscillation are also presented.

2.1 Cyclotron Motion and Landau Quantization

2.1.1 2D Electron System—Classical Theory

First, we consider the motion of an electron in 2D free space (xy plane) in classical theory. Hereafter we will denote the magnetic field as $\boldsymbol{B}$. As shown in Fig. 2.1, the Lorentz force acting on the electron with velocity $\boldsymbol{v}$ in B along the z-axis ($B \| z$) is written as

$$\boldsymbol{F} = -e\boldsymbol{v} \times \boldsymbol{B}. \tag{2.1}$$

Taking the mass of the electron as m_c, the above equation of motion is written as

$$m_c \frac{\mathrm{d}^2}{\mathrm{d}t^2} x = -eBv_y, m_c \frac{\mathrm{d}^2}{\mathrm{d}t^2} y = eBv_x, \tag{2.2}$$

S. Uji, *Fermiology*, Lecture Notes in Physics 1042,
https://doi.org/10.1007/978-981-95-0820-4_2

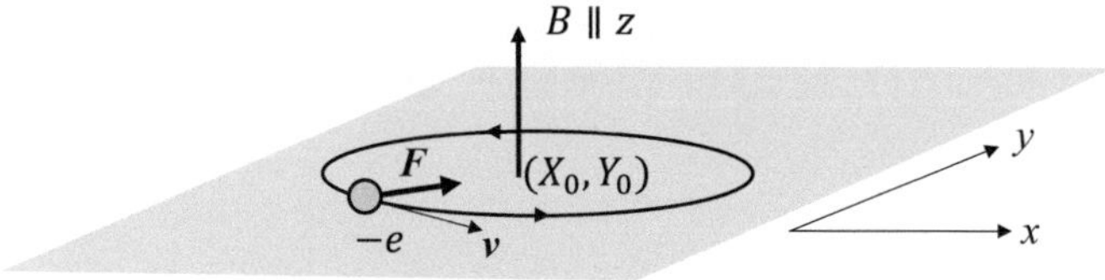

Fig. 2.1 Cyclotron motion of electron in 2D free space. (X_0, Y_0) is the center coordinate of the cyclotron motion

where $v_x = \mathrm{d}x/\mathrm{d}t$ and $v_x = \mathrm{d}y/\mathrm{d}t$ are the velocities of the electron in the x- and y-axes, respectively. The solutions are given by

$$v_x = v_0 \cos \omega_c (t - t_0),\ v_y = v_0 \sin \omega_c (t - t_0). \tag{2.3}$$

Here we take $v_x = v_0$ and $v_y = 0$ at $t = t_0$, and define

$$\omega_c = \frac{eB}{m_c}. \tag{2.4}$$

By integrating Eq. (2.3) over time, we obtain the coordinates of the electron (x, y),

$$\begin{aligned} x &= X_0 + \frac{v_0}{\omega_c} \sin \omega_c (t - t_0), \\ y &= Y_0 - \frac{v_0}{\omega_c} \cos \omega_c (t - t_0), \end{aligned} \tag{2.5}$$

where X_0 and Y_0 are integration constants. As shown in Fig. 2.1, the results show a circular motion with a radius v_0/ω_c whose center coordinate is (X_0, Y_0). This motion is called cyclotron motion and ω_c is the cyclotron (angular) frequency. Since the Lorentz force always acts in the direction perpendicular to both the velocity vector and the magnetic field, the magnetic field does not change the kinetic energy of the electrons: the kinetic energy is kept constant in this motion. This means that the electron always moves on an isoenergetic surface in k-space. The mass m_c in Eq. (2.4) is called the cyclotron effective mass or simply the effective mass. The effective mass is to be distinguished from the mass defined in Eq. (1.46).

Taking $m_c \boldsymbol{v} = \hbar \boldsymbol{k}$ in Eq. (2.1), we obtain $\hbar \partial \boldsymbol{k}/\partial t = -e\boldsymbol{v} \times \boldsymbol{B}$. Integrating this equation over time, we have

$$(\boldsymbol{k} - \boldsymbol{k}_0) = -(\boldsymbol{r} - \boldsymbol{r}_0) \times \frac{eB}{\hbar}, \tag{2.6}$$

where $\boldsymbol{k}_0$ and $\boldsymbol{r}_0$ are the integration constants. From the relation $\boldsymbol{k} \perp \boldsymbol{r}$, the orbital motion in k-space is obtained by rotating that in real space by 90° as shown in Fig. 2.2. As the magnetic field increases, the cyclotron frequency ω_c increases

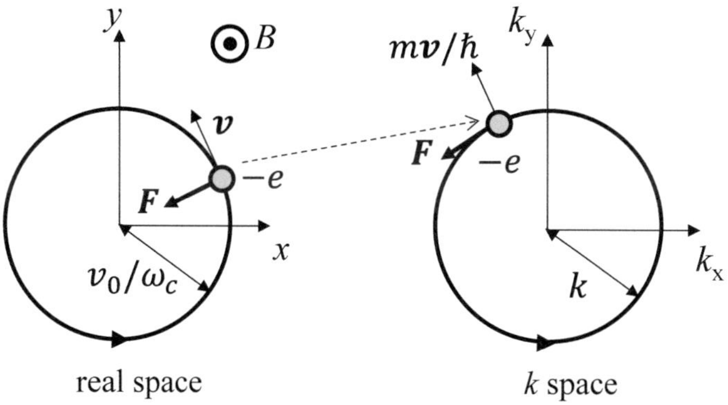

Fig. 2.2 Cyclotron motion in real space (left) and k-space (right). The orbital motion in k-space is obtained by rotating the motion in real space by 90°

in proportion to the magnetic field but the radius of cyclotron motion v_0/ω_c is inversely proportional to the magnetic field. Since Eq. (2.3) is written as $k_x = k_0 \cos\omega_c(t - t_0)$ and $k_y = k_0 \sin\omega_c(t - t_0)$, the radius of the cyclotron motion in k-space,

$$k_0 = m_c v_0/\hbar$$

is constant, independent of the magnetic field.

2.1.2 2D Electron System—Quantum Theory

We will discuss the electron motion in the framework of quantum theory. The momentum of the electrons in closed orbital motion must satisfy the Bohr quantization condition,

$$\oint \boldsymbol{p} \cdot \mathrm{d}\boldsymbol{r} = Nh,\ N = 0, 1, 2, \ldots \tag{2.7}$$

In a magnetic field, substituting the definition of the vector potential $\boldsymbol{B} = \mathrm{rot}\boldsymbol{A}$ into the Maxwell equation, we have $\mathrm{rot}\boldsymbol{E} + \partial\boldsymbol{B}/\partial t = \mathrm{rot}(\boldsymbol{E} + \partial\boldsymbol{A}/\partial t) = 0$. The electric field is written as $\boldsymbol{E} = -\nabla\phi$ where ϕ is the electrostatic potential. In a magnetic field, the Maxwell equation $\mathrm{rot}\boldsymbol{E} = 0$ in zero magnetic field is satisfied when the electric field is rewritten as $\boldsymbol{E} = -\nabla\phi - \partial\boldsymbol{A}/\partial t$. From the relations $\boldsymbol{F} = -e\boldsymbol{E} = e\nabla\phi + e\partial\boldsymbol{A}/\partial t$ and $\boldsymbol{F} = \hbar\partial\boldsymbol{k}/\partial t$, we have the equation of motion,

$$\frac{\partial}{\partial t}(\hbar\boldsymbol{k} - e\boldsymbol{A}) = e\nabla\phi. \tag{2.8}$$

Therefore, when the momentum of the electron is redefined as $\boldsymbol{p} = \hbar\boldsymbol{k} - e\boldsymbol{A}$, Eq. (2.8) is identical to the equation of motion in zero magnetic field, $\partial\boldsymbol{p}/\partial t = \boldsymbol{F} = e\nabla\phi$. The result shows that the Hamiltonian in a magnetic field can be obtained by replacing $\hbar\boldsymbol{k}$ by $\boldsymbol{p} + e\boldsymbol{A} = -i\hbar\nabla + e\boldsymbol{A}$ in the Hamiltonian in zero magnetic field,

$$\mathcal{H} = \frac{(\hbar\boldsymbol{k})^2}{2m} \rightarrow \mathcal{H} = \frac{(-i\hbar\nabla + e\boldsymbol{A})^2}{2m}.$$

This is the effect of the magnetic field on the orbital motion (kinetic energy) of the electron.

Substituting $\boldsymbol{p} = \hbar\boldsymbol{k} - e\boldsymbol{A}$ into Eq. (2.7), we have

$$\oint \boldsymbol{p} \cdot \mathrm{d}\boldsymbol{r} = \oint (\hbar\boldsymbol{k} - e\boldsymbol{A}) \cdot \mathrm{d}\boldsymbol{r}.$$

By using the relation $\boldsymbol{k} = -e\boldsymbol{r}\times\boldsymbol{B}/\hbar$ in Eq. (2.6), the first term of the parenthesis is rewritten as

$$\begin{aligned}\oint \hbar\boldsymbol{k} \cdot \mathrm{d}\boldsymbol{r} &= -e\oint \boldsymbol{r} \times \boldsymbol{B} \cdot \mathrm{d}\boldsymbol{r} \\ &= eB \cdot \oint \boldsymbol{r} \times \mathrm{d}\boldsymbol{r} = 2eBS = 2e\Phi. \end{aligned} \tag{2.9}$$

Here, S is the area of the cyclotron orbit and $\Phi = \mathrm{BS}$ is the total magnetic flux through S. Similarly, the second term of the parenthesis is rewritten as

$$\oint e\boldsymbol{A} \cdot \mathrm{d}\boldsymbol{r} = e\int \mathrm{rot}\boldsymbol{A} \cdot \mathrm{d}\boldsymbol{S} = e\int \boldsymbol{B} \cdot \mathrm{d}\boldsymbol{S} = eBS = e\Phi. \tag{2.10}$$

Therefore, the Bohr quantization condition is rewritten as,

$$\oint \boldsymbol{p} \cdot \mathrm{d}\boldsymbol{r} = Nh = e\Phi \rightarrow \Phi = N(h/e) = N\Phi_0,\ N = 0, 1, 2, \ldots \tag{2.11}$$

where $\Phi_0 = h/e = 4.13 \times 10^{-15}[\mathrm{Wb}] = 4.13 \times 10^{-7}\left[\mathrm{Gcm}^2\right]$ is the quantized magnetic flux called flux quantum. Equation (2.11) shows that an integer number of the flux quanta penetrate the cyclotron orbit in a magnetic field, showing that the closed area of the cyclotron orbit is quantized.

From the equation of motion for $B \| z$, $\hbar\partial\boldsymbol{k}/\partial t = \boldsymbol{F} = -e\mathrm{v} \times \boldsymbol{B}$, we obtain

$$\hbar\frac{\partial k_x}{\partial t} = -eBv_y, \hbar\frac{\partial k_y}{\partial t} = eBv_x. \tag{2.12}$$

By integrating over time, we have $\hbar k_x = -eB(y - Y_0)$ and $\hbar k_y = eB(x - X_0)$, where X_0 and Y_0 are the integration constants. These are rewritten as

$$x = X_0 + \frac{\hbar k_y}{eB}, y = Y_0 - \frac{\hbar k_x}{eB}. \tag{2.13}$$

Comparing these with Eq. (2.5), we note that (X_0, Y_0) is the center coordinate and $\left(\hbar k_y/eB, -\hbar k_x/eB\right)$ represents the orbital coordinate of the cyclotron motion. Using the Landau gauge $\boldsymbol{A}(0, Bx, 0)$ for $B\|z$, the relation $\hbar \boldsymbol{k} = \boldsymbol{p} + e\boldsymbol{A} = -i\hbar\nabla + e\boldsymbol{A}$ gives

$$\hbar k_x = -i\hbar\frac{\partial}{\partial x}, \hbar k_y = -i\hbar\frac{\partial}{\partial y} + eBx, \tag{2.14}$$

Substituting these into Eq. (2.13), we have

$$X_0 = \frac{i\hbar}{eB}\frac{\partial}{\partial y} = il_B^2\frac{\partial}{\partial y}, y = Y_0 + il_B^2\frac{\partial}{\partial x}. \tag{2.15}$$

Here, l_B is a quantity with the dimension of length defined as

$$l_B^2 = \hbar/eB \tag{2.16}$$

This is called the magnetic length. It is easy to verify the commutation relations, $\left[k_x, k_y\right] = -i/l_B^2, [k_x, x] = \left[k_y, y\right] = -i$, and $\left[k_x, y\right] = \left[k_y, x\right] = 0$. The center coordinates also have the relation, $[X_0, Y_0] = il_B^2$. These relations are gauge independent. Using Schwartz inequality, the uncertain relation is obtained as $\Delta X_0 \Delta Y_0 \geq l_B^2/2$, where $\Delta X_0(\Delta Y_0)$ is the uncertainty of $X_0(Y_0)$. This means that the central coordinates X_0 and Y_0 of the cyclotron motion are not simultaneously determined in quantum mechanics. As the magnetic field increases, the uncertainty l_B^2 decreases. This means that the electrons are confined to smaller regions in real space at higher magnetic fields.

The Hamiltonian in free space is written as

$$\mathcal{H} = \frac{\hbar^2}{2m_c}\left(k_x^2 + k_y^2\right). \tag{2.17}$$

The Zeeman effect is not taken into account for simplicity. Substituting Eq. (2.14) into Eq. (2.17) gives the Hamiltonian including the magnetic field effect,

$$\mathcal{H} = \frac{1}{2m_c}\left[\left(-i\hbar\frac{\partial}{\partial x}\right)^2 + \left(-i\hbar\frac{\partial}{\partial y} + eBx\right)^2\right]. \tag{2.18}$$

This Hamiltonian contains only the term $\partial/\partial y$ for y, we can write the solution as

$$\Psi(x, y) = \psi_N(x)\phi(y). \tag{2.19}$$

Since X_0 is commutative with $\mathcal{H}$, $i\hbar(\partial X_0/\partial t) = [X_0, \mathcal{H}] = 0$, X_0 is the conserved quantity (time-independent) but Y_0 is not. From Eq. (2.15), we have

$$X_0\Psi(x, y) = \frac{i\hbar}{eB}\frac{\partial}{\partial y}\Psi(x, y), \tag{2.20}$$

and thus

$$i\hbar\frac{\partial}{\partial y}\phi(y) = eBX_0\phi(y). \tag{2.21}$$

This solution is a plane wave,

$$\phi(y) = \exp\left(\frac{-ieBX_0y}{\hbar}\right) = \exp\left(\frac{-iX_0y}{l_B^2}\right). \tag{2.22}$$

Substituting $X_0 = (i\hbar/eB)(\partial/\partial y)$ into Eq. (2.18), we have

$$\mathcal{H} = -\frac{\hbar^2}{2m_c}\frac{\partial^2}{\partial x^2} + \frac{1}{2}m_c\omega_c^2(x - X_0)^2. \tag{2.23}$$

This is the Hamiltonian for a 1D harmonic oscillator. The solution is written as

$$\psi_N(x) = \frac{1}{\sqrt{2^N N! l_B\sqrt{\pi}}}H_N\left(\frac{x - X_0}{l_B}\right)\exp\left[-\frac{1}{2l_B^2}(x - X_0)^2\right], \tag{2.24}$$

where the coefficient $1/\sqrt{2^N N! l_B\sqrt{\pi}}$ is the normalized constant and $H_N(x)$ is the Nth order Hermite function. The wave function $\psi_N(x)$ is plotted for several N values in Fig. 2.3. $\psi_N(x)$ has N nodes (zero-crossing points). $\phi(y)$ is a plane wave, showing a constant electron density in the y-axis. In this gauge, X_0 is fixed, and thus Y_0 is completely uncertain from the uncertainty relation $\Delta X_0\Delta Y_0 \geq l_B^2/2$. In a different gauge, $\boldsymbol{A}(-By, 0, 0)$, we have a solution with x and y interchanged, the solution has N nodes in the y-axis.

If we define a reduced coordinate ξ

$$\xi = \sqrt{\frac{m_c\omega_c}{\hbar}}(x - X_0), \tag{2.25}$$

and operators,

$$a = \frac{1}{\sqrt{2}}\left(\xi + \frac{\partial}{\partial\xi}\right), a^+ = \frac{1}{\sqrt{2}}\left(\xi - \frac{\partial}{\partial\xi}\right), \tag{2.26}$$

the Hamiltonian is written as

$$\mathcal{H} - \hbar\omega_c\left(a^+a + \frac{1}{2}\right). \tag{2.27}$$

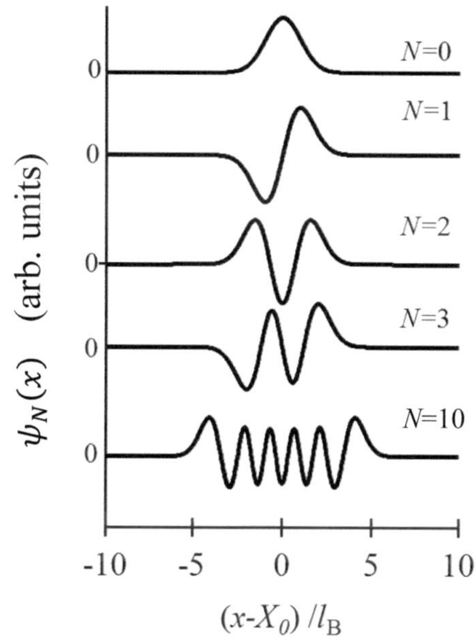

Fig. 2.3 Wave functions $\psi_N(x)$ for several N values. $\psi_N(x)$ has N nodes

The operators satisfy the commutation relation, $[a, a^+] = 1$. From the following relations,

$$a\psi_N(x) = \sqrt{N}\psi_{N-1}(x),\ a^+\psi_N(x) = \sqrt{N+1}\psi_{N+1}(x), \tag{2.28}$$

we note that a and a^+ are the annihilation and creation operators, respectively. The eigenvalue of the Schrödinger equation $\mathcal{H}\Psi(x, y) = E_{\mathrm{N}}\Psi(x, y)$ is quantized as

$$E_N = \hbar\omega_c\left(N + \frac{1}{2}\right), N = 0, 1, 2, \ldots \tag{2.29}$$

This is called Landau quantization, and the quantized energy level and its index N are called Landau level and Landau index, respectively. The ground state ($N = 0$) has the non-zero energy $E_{N=0} = (1/2)\hbar\omega_c$, which corresponds to zero-point energy of a harmonic oscillator. The Landau levels E_N are evenly spaced with $\hbar\omega_c$. Assuming that the lowest Landau level $E_{N=0}$ is equal to the kinetic energy of the cyclotron motion with the radius r_0,

$$\frac{\hbar\omega_c}{2} = \frac{m_c(r_0\omega_c)^2}{2}, \tag{2.30}$$

we obtain $r_0 = l_B$. This is the physical meaning of the magnetic length given in Eq. (2.16). For $B = 10$ T, we have $l_B = 8.1$nm.

Next we will see that the wave function $\psi_N(x)$ changes significantly depending on the gauge. For the symmetric gauge $\boldsymbol{A} = \boldsymbol{B}\times\boldsymbol{r}/2$, it may be easier to understand the relation to the classical cyclotron orbit. In cylindrical coordinates (r, φ), the Schrödinger equation can be divided into independent partial differential equations with respect to r and φ. By solving the Schrödinger equation, the eigenfunction is obtained, whose center coordinate (X_0, Y_0) has the relation,

$$X_0^2 + Y_0^2 = 2l_B^2\left(N + m_N + \frac{1}{2}\right) > 0, N = 0, 1, 2, \ldots \text{and } m_N \geq -N. \quad (2.31)$$

Here $R_0 = \sqrt{X_0^2 + Y_0^2}$ is the distance from the origin. Both X_0 and Y_0 are not the conserved quantities. The number m_N indicates the angular momentum along the z-axis. The total wave function is given by

$$\Psi_{Nm_N}(r, \varphi) = \psi_{Nm_N}(r)\exp(-im_{\mathrm{N}}\varphi), \quad (2.32)$$

and its radial function $\psi_{Nm_N}(r)$ is expressed as

$$\psi_{Nm_N}(r) = \frac{1}{l_B}\sqrt{\frac{N!}{2\pi(N + m_{\mathrm{N}})!}}\left(\frac{r}{\sqrt{2}l_B}\right)^{m_N}\exp\left(\frac{-r^2}{4l_B^2}\right)L_N^{m_N}\left(\frac{r^2}{2l_B^2}\right), \quad (2.33)$$

where $L_N^{m_N}(x)$ is the Nth order Laguerre polynomial. The eigenfunction in Eq. (2.32) is often called a vortex state because the rotation around (X_0, Y_0) changes the phase by $2\pi m_{\mathrm{N}}$. The eigenvalue is the same as Eq. (2.29). The radial function $\psi_{Nm_N}(r)$ is plotted for several N and m_N values in Fig. 2.4. Because of $L_0^0 = 1$, we have $\psi_{00}(r) \propto \exp(-r^2/4l_B^2)$, which extends over a range of $\sim 2l_B$ and its center coordinate is away from the origin by $R_0 = l_B$. As m_N increases for $N = 0$, $\psi_{0m_N}(r)$ spreads over a wider range and its center coordinate is also further away from the origin. As m_N increases significantly, the spreading wave function reaches the boundary of the space. This is the upper limit of m_N, which gives the degeneracy of the states. In this gauge, the distance R_0 is fixed but the center coordinate (X_0, Y_0) has uncertainty $\Delta X_0 \Delta Y_0 \geq l_B^2/2$. The function $\psi_{Nm_N}(r)$ has N nodal lines as shown in Fig. 2.4.

The above discussion has been made for the Landau quantization of electrons in free space. In a crystal, the Hamiltonian includes the periodic lattice potential, and thus it is difficult to solve the Schrödinger equation analytically. The energy bands in the crystal will generally have a complex functional form, which leads to a complex Fermi surface structure depending on the lattice potential.

Regardless of the Fermi surface structure, it should be noted that the electrons undergo orbital motion on the isoenergetic surface in a constant magnetic field as long as they are not scattered. When the orbits are closed, the Bohr quantization condition given by Eq. (2.7) must be satisfied; the electron energy levels are Landau-quantized. The Landau levels in the lattice potential are generally not equally spaced. As shown in Appendix 1.7, graphene has a linear energy dispersion (Dirac cone energy band) and the Landau levels are given by $E_N \propto \sqrt{N}$ as

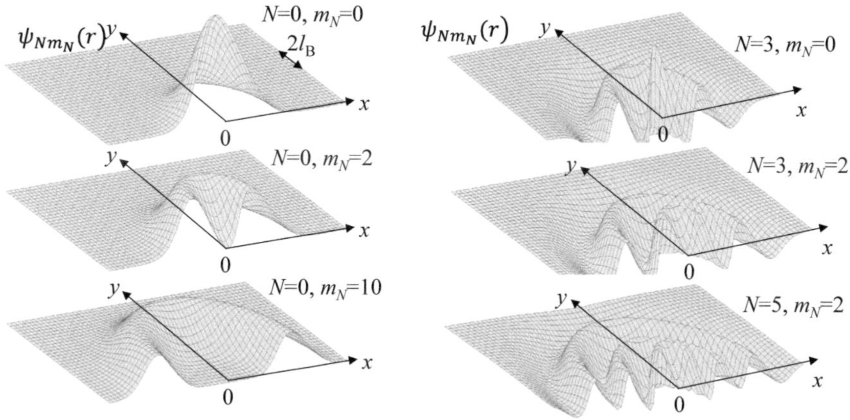

Fig. 2.4 Radial functions $\psi_{Nm_N}(r)$ for several N and m_N values in the symmetric gauge. $\psi_{Nm_N}(r)$ has N nodal lines

derived in Appendix 2.1. Although Landau quantization in a crystal is not as simple as in free space, a closer look at the eigenstates in free space will be helpful in understanding Landau quantization.

2.1.3 Degeneracy of Landau Level in 2D Electron System

Assuming the electronic state in the area of $L \times L$ and the periodic boundary condition $\phi(y) = \phi(y + L)$ for the plane wave in Eq. (2.22), we obtain $X_0 = 2n\pi l_B^2/L$ (n: integer). This relation means that the center coordinate X_0 of $\psi_N(x)$ in Eq. (2.24) is evenly spaced with $2\pi l_B^2/L$ in the x-axis but no restrictions on Y_0. The degeneracy of the state within the length L is given by $L/\left(2\pi l_B^2/L\right) = L^2/2\pi l_B^2$; there are $L^2/2\pi l_B^2$ states in the area $L \times L$. Therefore, the degeneracy D_N per unit area ($L^2 = 1$) is given by

$$D_N = \frac{1}{2\pi l_B^2} = \frac{m_c\omega_c}{2\pi\hbar} = \frac{B}{\Phi_0}. \tag{2.34}$$

This quantity is doubled when the spin degree of freedom is taken into account. Note that the degeneracy does not depend on the Landau index N. Since πl_B^2 is the area of the cyclotron orbit at the lowest Landau level in classical theory, the degeneracy is a measure of how many cyclotron orbits at the lowest Landau level can occupy the unit area. The relation $D_N = B/\Phi_0$ shows that the degeneracy is equal to the number of the flux quanta passing through the unit area.

All Landau levels, which are equally spaced with $\hbar\omega_c$, rise in proportion to the magnetic field. As shown in Fig. 2.5a, we assume that the Nth Landau level is just below the Fermi energy E_F. The density of states in the 2D electron system is $D(E) = m_c/2\pi\hbar^2$ from Eq. (1.8) without the spin degree of freedom. Therefore,

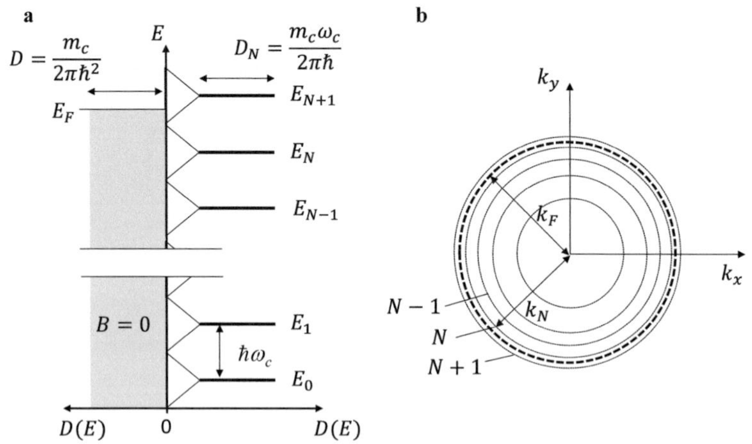

Fig. 2.5 **a** Landau quantization in 2D electronic state. All electronic states within $\hbar\omega_c$ (left) are condensed into the closest Landau levels (right). **b** Fermi surface (dashed circle) and quantized cyclotron orbits with indices $\ldots, N-1, N, N+1$ in magnetic field

the number of states in the range of $\hbar\omega_c$ is equal to the degeneracy of each Landau level with N,

$$\hbar\omega_c D(E) = \frac{m_c\omega_c}{2\pi\hbar} = D_N. \tag{2.35}$$

This result means that all electronic states within the range of $\hbar\omega_c$ are condensed into the closest Landau levels.

The area of the cyclotron orbit A_N in k-space is given by

$$A_N = \pi k_N^2, \tag{2.36}$$

where k_N is the radius of the Nth cyclotron orbit. From the relation $E_N = \hbar^2 k_N^2/2m_c = \hbar\omega_c(N+1/2)$, we have

$$A_N = \frac{2\pi eB}{\hbar}\left(N+\frac{1}{2}\right). \tag{2.37}$$

The area is proportional to the magnetic field. The radius of the cyclotron orbit increases with N and magnetic field, $k_N \propto \sqrt{(N+1/2)B}$, as shown in Fig. 2.5b. Since the area increases with magnetic field $A_N \propto B$, A_N becomes equal to the cross section of the cylindrical Fermi surface at a certain magnetic field. This situation corresponds to $E_N = E_F$. From the relation, $r = -\hbar k/eB$ in Eq. (2.6), we obtain the area of the cyclotron orbit in real space,

$$S_N = \left(\frac{\hbar}{eB}\right)^2 A_N = \frac{2\pi\hbar}{eB}\left(N+\frac{1}{2}\right). \tag{2.38}$$

S_N is inversely proportional to the magnetic field.

2.1.4 Landau Quantization in 3D Electron System

So far we have considered a 2D electron system. Now we will consider a 3D electron system. The Fermi surface of the 3D free electron system is a sphere as shown in Fig. 2.6a. All the electrons can take closed orbits in magnetic fields; all the states are Landau-quantized. In crystals, depending on the structure of the Fermi surface, the electrons may not take closed orbits. In such a case, the Bohr quantization condition need not be satisfied; the electronic states are not Landau-quantized. Returning to Fig. 2.6a, all the cyclotron orbits in the magnetic field ($\boldsymbol{B}\|z$) remain in the $k_x k_y$ plane, and the electron motion in the k_z-axis is not constrained by the magnetic field. The eigenvalues are given by

$$E_N = \hbar\omega_c\left(N + \frac{1}{2}\right) + \frac{\hbar^2 k_z^2}{2m_c}. \tag{2.39}$$

Using $E_F = \hbar^2\left(k_x^2 + k_y^2 + k_z^2\right)/2m_c$ in zero magnetic field, the difference between the Landau and Fermi levels is

$$E_F - E_N = \frac{\hbar^2}{2m_c}\left(k_x^2 + k_y^2\right) - \hbar\omega_c\left(N + \frac{1}{2}\right). \tag{2.40}$$

Figure 2.6b shows the isoenergetic surfaces (cylinders) for the Nth Landau level $\hbar\omega_c(N + 1/2)$, called the Landau tubes. The dashed curves in Fig. 2.6a show the intersection of the Landau tube and the spherical Fermi sphere, $E_F = E_N$. The cross section of the dashed orbit A_N is given by Eq. (2.37). There are other Landau tubes, for instance, ($N - 1$)th tube inside and ($N + 1$)th tube outside.

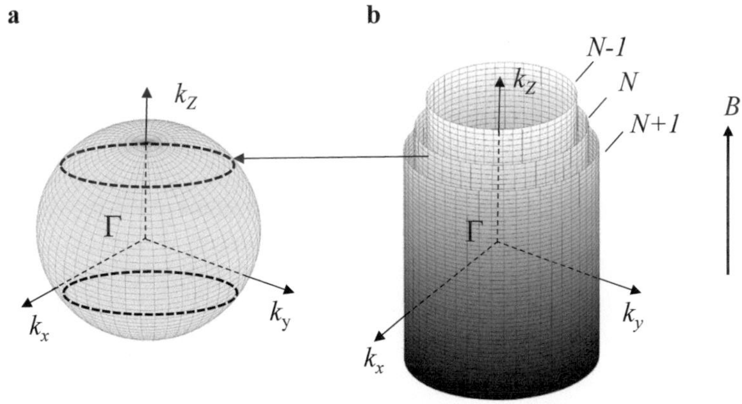

Fig. 2.6 **a** Spherical Fermi surface of 3D free electron system. The dashed curves show the intersection of the Landau tube and the spherical Fermi sphere, $E_F = E_N$ **b** Landau tubes with indices $N - 1, N$, and $N + 1$

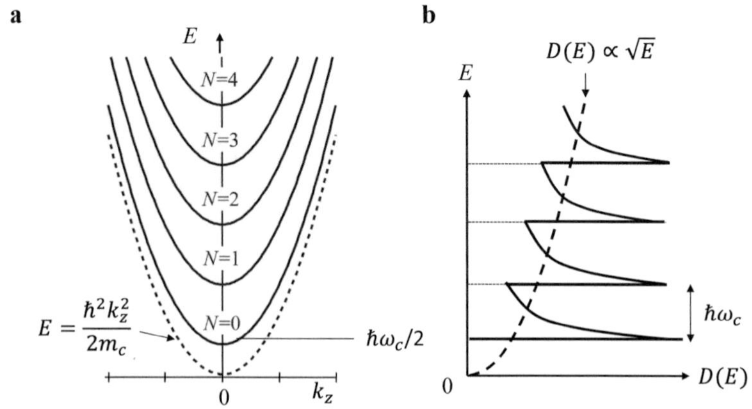

Fig. 2.7 **a** Landau levels of 3D free electron system, **b** Density of states in magnetic field. The dashed curve shows the density of states at zero magnetic field

As the magnetic field increases, the intersection on the Fermi surface expands and reaches the maximum cross section of the Fermi surface at a certain magnetic field, where $E_F = \hbar\omega_c(N + 1/2)$. The electrons inside the Fermi surface are condensed into the nearest Landau tube in a magnetic field.

Equation (2.39) shows that each Landau level has a parabolic dispersion with k_z as depicted in Fig. 2.7a. Using k_z from Eq. (2.39), we have

$$k_z = \left(\frac{2m_c}{\hbar^2}\right)^{1/2}\left[E_N - \hbar\omega_c\left(N + \frac{1}{2}\right)\right]^{1/2}, \tag{2.41}$$

We obtain the density of states by calculating the number of states in the $k_x k_y$ plane and in the k_z-axis independently,

$$D(E) = \frac{1}{4\pi^2}\left(\frac{2m_c}{\hbar^2}\right)^{3/2}\hbar\omega_c\sum_N\left[E - \left(N + \frac{1}{2}\right)\hbar\omega_c\right]^{-1/2}. \tag{2.42}$$

This equation is derived in Appendix 2.2. The resulting density of states is depicted in Fig. 2.7b. The density of states tends to diverge at $E = (N + 1/2)\hbar\omega_c$ and then gradually decreases with increasing E up to $(N + 3/2)\hbar\omega_c$, following $D(E) \propto 1/\sqrt{E}$. For comparison, the density of states at zero magnetic field is indicated by the dashed curve in Fig. 2.7b.

For reference, some parameters of the Fermi surfaces for two typical conductors are listed in Table 2.1, the copper (Cu) metal and organic conductor κ (BEDT-TTF)$_2$Cu(NCS)$_2$, where BEDT-TTF stands for *bis(ethylenedithio)tetrathiafulvalene*. The copper has a face-centered cubic cell with the lattice constant $a = 0.361$ nm as shown in Fig. 2.8a. κ-(BEDT-TTF)$_2$Cu(NCS)$_2$ has a monoclinic unit cell with the lattice constants $a = 1.63$ nm,

Table 2.1 Parameters of copper and organic conductor κ (BEDT-TTF)$_2$Cu(NCS)$_2$

Material	Cu	κ-(BEDT-TTF)$_2$Cu(NCS)$_2$
E_F (eV)	~7	~0.2
ω_c(rad/s)	~10^{12}	~10^{12}
$\hbar\omega_c$ (eV)	~10^{-3}	~10^{-3}
N	~7000	~300

The values ω_c and $\hbar\omega_c$ are calculated at a magnetic field of 10 T. N indicates the number of the Landau levels below E_F at 10 T

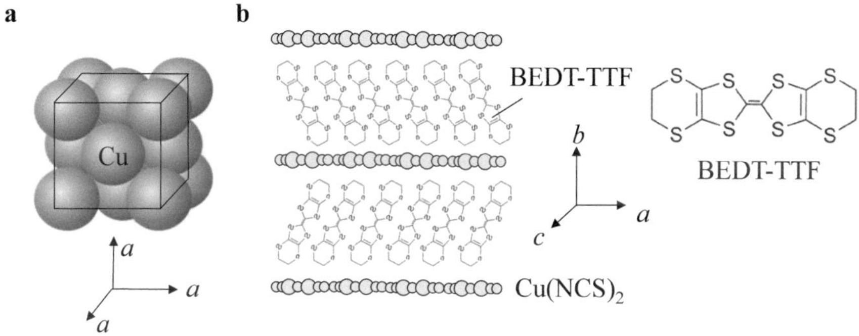

Fig. 2.8 Structures of **a** copper, and **b** organic conductor κ-(BEDT-TTF)$_2$Cu(NCS)$_2$ and BEDT-TTF molecule

$b = 0.844$ nm, $c = 1.31$ nm and $\beta = 110°$, whose schematic structure is shown in Fig. 2.8b. The first Greek letter represents the type of stacking structure of the BEDT-TTF molecule. The planar organic BEDT-TTF molecules are stacked in the a- and c-axes, and the Cu(NCS)$_2$ anions form insulating layers between the BEDT-TTF layers. Due to the layered structure, a highly 2D energy band is formed by the molecular orbital of BEDT-TTF; the conducting layer is in the ac plane and the least conducting axis is parallel to the b-axis. Note that the Fermi energy of κ-(BEDT-TTF)$_2$Cu(NCS)$_2$ is much smaller than that of Cu. The main reason is that the unit cell of κ-(BEDT-TTF)$_2$Cu(NCS)$_2$ is much larger and thus its low carrier density makes the Fermi surface and level much smaller.

2.2 Magnetic Field Dependence of Energy of Electronic State

2.2.1 Free Energy and Thermodynamic Grand Potential

We see how the total energy of the electronic state changes with a magnetic field [1]. The free energy F is defined as $F = E - TS$, where E is the internal energy and S is the entropy. We also define the thermodynamic grand potential as $\Omega =$

$F - \mu N_e$, where μ is the chemical potential and N_e is the number of electrons. Their differential expressions are respectively given by

$$dF = -SdT - MdB + \mu dN_e, \tag{2.43}$$

$$d\Omega = -SdT - MdB - N_e d\mu. \tag{2.44}$$

Here M is the magnetization and $B = \mu_0(H + M)$, where μ_0 is the permeability of the vacuum. For paramagnetic materials ($H \gg M$) with which we are concerned, we can take $B \approx \mu_0 H$. From the above equations, we obtain

$$M = -\left(\frac{\partial F}{\partial B}\right)_{N_e} = -\left(\frac{\partial \Omega}{\partial B}\right)_{\mu}, \tag{2.45}$$

$$\mu = \left(\frac{\partial F}{\partial N_e}\right)_B, \; N_e = -\left(\frac{\partial \Omega}{\partial \mu}\right)_B. \tag{2.46}$$

For simplicity, the above equations do not specify the constant temperature condition. In statistical mechanics, the thermodynamic grand potential is expressed as

$$\Omega = -k_B T \sum_i \ln\left(1 + e^{\frac{\mu - E_i}{k_B T}}\right), \tag{2.47}$$

where k_B is the Boltzmann constant and Σ indicates summing over all possible electronic states with the energy E_i. In a magnetic field, the sum of the states with k parallel to the magnetic field is replaced by its integral $\sum_k \to (1/2\pi) \int dk$ and those in a plane perpendicular to the magnetic field by the sum of the Landau-quantized states with index N. The thermodynamic grand potential per unit volume is given by

$$\Omega = -\frac{k_B T}{2\pi} \int_{-k_F}^{k_F} dk D_N \sum_N \ln\left(1 + e^{\frac{\mu - E_N}{k_B T}}\right). \tag{2.48}$$

The spin degree of freedom is not taken into account in the above sum. The Landau level E_N is given by Eq. (2.39) and the degeneracy of the Landau level D_N is given by Eq. (2.34), which depends only on the magnetic field.

Since the Fermi energy is equal to the chemical potential at $T = 0$ K, $\mu(T = 0) = E_F$, Eq. (2.48) for $T = 0$ K is written as

$$\Omega = \frac{1}{2\pi} \int_{-k_F}^{k_F} dk D_N \sum_N (E_N - E_F) \to \frac{1}{2\pi} \int_{-k_F}^{k_F} dk D_N \sum_N E_N. \tag{2.49}$$

For simplicity, the energy offset E_F is eliminated in the last expression. Rewriting Eq. (2.49) as

$$\Omega = \int \delta\Omega = \int_{-k_F}^{k_F} (\delta\Omega/\delta k)\mathrm{d}k,$$

we define the thermodynamic grand potential in the width δk parallel to the magnetic field,

$$\delta\Omega = \frac{\delta k D_N}{2\pi} \sum_N E_N. \tag{2.50}$$

The simplest case is an isolated 2D system, where the number of the electrons is conserved. The free energy is equal to the internal energy at $T = 0$ K, $F = E - TS \to E$ and the internal energy in a magnetic field perpendicular to the 2D system is simply written as

$$E = D_N \sum_N E_N. \tag{2.51}$$

2.2.2 2D Electron System

First, consider the case of a 2D free electron system with a constant number of electrons. In a magnetic field perpendicular to the 2D layer, the electronic states are Landau-quantized, as shown in Fig. 2.9a. We assume that the Fermi level lies in the range, $N\hbar\omega_c < E_F \le (N+1)\hbar\omega_c$. The E_{N-1} level is fully occupied but the E_N level is partially occupied by electrons whose number is given by $D_N(E_F - N\hbar\omega_c)/\hbar\omega_c$. As the magnetic field increases, all the Landau levels rise and some of the electrons occupying the E_N level fall to the E_{N-1} level. The chemical potential $\mu(B)$ is held at E_N [$\mu(B) = E_N$]; $\mu(B)$ linearly increases with magnetic field in this range as depicted in Fig. 2.9a. Once $N\hbar\omega_c$ exceeds E_F, the E_N level becomes empty, resulting in a discontinuous drop of $\mu(B)$ to E_{N-1}. As the magnetic field further increases, this process is repeated, resulting in a sawtooth oscillation of the chemical potential as shown in Fig. 2.10a.

In Fig. 2.9a, the Landau levels are fully occupied up to $N - 1$, and thus the internal energy given by Eq. (2.51) is calculated as

$$\begin{aligned} E(B) =& D_N \sum_{n=0}^{N} E_n = D_N \sum_{n=0}^{N-1} \left(n + \frac{1}{2}\right)\hbar\omega_c + D_N\left(N + \frac{1}{2}\right)(E_\mathrm{F} - N\hbar\omega_c) \\ =& -\frac{D(E_\mathrm{F})}{2}\left[\left(N + \frac{1}{2}\right)\hbar\omega_c - E_F\right]^2 + \frac{D(E_\mathrm{F})}{2}\left[E_F^2 + \frac{(\hbar\omega_c)^2}{4}\right], \end{aligned} \tag{2.52}$$

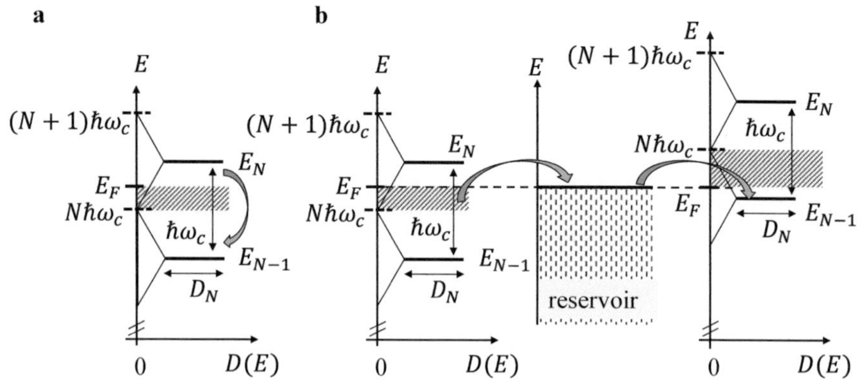

Fig. 2.9 Landau quantization and electron transfer between levels for **a** 2D and **b** 3D cases

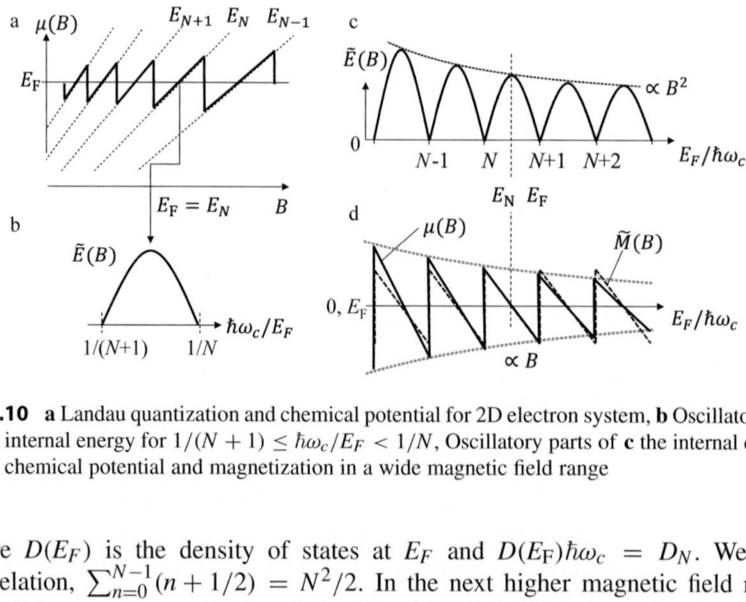

Fig. 2.10 **a** Landau quantization and chemical potential for 2D electron system, **b** Oscillatory part of the internal energy for $1/(N+1) \leq \hbar\omega_c/E_F < 1/N$, Oscillatory parts of **c** the internal energy, and **d** chemical potential and magnetization in a wide magnetic field range

where $D(E_F)$ is the density of states at E_F and $D(E_{\mathrm{F}})\hbar\omega_c = D_N$. We used the relation, $\sum_{n=0}^{N-1}(n+1/2) = N^2/2$. In the next higher magnetic field range, $(N-1)\hbar\omega_c < E_F \leq N\hbar\omega_c$, N in Eq. (2.52) is replaced by $N-1$. In this way, N decreases one by one as the magnetic field increases. Note that the first term $[(N+1/2)\hbar\omega_c - E_F]^2$ gives oscillatory behavior of the internal energy due to Landau quantization. The other magnetic-field-dependent term $D(E_F)(\hbar\omega_c)^2/8$ shows the orbital diamagnetic energy. From the relation $N \to E_F/\hbar\omega_c$ for $B \to 0$, we have the internal energy at $B = 0$, $E = D(E_F)E_F^2/2$, which is consistent with the result $E = \int_0^{E_F} D(E)E\mathrm{d}E$.

The magnetic field range for $\mu(B) = E_N$, $N\hbar\omega_c < E_F \leq (N+1)\hbar\omega_c$ is rewritten as

$$\frac{1}{N+1} \leq \frac{\hbar\omega_c}{E_F} < \frac{1}{N}. \tag{2.53}$$

Here we define the oscillatory term $\tilde{E}(B)$ of the internal energy in Eq. (2.52) as

$$\tilde{E}(B) = -\frac{D(E_F)E_F^2}{2}\left[\left(N+\frac{1}{2}\right)\frac{\hbar\omega_c}{E_F} - 1\right]^2 + \frac{D(E_F)}{8}(\hbar\omega_c)^2. \tag{2.54}$$

This is a quadratic function of B, $\tilde{E}(B)$ has a maximum at $E_F \approx E_N = (N+1/2)\hbar\omega_c$, as shown in Fig. 2.10b. In a wide magnetic field range, $\tilde{E}(B)$ oscillates (Fig. 2.10c): the energy $\tilde{E}(B)$ periodically has sharp minima as a function of the inverse magnetic field where $\mu(B)$ changes discontinuously. Since the magnetization is given by $M(B) = -(\partial E/\partial B)$, the oscillatory part of the magnetization $\tilde{M}(B)$ also forms a sawtooth with increasing magnetic field as shown in Fig. 2.10d, which is similar to $\mu(B)$. In this calculation, the electron spin is not taken into account, the magnetization is purely ascribed to the orbital magnetization. The amplitude of the oscillation of $\tilde{M}(B)$ is independent of the magnetic field. The effects of finite temperature, scattering and spin on the oscillations will be discussed in Sect. 3.2.

2.2.3 3D Electron System

It is not so likely that the electron systems we measure are completely 2D. Even for layered materials, there is a finite energy dispersion between the layers or they may be multiple-Fermi-surface systems. In these cases, the above argument for the 2D electron system should be modified. For simplicity, we assume an isotropic Fermi surface as shown in Fig. 2.6a and apply a magnetic field along the z (k_z)-axis. The electronic states in the k_xk_y-plane are Landau-quantized. As the magnetic field increases, the Landau levels (Landau tubes) with different quantum numbers N cross the Fermi level at different parts on the Fermi surface as already seen in Fig. 2.6. Here we see the Landau quantization in the range $E_{N-1} < E_F \leq E_N$ for the $k_Z = 0$ states as depicted in Fig. 2.9b. In the range $N\hbar\omega_c < E_F \leq E_N$ (left of Fig. 2.9b), the E_N level above E_F is empty; the electrons which occupied the states in the range between $N\hbar\omega_c$ and E_F at zero magnetic field (gray area) are transferred to the E_F level on the different parts of the Fermi surface ($k_Z \neq 0$) or other Fermi surfaces (if present), working as a reservoir. The number of the electrons N_L in the $k_Z = 0$ states is smaller than the number N_0 at zero magnetic field by $D(E_F)(E_F - N\hbar\omega_c)$; the electrons transferred to the reservoir have the energy $D(E_F)E_F(E_F - N\hbar\omega_c)$. Note that this transfer does not occur in a 2D electron system without the reservoir. When $N\hbar\omega_c$ reaches E_F, we have $N_L = N_0$ and thus the total energy of the $k_Z = 0$ states is equal to $D_N\sum_{n=0}^{N-1}(n+1/2)\hbar\omega_c$. In the

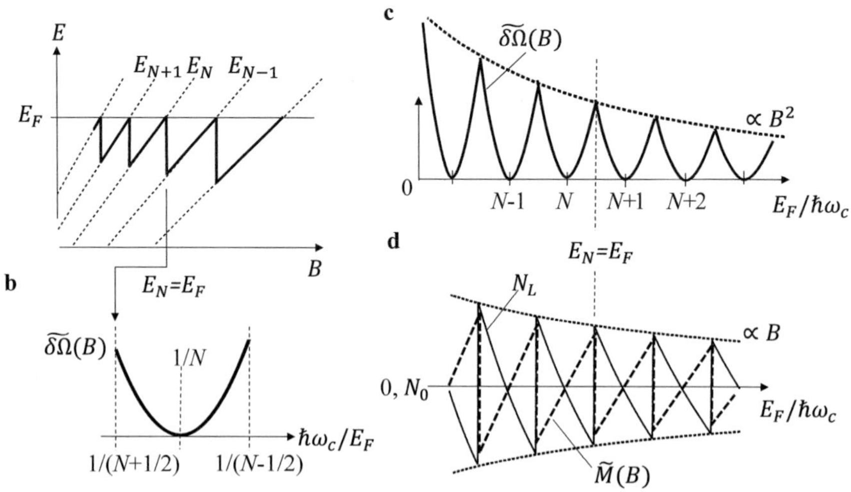

Fig. 2.11 **a** Landau quantization and highest Landau level occupied by electrons for 3D electron system, **b** Oscillatory part of the thermodynamic grand potential for $1/(N+1/2) \le \hbar\omega_c/E_F < 1/(N-1/2)$, **c** Oscillatory parts of the thermodynamic grand potential in a wide magnetic field range, **d** Number of electrons N_L in the $k_Z = 0$ states and oscillatory part of the magnetization in a wide magnetic field range

higher magnetic field range $E_{N-1} < E_F \le N\hbar\omega_c$ (right of Fig. 2.9b), the E_{N-1} level below E_F is still fully occupied by the electrons. This means that the electrons with the number of $D(E_F)(N\hbar\omega_c - E_F)$ (gray area) are transferred from the reservoir to the E_{N-1} level, and thus N_L is larger than N_0 by $D(E_F)(N\hbar\omega_c - E_F)$. As the magnetic field increases, all the Landau levels rise. Once E_{N-1} exceeds E_F, the E_{N-1} level becomes empty. As the magnetic field further increases, this process is repeated, and thus the highest Landau level occupied by the electrons in the $k_Z = 0$ states shows a sawtooth oscillation without exceeding E_F as shown in Fig. 2.11a.

In the range $E_{N-1} < E_F \le E_N$, the thermodynamic grand potential $\delta\Omega$ in Eq. (2.50) changes with magnetic field as follows,

$$\begin{aligned}\delta\Omega(B)\frac{2\pi}{\delta k} &= D_N\sum_{n=0}^{N} E_n = D_N\sum_{n=0}^{N-1}\left(n+\frac{1}{2}\right)\hbar\omega_c \\ &\quad + D(E_F)E_F(E_F - N\hbar\omega_c) \\ &= \frac{D(E_F)}{2}(N\hbar\omega_c - E_F)^2 + \frac{D(E_F)}{2}E_F^2. \end{aligned} \tag{2.55}$$

Since $E_F/\hbar\omega_c \to N$ for $B \to 0$, we have $\delta\Omega(B)(2\pi/\delta k) = D(E_F)E_F^2/2$ for $B = 0$, equal to the 2D case in Eq. (2.52). The range $E_{N-1} < E_F \le E_N$ is rewritten as

$$\frac{1}{N+1/2} \leq \frac{\hbar\omega_c}{E_F} < \frac{1}{N-1/2}. \tag{2.56}$$

The oscillatory term of the thermodynamic grand potential in Eq. (2.55) is defined as

$$\widetilde{\delta\Omega}(B)\frac{2\pi}{\delta k} = \frac{D(E_F)E_F^2}{2}\left[N\frac{\hbar\omega_c}{E_F} - 1\right]^2. \tag{2.57}$$

This is also a quadratic function of B; $\widetilde{\delta\Omega}(B)$ has a minimum at $E_F = N\hbar\omega_c$ as shown in Fig. 2.11b. In a wide magnetic field range, $\widetilde{\delta\Omega}(B)$ becomes a periodic function with inverse magnetic field (Fig. 2.11c). $\widetilde{\delta\Omega}(B)$ exhibits pronounced peaks at magnetic fields where $N_L(B)$ changes discontinuously. Since the magnetization is given by $M(B) = -(\partial\Omega/\partial B)$, the oscillatory part of the magnetization $\tilde{M}(B)$ also forms a sawtooth shape with increasing magnetic field (Fig. 2.11d). Again, $\tilde{M}(B)$ results purely from the orbital magnetization. For comparison, $N_L(B)$ is shown in Fig. 2.11d.

Both the internal energy of the 2D system and the thermodynamic grand potential of the electronic state with $k_Z = 0$ for the 3D system oscillate with increasing magnetic field. Therefore, the thermodynamic quantities, such as magnetization, heat capacity, which are derived from the derivative of the energies or the thermodynamic grand potential, also oscillate with the magnetic field. Since the density of states is derived from the thermodynamic grand potential, the resistance also oscillates as discussed in Chap. 3. These phenomena are commonly referred to as quantum oscillations. In particular, the quantum oscillations observed in resistance are called Shubnikov-de Haas (SdH) oscillations, and those in magnetization (or magnetic susceptibility) are called de Haas-van Alphen (dHvA) oscillations. Both oscillatory phenomena were first observed in the Bi metal and were published in 1930 [2, 3].

2.3 Frequency of Quantum Oscillation

We examine the relation between the frequency and the Fermi surface. Let A_F be the cross section of the 2D (cylindrical) Fermi surface or the cross section at $k_Z = 0$ of the 3D (spherical) Fermi surface in Fig. 2.6a. In the free electron system, the circular cross section is given by the Fermi wave number $A_F = \pi k_F^2$, and thus the Fermi energy is written as

$$E_F = \frac{\hbar^2 k_F^2}{2m_c} = \frac{\hbar^2}{2\pi m_c}A_F. \tag{2.58}$$

The number of the Landau levels $N(\gg 1)$ below E_F is inversely proportional to the magnetic field,

$$N = \frac{E_F}{\hbar\omega_c} = \frac{\hbar}{2\pi e} A_F \frac{1}{B}. \tag{2.59}$$

As the field increases, N decreases one by one. The magnetic field range $\Delta(1/B)$ that reduces N by one $\Delta N(= 1)$ can be defined as the frequency F_Q of the quantum oscillation,

$$F_Q = \frac{\Delta N}{\Delta(1/B)} = \frac{\hbar}{2\pi e} A_F, \; \frac{F_Q}{B} = \frac{E_F}{\hbar\omega_c}. \tag{2.60}$$

The frequency of the quantum oscillation is proportional to the cross section A_F, which is called the Onsager relation. Since the quantum oscillation is periodic with the inverse magnetic field, as already seen in Figs. 2.10 and 2.11, the frequency is expressed in the unit of the magnetic field, Tesla (T). Substituting the physical constants into Eq. (2.60), we obtain

$$F_Q[\mathrm{T}] = 1.05 \times 10^2 A_F\left[\mathrm{nm}^{-2}\right]. \tag{2.61}$$

Since the above discussion assumes energy bands in the free electron system one might think that the periodic oscillation with the inverse magnetic field is a special case. However, the 1/*B* periodicity is derived for more general cases.

As long as the Landau level E_N is a function of the product of the Landau index and the magnetic field, $E_N = f(NB)$, the 1/*B* periodic oscillation with the frequency given in Eq. (2.60) should emerge. For instance, the Landau levels of the Dirac cone energy band (Appendix 2.1) are given by $E_{N\pm} = \pm v_\mathrm{F}\sqrt{2e\hbar NB}$ in Eq. (2.80). This is clearly a function of *NB*, leading to a 1/*B* periodic oscillation.

For a general discussion of the periodicity, we expand the energy up to the second order near the Fermi wave number k_F,

$$E(k) = E(k_F) + \left.\frac{\partial E}{\partial k}\right|_{k_F}(k - k_F) + \frac{1}{2}\left.\frac{\partial^2 E}{\partial k^2}\right|_{k_F}(k - k_F)^2. \tag{2.62}$$

In the magnetic field $B \| z$, the electronic states are Landau-quantized in the $k_x k_y$ plane. The deviation of the wave number δk from k_F is written as

$$(k - k_F) = \delta k = \pm\sqrt{\delta k_x^2 + \delta k_y^2}. \tag{2.63}$$

Rewriting $\partial E/\partial k|_{k_F} = \hbar v_F$, the band in Eq. (2.62) is given by $E(k) = E(k_F) + \hbar v_F \delta k$ up to the first order (linear) term. This is equivalent to Eq. (2.80) and thus the Landau level E_N is a function of *NB*, $E_N = f(NB)$, as seen later. If the second order (parabolic) term in Eq. (2.62) is dominant, we obtain $E(k) \propto \delta k^2$, equivalent to the free electron case discussed in Sect. 2.1. Either case results in a 1/*B* periodic oscillation.

Since the frequency of the quantum oscillation satisfies $F_Q \gg B(N \gg 1)$ in most conductors, the energy dispersion in a narrow energy range near E_F determines the periodicity of the quantum oscillations. Since the band dispersion is generally approximated as $E(k) \propto k$ near E_F for $F_Q \gg B$; the first order term in Eq. (2.62) is dominant, the 1/B periodic quantum oscillation is a common feature.

Figure 2.12a shows the quantum oscillation observed in the magnetic torque for the quasi-2D organic conductor κ-(BEDT-TTF)$_2$Cu(NCS)$_2$ in magnetic fields perpendicular to the conduction plane. The torque is given by $\boldsymbol{\tau}_{\text{torque}} = \boldsymbol{M} \times \boldsymbol{B}$, as will be seen in Chap. 3. The schematic structure of this organic conductor is already shown in Fig. 2.8b. A closer look at the structure of this conductor shows that the neighboring BEDT-TTF molecules are dimerized, as shown in Fig. 2.12b. The overlapping molecular orbitals form bonding and antibonding orbitals. The bonding orbital occupied by two electrons does not contribute to the conduction. The antibonding orbital occupied by one electron overlaps with that of the neighboring dimerized molecules in the *ac* plane and forms a 2D conduction band in the BEDT-TTF molecular layer. Since the antibonding orbitals are localized in the BEDT-TTF dimer, the energy band is well described in the tight binding approximation. The band dispersion is not necessarily given by $E(k) \propto k^2$ at E_F; the Landau levels are not equally spaced in a wide energy range. However, the 1/B periodic oscillation is evident in the wide magnetic field range as shown in Fig. 2.12a.

From the relation $\boldsymbol{\tau}_{\text{torque}} = \boldsymbol{M} \times \boldsymbol{B}$, the quantum oscillation observed in the magnetic torque measurement is often called d*HvA* oscillation. From the wavelength of the oscillation $\Delta(1/B) = 0.0016[1/\text{T}]$, the frequency is calculated as $F_Q = 1/\Delta(1/B) = 625$ T from Eq. (2.60). The experiments are performed at a finite temperature and the electrons are scattered due to the inhomogeneity of the sample. Due to these effects, the sawtooth wave of the magnetization depicted in Fig. 2.11d is smoothed and a sinusoidal waveform is observed. The details of the finite temperature and scattering effects are discussed in Chap. 3.

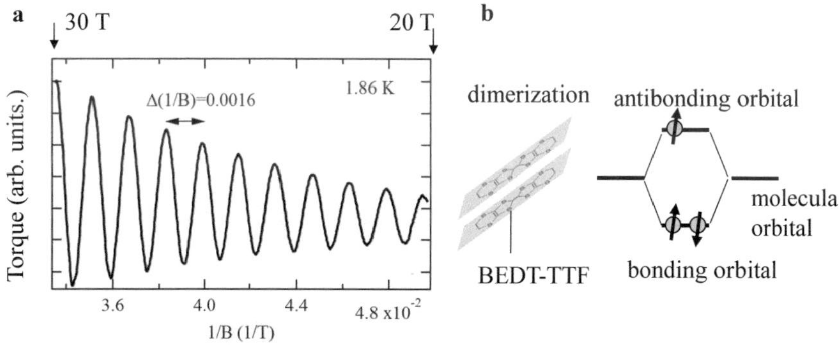

Fig. 2.12 **a** Quantum oscillation observed in magnetic torque for quasi-2D organic conductor κ-(BEDT-TTF)$_2$Cu(NCS)$_2$, **b** Dimerized BEDT-TTF molecules, and bonding and antibonding orbitals formed by the dimerized molecules

2.4 Quantum Oscillation and Fermi Surface Structure

We have seen that the frequency of the quantum oscillations is proportional to the cross section of the Fermi surface. We will show below that the observed frequency corresponds to the extremal cross section of the Fermi surface. We return to the thermodynamic grand potential of the 3D electron system shown in Fig. 2.11c. Since the quantum oscillation has the wavelength $1/F_Q = \Delta(1/B)$, the functional form of the quantum oscillation can be expanded by Fourier series, $\cos(2\pi p F_Q/B), p = 1, 2, 3, \ldots,$ Taking into account of the spin degree of freedom, $\delta\widetilde{\Omega}$ in Eq. (2.57), except for the diamagnetic energy, is written as

$$\delta\widetilde{\Omega}(B) = \delta k \frac{e^2}{4\pi^2 m_c} \sum_{p=1}^{\infty} \frac{B^2}{\pi^2 p^2} \cos\left[2\pi \mathrm{p}\left(\frac{F_Q}{B} - \frac{1}{2}\right)\right]. \tag{2.64}$$

This is the quantum oscillation for the 2D electron system connected to a reservoir, which is derived in Appendix 2.3.

Assuming the spherical Fermi surface of the free electron system in Fig. 2.6a, we take $F_Q(k_z)$ as the frequency corresponding to the cross section in the k_z plane for $B \| k_z$. Since the cross section has a maximum at $k_z = 0$, the frequency can be expanded around $k_z = 0$ as follows,

$$F_Q(k_z) = F_Q(0) + \frac{1}{2} F_Q^{''} k_z^2, \tag{2.65}$$

where $F_Q^{''} = \partial^2 F_Q(k_z)/\partial k_z^2 < 0$. The total thermodynamic grand potential is obtained by integrating $\delta\widetilde{\Omega}$ in Eq. (2.64) over the Fermi surface. What we need to calculate is

$$I_{\text{osc}} = \int \mathrm{d}k_z \cos\left[2\pi p\left(\frac{F_Q(k_z)}{B} - \frac{1}{2}\right)\right]. \tag{2.66}$$

Under normal conditions, the cosine function oscillates rapidly as a function of k_z and thus we extend the integration range to infinity $\pm\infty$. As long as this approximation holds, the following argument can be applied to any Fermi surface structure. Substituting Eq. (2.65) into Eq. (2.66), we obtain

$$I_{\text{osc}} = \sqrt{\frac{B}{p\left|F_Q^{''}\right|}} \cos\left[2\pi p\left(\frac{F_Q(0)}{B} - \frac{1}{2}\right) \pm \frac{\pi}{4}\right]. \tag{2.67}$$

This equation is derived in Appendix 2.4. We note that the phase $-\pi/4$ ($\pi/4$) is added for a maximum (minimum) cross section. Quantum oscillations from non-extremal cross sections of the Fermi surface cancel each other out and are therefore not observed. Only from the extremal cross sections, the quantum oscillation is observed. We can conclude that "quantum oscillations arise from the

Landau quantization of the electronic states in the extremal cross sections of the Fermi surface perpendicular to the magnetic field."

From the relation $F_Q = (\hbar/2\pi e)A_F$ in Eq. (2.60), we can define the curvature factor of the Fermi surface cross section,

$$A_F'' = \left|\frac{\partial^2 A_F(k_z)}{\partial k_z^2}\right| = \frac{2\pi e}{\hbar}\left|F_Q''\right|. \tag{2.68}$$

From Eq. (2.67), the oscillation amplitude is written as $I_{\text{osc}} \propto 1/\sqrt{A_F''}$; the smaller A_F'', the larger the oscillation amplitude. This is intuitive and easy to understand. As seen in Fig. 2.13, even for the same extremal cross section A_F, more electronic states near the extremal cross section can contribute to quantum oscillations for the smaller curvature factor A_F''; larger oscillation amplitude for smaller A_F''. The curvature factor for a quasi-2D Fermi surface is derived in Appendix 2.5. From Eqs. (2.67) and (2.68), the quantum oscillation of the thermodynamic grand potential is calculated as

$$\begin{aligned}\tilde{\Omega}(B) &= \int \mathrm{d}\widetilde{\Omega} = \int \frac{\delta\widetilde{\Omega}}{\delta k}\mathrm{d}k_z = \left(\frac{e}{2\pi\hbar}\right)^{3/2}\frac{e\hbar}{m_c}\frac{1}{\left(A_F''\right)^{1/2}} \\ &\times \sum_{p=1}^{\infty}\frac{B^{5/2}}{\pi^2 p^{5/2}}\cos\left[2\pi p\left(\frac{F_Q(0)}{B}-\frac{1}{2}\right)\pm\frac{\pi}{4}\right].\end{aligned} \tag{2.69}$$

This is the quantum oscillation for the 3D electron system. The amplitude of the oscillation increases rapidly with magnetic field, $\tilde{\Omega} \propto B^{5/2}$.

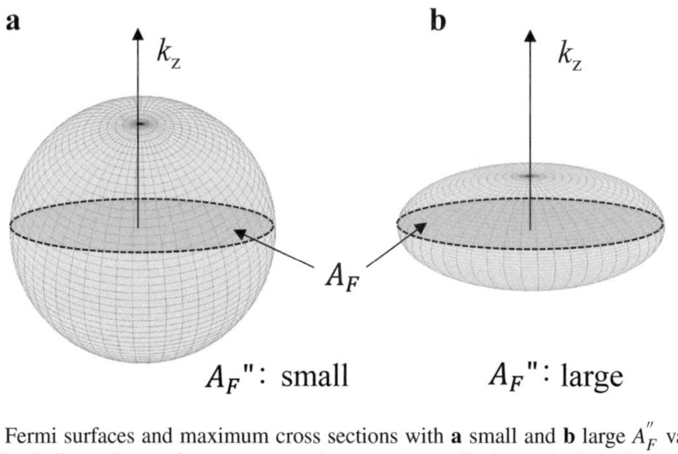

Fig. 2.13 Fermi surfaces and maximum cross sections with **a** small and **b** large A_F'' values. The dotted circles indicate the maximum cross sections A_F perpendicular to the k_z-axis

For comparison, we show the oscillatory part of the internal energy in Eq. (2.54), except for the diamagnetic energy,

$$\tilde{E}(B) = -\frac{e^2}{2\pi m_c}\sum_{p=1}^{\infty}\frac{B^2}{\pi^2 p^2}\cos\left[2\pi p\frac{F_Q}{B}\right]. \tag{2.70}$$

This is also derived in Appendix 2.3. The oscillation has no phase factor unlike Eq. (2.64). Equation (2.70) is applicable to the 2D electron system completely isolated from a reservoir. If the 2D electron system is in contact with a reservoir or has multiple Fermi surfaces, Eq. (2.64) is applicable. For $p = 1$, Eq. (2.70) is written as

$$\tilde{E}(B) \propto -\cos\left[2\pi\frac{F_Q}{B}\right] = \cos\left[2\pi\left(\frac{F_Q}{B} - \frac{1}{2}\right)\right].$$

Note that this is equivalent to Eq. (2.64), no difference in the oscillatory part between the two cases for $p = 1$.

In general, the Fermi surface structures of 3D materials are complicated and vary from material to material. The complicated Fermi surfaces will have many extremal cross sections. When there are multiple extremal cross sections perpendicular to the magnetic field, the quantum oscillations summed over all the extremal cross sections are observed. By observing the quantum oscillations at various magnetic field orientations, we can construct the overall picture of the Fermi surface.

We see the angular dependence of the frequency for two simple examples, ellipsoidal and cylindrical (perfect 2D) Fermi surfaces as depicted in Fig. 2.14a. Let F_{Q0} be the frequency observed for these Fermi surfaces in a magnetic field parallel to the k_z-axis. The ellipsoidal Fermi surface has the maximum cross section in any magnetic field direction. As the magnetic field is tilted from the k_z-axis to the k_x-axis, the frequency $F_Q(\theta)$ increases as shown in Fig. 2.14b. For the cylindrical Fermi surface, all the cross sections have the same area; all the electronic states contribute to the quantum oscillation. The frequency increases as the magnetic field is tilted from the k_z-axis to the k_x-axis, and then diverges in the magnetic field parallel to the k_x-axis, following $F_Q(\theta) = F_{Q0}/\cos(\theta)$. At semiconductor interfaces such as GaAs [4, 5] or in layered transition metal dichalcogenides [6, 7], an ideal 2D electron system is formed, which actually exhibits $F_Q(\theta) = F_{Q0}/\cos(\theta)$. Due to the extremely high mobility, integer and fractional quantum Hall effects have been found at high magnetic fields, whose electronic states are of great interest.

2.5 Observation Conditions of Quantum Oscillation

In this section, we discuss the experimental conditions under which quantum oscillations can be observed. First, the Fermi surface must be closed, where the electronic states arc Landau-quantized in magnetic fields. Second, the closed Fermi

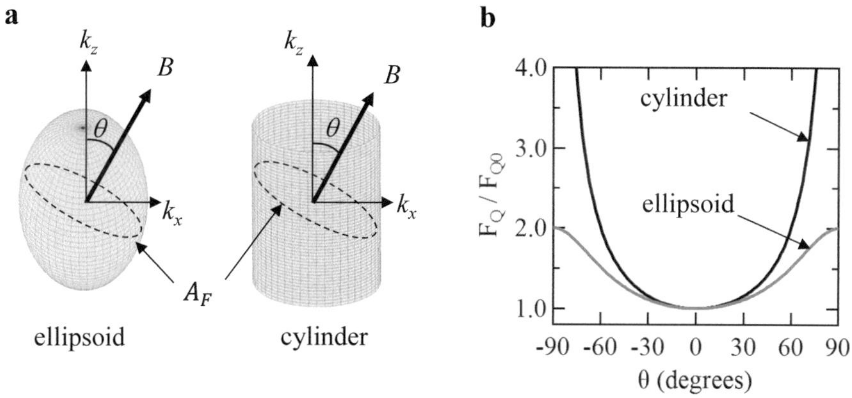

Fig. 2.14 **a** Ellipsoidal and cylindrical Fermi surfaces. The dotted ellipses indicate the cross sections perpendicular to the magnetic field. **b** Frequencies of quantum oscillations as a function of the magnetic field angle for the ellipsoidal and cylindrical Fermi surfaces shown in **a**

surface must have an extremal cross section perpendicular to the magnetic field. In addition to the above, low temperatures and strong magnetic fields are required to clearly observe quantum oscillations. The reasons are explained qualitatively below.

2.5.1 Effect of Finite Temperature

At finite temperatures, the Fermi edge, an abrupt change of the Fermi distribution function $f(E)$ at the Fermi level, is smeared due to thermal excitation in the range of $k_B T$, as depicted in Fig. 2.15a. Since the Landau level spacing is $\hbar\omega_c$, a sufficiently low temperature as compared to $\hbar\omega_c$ is required to observe the quantum oscillation,

$$k_B T \ll \hbar\omega_c. \tag{2.71}$$

If the effective mass of the electrons m_c is equal to the free electron mass m_0, $\hbar\omega_c/k_B \approx 13.4$ K at $B = 10$ T. Therefore, we need low temperatures $T \ll 13.4$ K to observe the quantum oscillation at ~10 T. If $m_c \gg m_0$, we may need a special cooling system such as a dilution refrigerator to cool the sample. By contrast, since $m_c \ll m_0$ for some semimetals, quantum oscillations can be observed even at a few tens of Kelvins.

2.5.2 Effect of Scattering

For Landau quantization of the electronic states in magnetic fields, the Bohr quantization condition in Eq. (2.7) needs to be satisfied. If the electrons are scattered

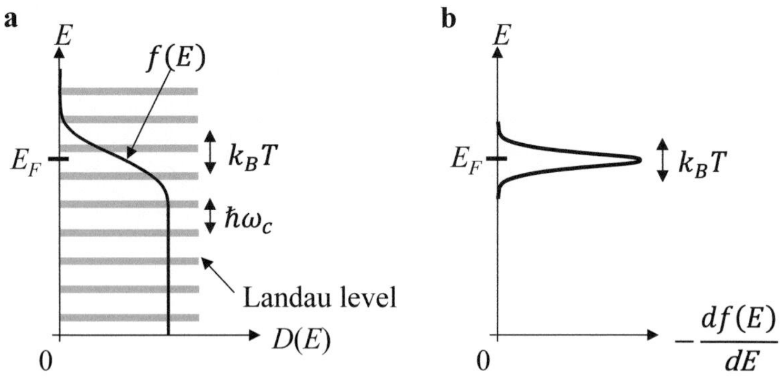

Fig. 2.15 **a** Landau levels and Fermi distribution function $f(E)$, **b** $-\mathrm{d}f(E)/\mathrm{d}E$

with other electrons, phonons, or impurities, the electrons cannot complete the cyclotron motion, as illustrated in Fig. 2.16a. Therefore, the scattering time of the electrons must be sufficiently longer than the period of cyclotron motion $1/\omega_c$ for the observation of the quantum oscillations,

$$\tau \gg \frac{1}{\omega_c} \quad \text{or} \quad \frac{\hbar}{\tau} \ll \hbar\omega_c. \tag{2.72}$$

From the uncertainty relation for time and energy $\Delta E \Delta t \approx \hbar$, the quantity $\hbar/\tau$ in the second inequality corresponds to the uncertainty of the energy ΔE due to the life (scattering) time $\tau \approx \Delta t$. It means that the discrete Landau levels would have a width of $\hbar/\tau$ due to the scattering (Fig. 2.16b). The width $\hbar/\tau$ needs to

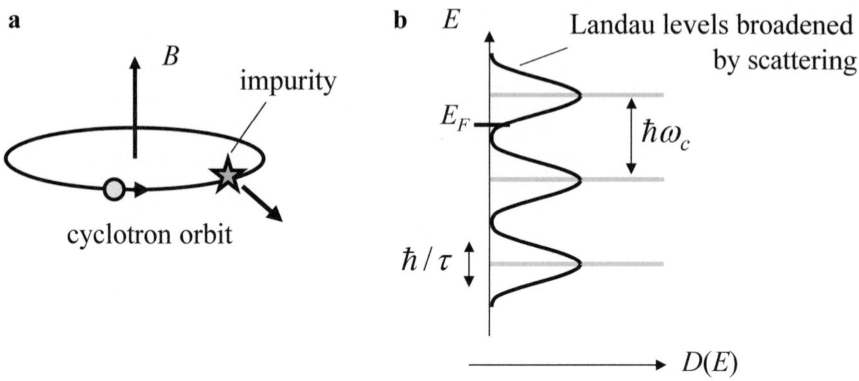

Fig. 2.16 **a** Cyclotron orbit and scattering with impurity, **b** Landau levels without scattering and those broadened by scattering

be sufficiently smaller than the Landau level spacing $\hbar\omega_c$ to observe the quantum oscillation.

In general, the resistance of a conductor such as a metal decreases with decreasing temperature. This is because electron–phonon and electron–electron scattering are more suppressed at lower temperatures. On the other hand, scattering by irregular potentials such as impurities or lattice defects results in a constant resistance at low temperatures. This is the origin of the residual resistivity at ~0 K, and for this reason, the residual resistivity is a good indicator of the sample quality. Since the absolute value of the residual resistivity also depends on the carrier concentration and the Fermi surface structure, the ratio $R_{\mathrm{RT}}/R_{\mathrm{res}}$, the room temperature resistance R_{RT} divided by the residual resistance R_{res} at a low temperature, may be a better indicator. It is called the residual resistance ratio (RRR). The higher the RRR, the less impurities and lattice defects are present. It is often found that the RRR exceeds 1000 for pure metals.

Figure 2.17a presents the temperature dependence of the resistance for an organic conductor $(\mathrm{BEDT\text{-}TTF})_2\mathrm{Br(DIA)}$. This conductor has a layered structure, where the BEDT-TTF molecules form conducting layers, similar to κ-$(\mathrm{BEDT\text{-}TTF})_2\mathrm{Cu(NCS)}_2$ shown in Fig. 2.8b. This sample has a large RRR of ~380, showing that the sample is of very good quality. The quantum oscillations (SdH oscillations) are clearly observed in the resistance at high magnetic fields as shown in Fig. 2.17b. In organic conductors, quantum oscillations have often been observed for RRR > 100. However, it should be noted that this is merely an approximate threshold. When the conductor undergoes a phase transition, where the carrier concentration, shape and size of the Fermi surface, etc. change, the RRR cannot be relied upon as the indicator.

2.6 Summary

In this chapter we have seen that the electrons in free space undergo cyclotron motion in magnetic fields and the electronic states are Landau-quantized due to the Bohr quantization condition. The Landau levels in free space are evenly spaced with $\hbar\omega_c$. The Landau quantization leads to periodic oscillations in the thermodynamic potential or the internal energy as a function of the inverse magnetic field. In real conductors the energy bands generally form a complex Fermi surface structure depending on the lattice potential. Only when the electrons can form closed orbits on the Fermi surfaces, the electronic states are Landau-quantized. In lattice potentials, it is difficult to solve the Schrödinger equation analytically and thus the Landau levels are not explicitly expressed. Even in such a case, since the energy dispersion is approximated as linear, $E(k) \propto k$ in a narrow energy range near E_F, the periodic quantum oscillation with the inverse magnetic field is a common feature. For 3D Fermi surfaces, the quantum oscillation has a frequency proportional to the extremal cross section of the Fermi surface, perpendicular to the magnetic field. When there exist multiple extremal cross sections, the quantum oscillations arise from all of them. The quantum oscillations are observable if the temperature

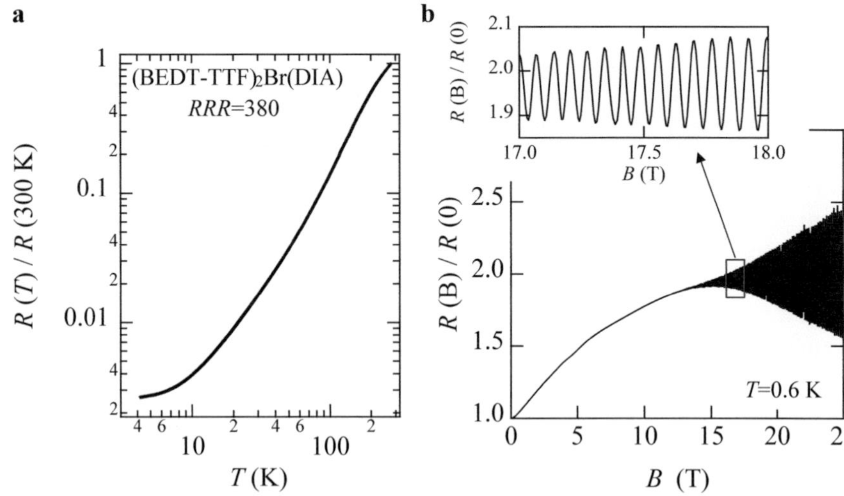

Fig. 2.17 **a** Temperature dependence of the interlayer resistance for an organic conductor (BEDT-TTF)$_2$Br(DIA), **b** Magnetic field dependence of the resistance. SdH oscillations are clearly observed. Inset: close-up of the SdH oscillation at ~17.5 T

is low enough $k_B T \ll \hbar\omega_c$ and the sample quality or the magnetic field is high enough $\omega_c \tau \gg 1$.

Appendix 2.1

Graphene has the Dirac cone energy bands as shown in Appendix 1.7. Here we see how the electronic state is Landau-quantized in a magnetic field. As shown in Eq. (1.53), the energy bands of the graphene have the linear dispersion at the K and K' points. The simplest model Hamiltonian describing a linear dispersion band is expressed as

$$\mathcal{H} = \hbar v_{\mathrm{F}} \begin{pmatrix} 0 & k_x + \mathrm{i}k_y \\ k_x - \mathrm{i}k_y & 0 \end{pmatrix}. \tag{2.73}$$

As given in Sect. 1.2.3, the Schrödinger equation in the matrix representation is written by

$$\hbar v_{\mathrm{F}} \begin{pmatrix} 0 & k_x + \mathrm{i}k_y \\ k_x - \mathrm{i}k_y & 0 \end{pmatrix} \begin{pmatrix} C_A \\ C_B \end{pmatrix} = E(k) \begin{pmatrix} C_A \\ C_B \end{pmatrix}.$$

For simplicity, we take $\varepsilon = 0$. By solving the secular equation,

$$\begin{vmatrix} -E(k) & \hbar v_{\mathrm{F}}(k_x + \mathrm{i}k_y) \\ \hbar v_{\mathrm{F}}(k_x - \mathrm{i}k_y) & -E(k) \end{vmatrix} = 0, \tag{2.74}$$

we obtain the eigenvalues,

$$E_{\pm}(k) = \pm\hbar v_{\mathrm{F}} k, k^2 = k_x^2 + k_y^2. \tag{2.75}$$

This is equivalent to the energy dispersion of the graphene at the K and K' points. In this case, the Dirac cone is formed at the Γ point as shown in Fig. 2.18a. The velocity of the electrons is given by $v = (1/\hbar)\partial E_{\pm}(k)/\partial k = \pm v_{\mathrm{F}}$. For the Landau gauge $\boldsymbol{A}(0, Bx, 0)$ in a magnetic field $\boldsymbol{B}(0, 0, B)$, we have $\hbar k_x = -i\hbar\partial\big/\partial x, \hbar k_y = i\hbar\partial\big/\partial y + eBx,$, as given in Eq. (2.14). Here we define the operators a and a^+, respectively,

$$a = \frac{l_B}{\sqrt{2}}\left(k_x - \mathrm{i}k_y\right), a^+ = \frac{l_B}{\sqrt{2}}\left(k_x + \mathrm{i}k_y\right). \tag{2.76}$$

From $\left[k_x, k_y\right] = -ieB/\hbar = -\mathrm{i}/l_B^2$, we can confirm the commutation relation,

$$\left[a, a^+\right] = il_B^2\left[k_x, k_y\right] = 1. \tag{2.77}$$

This relation shows that a and a^+ are the annihilation and creation operators, respectively. Substituting Eq. (2.76) into Eq. (2.73), we have the Hamiltonian in a magnetic field,

$$\mathcal{H} = \frac{\sqrt{2}\hbar v_{\mathrm{F}}}{l_B}\begin{pmatrix} 0 & a^+ \\ a & 0 \end{pmatrix}. \tag{2.78}$$

We specify the eigenstate $|N\rangle$ with the Landau index N. The eigenvalue E can be obtained from the secular equation,

$$\begin{vmatrix} -E & \sqrt{2}\hbar v_{\mathrm{F}} a^+/l_B \\ \sqrt{2}\hbar v_{\mathrm{F}} a/l_B & -E \end{vmatrix} = 0. \tag{2.79}$$

By using the relation $a^+a|N\rangle = N|N\rangle$, we obtain the eigenvalues (Landau levels),

$$E_{N\pm} = \pm v_{\mathrm{F}}\sqrt{2e\hbar NB}, N = 0, 1, 2, \ldots \tag{2.80}$$

which are depicted in in Fig. 2.18b. As we have already seen, all the k states are condensed into the Landau levels. The two states with $E_{N\pm}$ are degenerate for $N = 0$. In contrast to the free space case in Eq. (2.29), the eigenstate has no zero-point energy and the Landau levels are not equally spaced. As N increases, the energy spacing between the Nth and $(N - 1)$th levels asymptotically approaches

$$\Delta E_N = v_{\mathrm{F}}\sqrt{2e\hbar NB} - v_{\mathrm{F}}\sqrt{2e\hbar(N-1)B} \rightarrow v_{\mathrm{F}}\sqrt{e\hbar B/2N}. \tag{2.81}$$

The energy spacing decreases with increasing N.

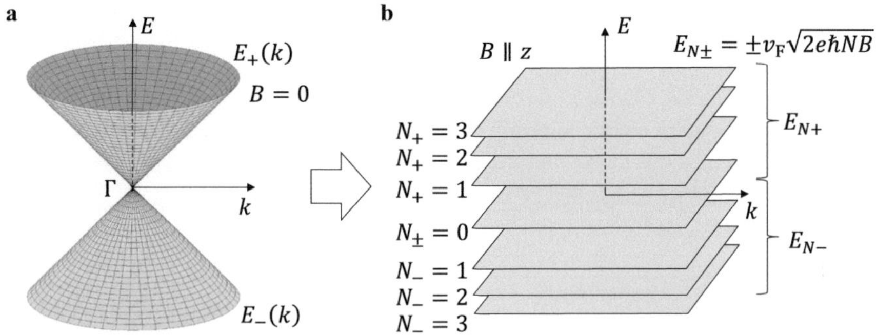

Fig. 2.18 **a** Dirac cone energy bands at zero magnetic field, **b** Landau levels in magnetic field ($B\|z$)

Appendix 2.2

The density of states in Eq. (2.42) can be driven as follows. The number of states n per unit volume is obtained by summing the degeneracy of the Landau levels $1/2\pi l_B^2$,

$$n = \sum_{N,k_z} \frac{1}{2\pi l_B^2}. \tag{2.82}$$

The sum of k_z can be replaced by the integral,

$$\sum_{k_z} \to \frac{2}{\pi} \int_0^{k_F} dk_z \to \frac{2}{\pi} \int_0^{E_F} \frac{m_c}{\hbar^2 k_z} dE. \tag{2.83}$$

The spin degree of freedom is taken into account in Eq. (2.83). Substituting Eq. (2.41) into Eq. (2.83), we obtain

$$n = \int_0^{E_F} \frac{1}{4\pi^2} \left(\frac{2m_c}{\hbar^2}\right)^{3/2} \hbar\omega_c \sum_N \left[E - \left(N + \frac{1}{2}\right)\hbar\omega_c\right]^{-1/2} dE \equiv \int_0^{E_F} D(E) dE, \tag{2.84}$$

and thus the density of states $D(E)$ is given by

$$D(E) = \frac{1}{4\pi^2} \left(\frac{2m_c}{\hbar^2}\right)^{3/2} \hbar\omega_c \sum_N \left[E - \left(N + \frac{1}{2}\right)\hbar\omega_c\right]^{-1/2}.$$

Appendix 2.3

We derive Eq. (2.64) from the oscillatory part of the thermodynamic grand potential in Eq. (2.55),

$$\widetilde{\delta\Omega}'(B)\frac{2\pi}{\delta k} = \frac{D(E_F)}{2}(N\hbar\omega_c - E_F)^2. \tag{2.85}$$

Taking $x = E_F/\hbar\omega_c = F_Q/B$ in Eq. (2.85), we obtain

$$\frac{\widetilde{\delta\Omega}'(B)}{\delta k}\frac{1}{(\hbar\omega_c)^2}\frac{4\pi}{D(E_F)} = (N - x)^2. \tag{2.86}$$

Further taking $x = y/2\pi + N - 1/2$, Eq. (2.86) is rewritten as

$$\frac{\widetilde{\delta\Omega}'(B)}{\delta k}\frac{1}{(\hbar\omega_c)^2}\frac{16\pi^3}{D(E_F)} = (\pi - y)^2, \tag{2.87}$$

where the range of y, corresponding to Eq. (2.56), is $0 < y \leq 2\pi$. This is a periodic function with the period of 2π and its Fourier series expansion is given by

$$(\pi - y)^2 \rightarrow 4\sum_{p=1}^{\infty}\frac{\cos(py)}{p^2} + \frac{\pi^2}{3}. \tag{2.88}$$

From $D(E_F) = m_c/\pi\hbar^2$ in Eq. (1.8), including the spin degree of freedom, Eq. (2.85) is written as

$$\frac{\widetilde{\delta\Omega}'(B)}{\delta k} = \frac{e^2}{4\pi^2 m_c}\sum_{p=1}^{\infty}\frac{B^2}{\pi^2 p^2}\cos\left[2\pi p\left(\frac{F_Q}{B} - \frac{1}{2}\right)\right] + \frac{e^2B^2}{48\pi^2 m_c}. \tag{2.89}$$

The first term is the quantum oscillation given by Eq. (2.64) and the non-oscillatory second term $\left(\propto B^2\right)$ expresses the orbital diamagnetic energy. The orbital diamagnetic energy Ω_L per unit volume is calculated as

$$\Omega_L(B) = \int \delta\Omega_L = \int_{-k_F}^{k_F} dk\frac{e^2B^2}{48\pi^2 m_c} = \frac{e^2 k_F B^2}{24\pi^2 m_c}.$$

Therefore, we obtain the orbital diamagnetic moment M_L and its susceptibility χ_L, respectively,

$$M_L(B) = -\frac{\partial\Omega_L}{\partial B} = -\frac{e^2 k_F B}{12\pi^2 m_c},\ \chi_L = \frac{\partial M_L}{\partial B} = -\frac{e^2 k_F}{12\pi^2 m_c}. \tag{2.90}$$

The quantity χ_L is called Landau diamagnetism. For the free electrons with the spin $S = 1/2$ and the g-factor $g = 2$, the magnetic susceptibility (Pauli paramagnetism) is given by $\chi_{\text{Pauli}} = \mu_B^2 D(E_F)$, where μ_B is the Bohr magneton, $\mu_B = e\hbar/2m_c$. Using $D(E_F) = m_c k_F/\pi^2\hbar^2$ in Eq. (1.7) with the spin degree of freedom, we obtain

$$\chi_{\text{Pauli}} = e^2 k_F/4\pi^2 m_c,$$

and thus

$$\chi_L = -\chi_{\text{Pauli}}/3.$$

Next, we derive Eq. (2.70) from Eq. (2.54). Taking $E_F/\hbar\omega_c = F_Q/B = x$ and $x = y/2\pi + N$, Eq. (2.54) is written as

$$\tilde{E}(B)\frac{2}{D(E_F)(\hbar\omega_c)^2} = \frac{-1}{4\pi^2}(\pi - y)^2 + \frac{1}{4} \text{ for } 0 < y \le 2\pi. \tag{2.91}$$

Using the Fourier series expansion of Eq. (2.88), $\tilde{E}(B)$ is given by

$$\tilde{E}(B) = -\frac{e^2}{2\pi m_c}\sum_{p=1}^{\infty}\frac{B^2}{\pi^2 p^2}\cos\left[2\pi p\frac{F_Q}{B}\right] + \frac{e^2 B^2}{12\pi m_c}. \tag{2.92}$$

The first term represents the quantum oscillation and the second term $\left(\propto B^2\right)$ corresponds to the orbital diamagnetic energy.

Appendix 2.4

We derive Eq. (2.67) [1]. Substituting Eq. (2.65) into Eq. (2.66), we have

$$\begin{aligned} I_{\text{osc}} &= \int \mathrm{d}k_z \cos\left[2\pi p\left(\frac{F_Q(k_z)}{B} - \frac{1}{2}\right)\right] \\ &= \int \mathrm{d}k_z \cos\left[2\pi p\left(\frac{F_Q(0)}{B} - \frac{1}{2}\right) + \pi p\frac{F_Q'' k_z^2}{B}\right] \\ &= \cos X \int \mathrm{d}k_z \cos\left(\pi p\frac{F_Q'' k_z^2}{B}\right) - \sin X \int \mathrm{d}k_z \sin\left(\pi p\frac{F_Q'' k_z^2}{B}\right), \end{aligned} \tag{2.93}$$

where $X = 2\pi p\left[F_Q(0)/B - 1/2\right]$. Assuming a spherical Fermi surface as shown in Fig. 2.6a, we obtain $\left|F_Q''\right| = \hbar/e$ (see Appendix 2.5). The integration of k_z ranges from $-\pi/c$ to π/c, where c is the lattice constant. Assuming $k_z = \pi/c$, we have

$$\frac{\left|F_Q''\right| k_z^2}{B} \approx \frac{\Phi_0}{Bc^2}. \tag{2.94}$$

Substituting typical values of conductors, $c = 1$ nm and $B = 5$ T, we have $\Phi_0/Bc^2 \approx 10^3$. Since $\cos\left(\pi p F_Q'' k_z^2/B\right)$ and $\sin\left(\pi p F_Q'' k_z^2/B\right)$ oscillate rapidly in the integration range, we can replace the integration in Eq. (2.93) by $\int_{-\infty}^{\infty} \mathrm{d}k_z$. Taking $Y^2 = \pi p \left|F_Q''\right| k_z^2/B$, we have

$$\int_{-\infty}^{\infty} \mathrm{d}k_z \cos\left(\pi p \frac{F_Q'' k_z^2}{B}\right) = 2\sqrt{\frac{B}{\pi p \left|F_Q''\right|}} \int_0^{\infty} \mathrm{d}Y \cos\left(Y^2\right) = \sqrt{\frac{B}{2p\left|F_Q''\right|}}. \tag{2.95}$$

Note that F_Q'' is an even function with respect to k_z and $\int_0^{\infty} \mathrm{d}Y \cos\left(Y^2\right) = \sqrt{\pi/8}$. Using another relation, $\int_0^{\infty} \mathrm{d}Y \sin\left(Y^2\right) = \sqrt{\pi/8}$, we obtain

$$\int_{-\infty}^{\infty} \mathrm{d}k_z \sin\left(\pi p \frac{F_Q'' k_z^2}{B}\right) = \mathrm{sgn}\left(F_Q''\right)\sqrt{\frac{B}{2p\left|F_Q''\right|}}, \tag{2.96}$$

where sgn(Z) indicates the sign of the value Z. Using $\cos(\pi/4) = \sin(\pi/4) = 1/\sqrt{2}$, Eq. (2.93) is written as,

$$\begin{aligned} I_{\mathrm{osc}} &= \sqrt{\frac{B}{p\left|F_Q''\right|}} \left(\frac{1}{\sqrt{2}} \cos X - \mathrm{sgn}\left(F_Q''\right) \frac{1}{\sqrt{2}} \sin X\right) \\ &= \sqrt{\frac{B}{p\left|F_Q''\right|}} \left[\cos\left(\frac{\pi}{4}\right) \cos X - \mathrm{sgn}\left(F_Q''\right) \sin\left(\frac{\pi}{4}\right) \sin X\right] \\ &= \sqrt{\frac{B}{p\left|F_Q''\right|}} \cos\left[2\pi p\left(\frac{F_Q(0)}{B} - \frac{1}{2}\right) \pm \frac{\pi}{4}\right]. \end{aligned} \tag{2.97}$$

The phase factor is $-\pi/4$ for the maximum cross section $\left(F_Q'' < 0\right)$ and is $\pi/4$ for the minimum cross section $\left(F_Q'' > 0\right)$.

Appendix 2.5

We derive the curvature factor for a quasi-2D Fermi surface. We assume free electron layers in the xy plane and a tight binding energy band in the z-axis,

$$E(\boldsymbol{k}) = \frac{\hbar^2}{2m_c}\left(k_x^2 + k_y^2\right) - 2t_c\cos(k_z c). \tag{2.98}$$

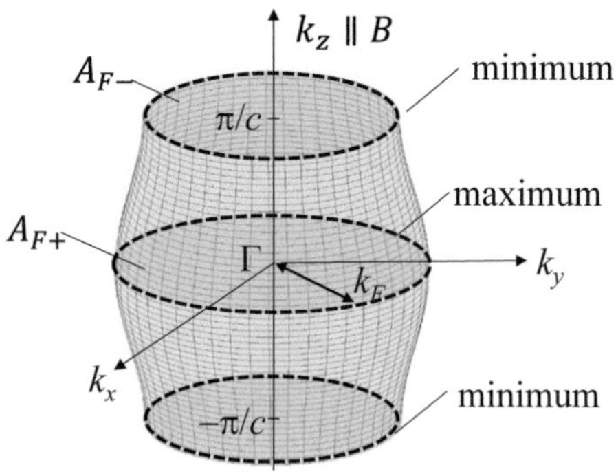

Fig. 2.19 Quasi-2D Fermi surface, and its minimum and maximum cross sections perpendicular to the magnetic field ($B \| k_z$)

If the Fermi energy is predominantly determined by the first term, $E_F = \hbar^2 k_F^2/2m_c \gg 2t_c$, this band forms a quasi-2D Fermi surface, cylindrical Fermi surface corrugated in the k_z-axis as depicted in Fig. 2.19. Since the Fermi wave number k_F is defined as the radius of the cross section, the cross section is written as $A_F = \pi k_F^2 = (2m_c\pi/\hbar^2)[E_F + 2t_c\cos(k_z c)]$, and thus

$$A_F^{''} = \left|\frac{\partial^2 A_F(k_z)}{\partial k_z^2}\right| = \left|\frac{4m_c\pi}{\hbar^2}c^2 t_c \cos(k_z c)\right|. \tag{2.99}$$

For $k_z = 0, \pm\pi/c$, we have $A_F^{''} = 4m_c\pi c^2 t_c/\hbar^2$. From the relation $\tilde{\Omega} \propto 1/\sqrt{A_F''}$, the oscillation amplitude increases as t_c decreases. The maximum and minimum cross sections are respectively,

$$A_{F\pm} = \frac{2m_c\pi}{\hbar^2}(E_F \pm 2t_c). \tag{2.100}$$

Two different quantum oscillations with the frequencies, $F_{Q\pm} = A_{F\pm}(\hbar/2\pi e)$ can be observed for this Fermi surface. For a highly 2D case such as $2t_c < \hbar\omega_c$, it is impossible to observe $F_{Q\pm}$ separately. For the perfect 2D Fermi surface ($t_c = 0$), note that $A_F^{''} = 0$ and Eq. (2.69) is not applicable. Instead, Eq. (2.64) should be used. For spherical Fermi surface in Fig. 2.6a, the energy band is given by $E(\boldsymbol{k}) = (\hbar^2/2m_c)\left(k_x^2 + k_y^2 + k_z^2\right)$ and thus $A_F = \pi k_F^2 = (2m_c\pi/\hbar^2)E_F - \pi k_z^2$. It gives $A_F^{''} = 2\pi$ and $\left|F_Q''\right| = \hbar/e$.

References

1. Shoenberg, D.: Magnetic Oscillations in Metals, pp. 22–82. Cambridge Univ. Press (1984)
2. Schubnikow, L., de Haas, W.J.: A new phenomenon in the change of resistance in a magnetic field of single crystals of bismuth. Nature **126**, 500 (1930). https://doi.org/10.1038/126500a0
3. de Haas, W.J., van Alphen, P.M.: Proc. Acad. Sci. Amst. **33**, 1106–1118 (1930)

4. Yoshioka, D.: The Quantum Hall Effect. Springer (2002)
5. Chakraborty, T., Pietiläinen, P.: The Quantum Hall Effects, Integral and Fractional, Springer Series in Solid-State Sciences (1995)
6. Wang, H., Liu, F., Fu, W., Fang, Z., Zhoue, W.: Liu, Z/: Two-dimensional heterostructures: fabrication, characterization, and application. Nanoscale **6**, 12250–12272 (2014). https://doi.org/10.1039/c4nr03435j
7. Li, M.-Y., Chen, C.-H., Shi, Y., Li, L.-J.: Heterostructures based on two-dimensional layered materials and their potential applications. Mater. Today **19**, 322–335 (2016). https://doi.org/10.1016/j.mattod.2015.11.003

Quantum Oscillation and Related Phenomena

3

Abstract

This chapter starts with the Lifshitz–Kosevich formula, which describes quantum oscillations and their reduction factors due to the effects of finite temperature, scattering and spin. Although quantum oscillations can generally be observed only for closed Fermi surfaces, quantum oscillations can sometimes be observed even for open Fermi surfaces, where the interband tunneling effect, called magnetic breakdown, plays an essential role. When the oscillatory part of the magnetization is sufficiently large, the waveform of the quantum oscillation is strongly deformed, which gives rise to a variety of frequencies. In contrast to quantum oscillations, the resistance can oscillate with magnetic field for open orbits without Landau quantization, which arises from quantum interference of the electron wave function. In materials with characteristic energy bands, the electrons acquire a phase as they move along an orbit on the Fermi surface. This phase is called the geometric phase or Berry phase. We will also look at how the Berry phase appears in the quantum oscillations and how it is experimentally determined.

3.1 Lifshitz–Kosevich (L–K) Formula

3.1.1 Thermodynamic Grand Potential

The thermodynamic grand potential in magnetic fields is expressed as

$$\Omega = -\frac{k_B T}{2\pi}\int \mathrm{d}k\, D_N \sum_N \ln\left(1 + e^{\frac{\mu - E_N}{k_B T}}\right). \tag{3.1}$$

In Chap. 2, we obtained the quantum oscillation at 0 K without scattering and spin effects. By taking into account the effects of finite temperature, scattering,

S. Uji, *Fermiology*, Lecture Notes in Physics 1042,
https://doi.org/10.1007/978-981-95-0820-4_3

and spin, the quantum oscillation of the thermodynamic grand potential per unit volume [1] for a 3D electron system is expressed as

$$\tilde{\Omega}(B) = \left(\frac{e}{2\pi\hbar}\right)^{3/2}\left(\frac{e\hbar}{m_c}\right)\frac{B^{5/2}}{\pi^2\left(A_F''\right)^{1/2}} \times \sum_{p=1}^{\infty}\frac{1}{p^{5/2}}R_T R_D R_S \cos\left[2\pi p\left(\frac{F_Q}{B}-\frac{1}{2}\right)+\varphi_0\right]. \tag{3.2a}$$

This calculation was performed by Lifshitz and Kosevich, and the formula is called Lifshitz–Kosevich formula, or L–K formula simply. The cosine term is the oscillatory part and its phase factor is

$$\varphi_0 = \pi/4 \text{ for minimum cross section,}$$

$$\varphi_0 = -\pi/4 \text{ for maximum cross section.}$$

For comparison, we show the result for a 2D electron system connected to a reservoir,

$$\tilde{\Omega}_{2D}(B) = \frac{e}{4\pi^2\hbar}\left(\frac{e\hbar}{m_c}\right)\frac{B^2}{\pi^2}\sum_{p=1}^{\infty}\frac{1}{p^2}R_T R_D R_S \cos\left[2\pi p\left(\frac{F_Q}{B}-\frac{1}{2}\right)\right], \tag{3.2b}$$

where we take $\delta k = 1$ for simplicity. These formulae include three reduction factors, R_T, R_D, and R_S, which will be given later. Without these reduction factors, Eqs. (3.2a) and (3.2b) are equivalent to Eqs. (2.69) and (2.64), respectively. The curvature factor of the Fermi surface is given by Eq. (2.68), $A_F'' = \left|\partial^2 A_F/\partial k_{\parallel}^2\right|$, where $k_{\parallel}$ is the wave number in the magnetic field direction. The oscillation with F_Q is called the fundamental oscillation and those with pF_Q ($p = 2, 3, 4, \ldots$) are called its harmonics.

Here we roughly estimate the amplitude of the quantum oscillation. For a spherical Fermi surface of a 3D free electron system in Fig. 2.6a, we have $A_F'' = 2\pi$ (Appendix 2.5). Using $E_F = \hbar^2 A_F/2\pi m_c = e\hbar F_Q/m_c$, and $n(E_F) = \left(2m_c E_F/\hbar^2\right)^{3/2}/\left(3\pi^2\right) = \left(2eF_Q/\hbar\right)^{3/2}/\left(3\pi^2\right)$, we obtain

$$\tilde{\Omega}(B) = \frac{3}{2^{7/2}\pi^2}n(E_F)E_F\left(\frac{B}{F_Q}\right)^{5/2} \times \sum_{p=1}^{\infty}\frac{1}{p^{5/2}}R_T R_D R_S \cos\left[2\pi p\left(\frac{F_Q}{B}-\frac{1}{2}\right)+\varphi_0\right]. \tag{3.3}$$

The total internal energy of the electrons is given by $E_{\text{total}} \approx n(E_F)E_F/4$. Assuming that all the reduction factors are unity, and typical values, $F_Q = 1000$ T and $B = 10$ T, we obtain $\tilde{\Omega}(B)/E_{\text{total}} \approx 10^{-6}$; the oscillation amplitude is only a small fraction of the total energy.

3.1.2 Magnetization and Magnetic Torque

From the relation $\Omega = -\boldsymbol{M} \cdot \boldsymbol{B} = -MB\cos(\theta)$, where θ is the angle between the magnetization and the magnetic field, we have the magnetization parallel to the magnetic field $M_{\parallel} = -\partial\Omega/\partial B = M\cos(\theta)$. Since the condition $F_Q \gg B (N \gg 1)$ is satisfied in most cases, the derivative of F_Q/B is dominant in $\partial\tilde{\Omega}(B)/\partial B$, and thus we obtain

$$\tilde{M}_{\parallel}(B) = -\left(\frac{e}{2\pi\hbar}\right)^{3/2}\left(\frac{e\hbar}{m_c}\right)\frac{2F_Q B^{1/2}}{\pi\left(A_F''\right)^{1/2}} \times \sum_{p=1}^{\infty}\frac{1}{p^{3/2}}R_T R_D R_S \sin\left[2\pi p\left(\frac{F_Q}{B}-\frac{1}{2}\right)+\varphi_0\right], \tag{3.4a}$$

$$\tilde{M}_{2D\parallel}(B) = -\frac{e}{2\pi^2\hbar}\left(\frac{e\hbar}{m_c}\right)\frac{F_Q}{\pi}\sum_{p=1}^{\infty}\frac{1}{p}R_T R_D R_S \sin\left[2\pi p\left(\frac{F_Q}{B}-\frac{1}{2}\right)\right]. \tag{3.4b}$$

The magnetization perpendicular to the magnetic field is given by $M_{\perp} = M\sin(\theta) = (1/B)\partial\Omega(\theta)/\partial\theta$. Similarly, assuming that the dominant term of the derivative arises from $\partial F_Q/\partial\theta$, we have

$$\frac{\partial\tilde{\Omega}}{\partial\theta} \approx \frac{\partial\tilde{\Omega}}{\partial F_Q}\frac{\partial F_Q}{\partial\theta}, \quad \frac{\partial\tilde{\Omega}}{\partial F_Q} \approx -\frac{B}{F_Q}\frac{\partial\tilde{\Omega}}{\partial B} = \frac{B}{F_Q}\tilde{M}_{\parallel},$$

and thus the perpendicular magnetizations are written as

$$\tilde{M}_{\perp}(B) = \frac{1}{F_Q}\left(\frac{\partial F_Q(\theta)}{\partial\theta}\right)\tilde{M}_{\parallel},$$
$$\tilde{M}_{2D\perp}(B) = \frac{1}{F_Q}\left(\frac{\partial F_Q(\theta)}{\partial\theta}\right)\tilde{M}_{2D\parallel}. \tag{3.5}$$

Note that only the oscillatory part is discussed above. For instance, the Pauli paramagnetic moment is induced by the spin polarization in magnetic fields. Using the Bohr magneton $\mu_B = e\hbar/2m_c$, the Pauli paramagnetic moment M_{Pauli} and the paramagnetic susceptibility χ_{Pauli} are written as

$$M_{\mathrm{Pauli}} = \chi_{\mathrm{Pauli}}B, \quad \chi_{\mathrm{Pauli}} = \mu_B^2 D(E_F). \tag{3.6}$$

In addition, orbital paramagnetism and orbital diamagnetism (Landau diamagnetism) are present. The quantum oscillation $\tilde{M}$ is superimposed on these non-oscillatory backgrounds.

The magnetic torque is defined as $\boldsymbol{\tau}_{\mathrm{torque}} = \boldsymbol{M} \times \boldsymbol{B} = MB\sin(\theta) = -\partial\Omega/\partial\theta$. As Eq. (3.5) is derived above, the oscillatory parts of the magnetic torque are written as

$$\tilde{\tau}_{\mathrm{torque}}(B) = -\frac{B}{F_Q}\left(\frac{\partial F_Q}{\partial\theta}\right)\tilde{M}_{\parallel}, \; \tilde{\tau}_{2D\ \mathrm{torque}}(B) = -\frac{B}{F_Q}\left(\frac{\partial F_Q}{\partial\theta}\right)\tilde{M}_{2D\parallel}. \tag{3.7}$$

Since $\tilde{M}_{\perp}(B)$ and $\tilde{\tau}_{\text{torque}}$ are proportional to $\partial F_Q/\partial\theta$, the quantum oscillations in these quantities are significantly reduced for isotropic Fermi surfaces. Even for anisotropic Fermi surfaces, as shown in Fig. 2.14, the quantum oscillations are reduced when the magnetic field is parallel to the k_z-axis, where $\partial F_Q/\partial\theta = 0$.

3.1.3 Density of States

The number of the electrons is written as $N_e = -\partial\Omega/\partial\mu$ from Eq. (2.46). At low temperatures, μ can be approximated by $\mu \approx E_{\text{F}} = (e\hbar/m_c)F_Q$, and thus the oscillatory part of the electron number $\tilde{n}$ per unit volume is given by

$$\tilde{n} = -\frac{\partial\tilde{\Omega}}{\partial\mu} \approx -\frac{\partial\tilde{\Omega}}{\partial E_{\text{F}}} = -\frac{m_c}{e\hbar}\frac{\partial\tilde{\Omega}}{\partial F_Q}. \tag{3.8}$$

The oscillatory part of the density of states is written as

$$\tilde{D}(B) = \frac{\partial\tilde{n}}{\partial E_{\text{F}}} = -\left(\frac{m_c}{e\hbar}\right)^2 \frac{\partial^2\tilde{\Omega}(B)}{\partial F_Q^2}, \tag{3.9}$$

and thus

$$\tilde{D}(B) = \left(\frac{e}{2\pi\hbar}\right)^{3/2}\left(\frac{m_c}{e\hbar}\right)\frac{4B^{1/2}}{\left(A_F''\right)^{1/2}} \times \sum_{p=1}^{\infty}\frac{1}{p^{1/2}}R_D R_S \cos\left[2\pi p\left(\frac{F_Q}{B}-\frac{1}{2}\right)+\varphi_0\right], \tag{3.10a}$$

$$\tilde{D}_{2D}(B) = \frac{e}{\pi^2\hbar}\left(\frac{m_c}{e\hbar}\right)\sum_{p=1}^{\infty} R_D R_S \cos\left[2\pi p\left(\frac{F_Q}{B}-\frac{1}{2}\right)\right]. \tag{3.10b}$$

Note that the density of states does not include R_T. This is because, as we will see in Sect. 3.2, the R_T factor is the finite temperature effect but the density of states is defined as a temperature-independent quantity.

3.1.4 Electrical Resistivity and Conductivity

Let us now consider the electrical resistivity and conductivity in the framework of classical theory based on the Drude model. For simplicity, we assume a 2D free electron system in the xy plane with the electric field $\boldsymbol{E}\left(E_x, E_y, 0\right)$ and the magnetic field $\boldsymbol{B}(0, 0, B)$. From the equation of motion, $\boldsymbol{F} = m_c\dot{\boldsymbol{v}} = -e(\boldsymbol{E} + \boldsymbol{v}\times\boldsymbol{B})$, we have

$$m_c\dot{v}_x = -eE_x - ev_yB, m_c\dot{v}_y = -eE_y + ev_xB. \tag{3.11}$$

The momentum of the electron $\boldsymbol{p} = m_c\boldsymbol{v}$ gained by $\boldsymbol{E}$ and $\boldsymbol{B}$ is lost by scattering. Let τ be the scattering time, which is defined as the time that the electrons can move without being scattered. Assuming that the electrons are in a steady state, we replace the electron velocity v by the average value $\langle v \rangle$ in the above equations. The average momentum $\langle p \rangle = m_c\langle v \rangle$ that the electron acquires in the interval τ is written as

$$\begin{aligned} \langle p_x \rangle &= m_c\langle v_x \rangle = \left\langle \int_0^\tau m_c\dot{v}_x \mathrm{d}t \right\rangle = -e\left(E_x + \langle v_y \rangle B\right)\tau, \\ \langle p_y \rangle &= m_c\langle v_y \rangle = \left\langle \int_0^\tau m_c\dot{v}_y \mathrm{d}t \right\rangle = -e\left(E_y - \langle v_x \rangle B\right)\tau. \end{aligned} \tag{3.12}$$

The current density is given by $\boldsymbol{j} = -N_e e\boldsymbol{v}$, where N_e is the number of electrons per unit volume. From Eq. (3.12), the electrical resistivity tensor $\boldsymbol{\rho}$ defined as $\boldsymbol{E} = \boldsymbol{\rho}\boldsymbol{j}$ is written as

$$\begin{pmatrix} E_x \\ E_y \end{pmatrix} = \begin{pmatrix} \rho_{xx} & \rho_{xy} \\ \rho_{yx} & \rho_{yy} \end{pmatrix}\begin{pmatrix} j_x \\ j_y \end{pmatrix} = \begin{pmatrix} \rho_0 & B/N_e e \\ -B/N_e e & \rho_0 \end{pmatrix}\begin{pmatrix} j_x \\ j_y \end{pmatrix}. \tag{3.13}$$

The diagonal term is

$$\rho_0 = \frac{m_c}{N_e e^2 \tau}. \tag{3.14}$$

For example, when the electric field E_x in the x direction is measured for a current j_y in the y direction, the Hall resistivity is defined as $\rho_{xy} = E_x/j_y = B/N_e e$. Because of the isotropic xy plane, we have $\rho_{xx} = \rho_{yy} = \rho_0$. Similarly, we define the electrical conductivity tensor $\boldsymbol{\sigma}$ as $\boldsymbol{j} = \boldsymbol{\sigma}\boldsymbol{E}$. Since $\boldsymbol{\sigma}$ is the inverse matrix of $\boldsymbol{\rho}$, we obtain

$$\rho_{xx} = \frac{\sigma_{yy}}{\sigma_{xx}\sigma_{yy} + \sigma_{xy}^2}, \ \rho_{xy} = \frac{-\sigma_{xy}}{\sigma_{xx}\sigma_{yy} + \sigma_{xy}^2}, \tag{3.15}$$

and

$$\begin{pmatrix} j_x \\ j_y \end{pmatrix} = \begin{pmatrix} \sigma_{xx} & \sigma_{xy} \\ \sigma_{yx} & \sigma_{yy} \end{pmatrix}\begin{pmatrix} E_x \\ E_y \end{pmatrix} = \sigma_0\begin{pmatrix} \frac{1}{1+\omega_c^2\tau^2} & -\frac{\omega_c\tau}{1+\omega_c^2\tau^2} \\ \frac{\omega_c\tau}{1+\omega_c^2\tau^2} & \frac{1}{1+\omega_c^2\tau^2} \end{pmatrix}\begin{pmatrix} E_x \\ E_y \end{pmatrix}, \tag{3.16}$$

$$\sigma_0 = \frac{1}{\rho_0} = \frac{N_e e^2 \tau}{m_c}. \tag{3.17}$$

These equations can be derived more generally by Chambers formula, as will be discussed in Chap. 4.

In Eq. (3.13), ρ_{xx} is independent of ω_c, no magnetoresistance. This is due to the assumption that there is only one type of charge carrier. In a real conductor, the mass and scattering time of the electrons will be different at different parts of the Fermi surface or at each Fermi surface for a multi Fermi surface system. For example, if there are two different carriers A and B with different N_e, m_c, and τ, the conductivity is given by the sum of the individual term, $\sigma_{xx} = \sigma_{xx}^A + \sigma_{xx}^B$ and $\sigma_{xy} = \sigma_{xy}^A + \sigma_{xy}^B$. Substituting these into Eq. (3.15), we find that ρ_{xx} depends on ω_c; magnetoresistance appears.

We assume that the electrons are scattered by impurities or defects and then transferred from the initial state $i\rangle$ to a final state $f\rangle$. According to the Fermi's golden rule, the transition probability is given by

$$T_{i\to f} = \frac{2\pi}{\hbar}\left|\langle f|V'|i\rangle\right|^2 D\,\delta\left(E_f - E_i\right), \tag{3.18}$$

where V' is the scattering potential, E_i and E_f are the eigenvalues of the initial and final states, respectively, D is the density of states at E_f, and δ is the delta function. By summing over the final state, the scattering time τ can be defined as $1/\tau = \sum_{f(\neq i)} T_{i\to f}$. Taking D_0 as the non-oscillatory density of states, we have

$$D = D_0\left[1 + \frac{\tilde{D}(\mathrm{B}\,)\,R_T}{D_0}\right]. \tag{3.19}$$

Substituting this equation into Eq. (3.18), the scattering time will be given by,

$$\frac{1}{\tau} = \sum_{f(\neq i)} T_{i\to f} = \frac{1}{\tau_0}\left[1 + \frac{\tilde{D}(\mathrm{B}\,)\,R_T}{D_0}\right], \tag{3.20}$$

where τ_0 is non-oscillatory part of the scattering time,

$$\frac{1}{\tau_0} = \frac{2\pi}{\hbar}\sum_{f(\neq i)}\left|\langle f|V'|i\rangle\right|^2 D_0\delta\left(E_f - E_i\right). \tag{3.21}$$

If the quantum oscillation is observed, we expect $\omega_c\tau \gg 1$ from Eq. (2.72). Then the conductivity is given by

$$\sigma_{xx}(B) = \frac{\sigma_0}{1+\omega_c^2\tau^2} \approx \frac{\sigma_0}{\omega_c^2\tau^2} \propto \frac{1}{\tau}. \tag{3.22}$$

Substituting Eq. (3.20) into Eq. (3.22), we will obtain

$$\sigma_{xx}(B) = \sigma_0(B)\left(1 + \alpha_{\mathrm{AH}}\frac{\tilde{D}(B)R_T}{D_0}\right) = \sigma_0(B) + \tilde{\sigma}_{xx}(B), \tag{3.23}$$

where $\sigma_0(B)$ and $\tilde{\sigma}_{xx}(B)$ are the non-oscillatory and oscillatory conductivities, respectively. This is a general expression of the conductivity in the magnetic field.

By considering various phonon modes and scattering mechanisms for an isotropic Fermi surface, it is derived that the factor α_{AH} is of the order of unity [2].

What we experimentally observe is resistivity (not conductivity). From Eq. (3.14), we similarly obtain $\rho_{xx} \propto 1/\tau$ and thus

$$\rho_{xx}(B) = \rho_0(B)\left(1 + \alpha_{AH}\frac{\tilde{D}(B)R_T}{D_0}\right) = \rho_0(B) + \tilde{\rho}_{xx}(B), \tag{3.24}$$

where $\rho_0(B)$ and $\tilde{\rho}_{xx}(B)$ are the non-oscillatory and oscillatory resistivities, respectively. Equation (3.24) is rewritten as,

$$\frac{\tilde{\rho}_{xx}(B)}{\rho_0(B)} = \frac{\rho_{xx}(B)}{\rho_0(B)} - 1 = \alpha_{AH}\frac{\tilde{D}(B)R_T}{D_0}. \tag{3.25}$$

Therefore, if $\rho_{xx}(B)/\rho_0(B)-1$ is plotted as a function of magnetic field, the SdH oscillations can be analyzed by the function $\tilde{D}(B)R_T$. For a 2D electron system, $\tilde{D}(B)$ should be replaced by $\tilde{D}_{2D}(B)$.

When the Fermi surface has multiple extremal cross sections, the oscillations arising from all the cross sections can be observed simultaneously. Even for a single cross section, multiple frequencies are often observed as shown below. To find the frequencies contained in the signal, (1) we first plot the oscillation $\rho_{xx}(B)/\rho_0(B) - 1$ with 1/*B* so that the oscillation is periodic, and then (2) we compute its Fourier spectrum. Figure 3.1a shows the resistivity of a quasi-2D organic conductor, (BEDT-TTF)$_2$KHg(SCN)$_4$ in the magnetic field perpendicular to the conduction layers [3]. This conductor has a layered structure similar to κ-(BEDT-TTF)$_2$Cu(NCS)$_2$ shown in Fig. 2.8b. In a high magnetic field range, the quantum oscillation is evident. The dashed curve in Fig. 3.1a indicates the non-oscillatory background $\rho_0(B)$. The inset shows $\rho_{xx}(B)\rho_0(B) - 1$ as a function of 1/*B*, where the periodic SdH oscillation is evident. The wavelength is $\Delta(1/B) = 1.5 \times 10^{-3}$, giving the frequency $F_Q = 670$T. The oscillatory part of the density of states in Eq. (3.10b) is given by

$$\tilde{D}_{2D}(B) \propto \sum_{p=1}^{\infty} R_D R_S \cos\left[2\pi p\left(\frac{F_Q}{B} - \frac{1}{2}\right)\right].$$

In addition to the fundamental oscillation with F_Q, the harmonics with pF_Q (p = 2, 3, 4,…) can be observed although the harmonics are generally smaller than the fundamental one as discussed in Sect. 3.2. The Fourier spectrum of the oscillation is shown in Fig. 3.1b, where the second harmonic $2F_Q$ ($p = 2$) is evident in addition to F_Q. The peak height of the Fourier spectrum shows the oscillation amplitude averaged over the Fourier transform range. The origin of the rather large second harmonic oscillation is discussed in Sect. 3.2.

The L–K formula is a good approximation when the magnetic field is not very large, i.e., for $F_Q/B \gg 1$. This condition is equivalent to the number of Landau levels below the Fermi level being very large, $N = E_F/\hbar\omega_c = F_Q/\mathrm{B} \gg 1$ in

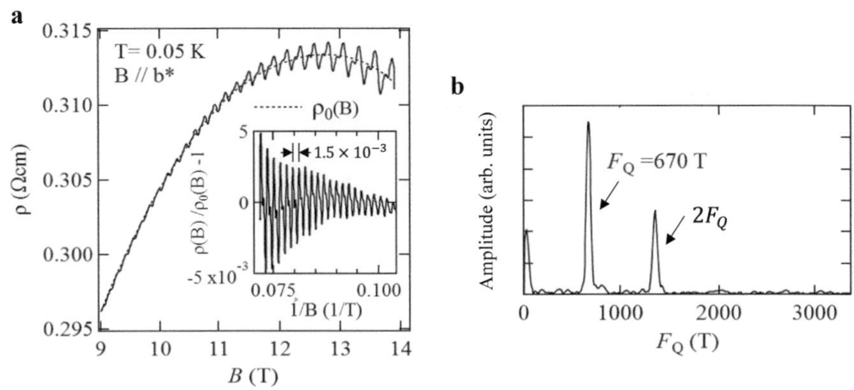

Fig. 3.1 **a** Resistivity of quasi-2D organic conductor $(BEDT\text{-}TTF)_2KHg(SCN)_4$ [3]. Inset: oscillatory part of the resistance. **b** Fourier spectrum of the oscillatory part of the resistivity $\rho_{xx}(B)/\rho_0(B) - 1$. The fundamental ($F_Q$) and second harmonic ($2F_Q$) oscillations are observed

Eq. (2.60). For most conductors, the frequency of the quantum oscillation has a large value, $F_Q = 100$–5000 T, and thus the condition $N \gg 1$ is satisfied in a normal experimental condition $B \approx 10$ T. In some conductors, however, quantum oscillations with low frequencies $F_Q < 10$ T are observed, in which case $E_F \approx \hbar\omega_c$, and the L–K formula is not strictly applicable.

In particular, there is only one Landau level allowed for the electrons for $E_F < \hbar\omega_c$; all electrons occupy the lowest Landau level with $N = 0$. This situation is called the quantum limit. In the quantum limit, one would classically expect all electrons to be in the cyclotron motion with the same orbital radius, $l_B = \sqrt{\hbar/eB}$ in Eq. (2.16). The cyclotron orbits are separated from each other in real space, showing that each electron is localized in a restricted area. For a particular 2D electron system, it has been reported that the electrical resistance diverges in the quantum limit; the electronic state becomes insulating. The origin of the insulating state, called Wigner crystal, is an interesting issue, which has been discussed in terms of the Coulomb repulsion between the electrons [4, 5].

3.2 Reduction Factors of Quantum Oscillation

Let us look specifically at the three damping factors R_T, R_D, and R_S of the quantum oscillation in the L–K formula. The absolute values of these factors have a maximum value of 1 and become smaller than 1 depending on the temperature, magnetic field, g-factor, etc.; the oscillation amplitude is reduced by these factors.

3.2.1 Temperature Reduction Factor

The Fermi edge is smeared out as the temperature increases as seen in Fig. 2.15. The temperature reduction factor R_T results from this finite temperature effect on the oscillation amplitude, given by

$$R_T = \frac{Kp\mu_c T/B}{\sinh(Kp\mu_c T/B)} = \frac{z}{\sinh(z)}, \quad z = \frac{Kp\mu_c T}{B}. \tag{3.26}$$

This factor is derived in Appendix 3.1. The function $R_T(z)$ has a maximum value of 1 at $z = 0$ and then monotonically decreases with increasing z as shown in Fig. 3.10b. The ratio of the effective mass m_c to the free electron mass m_0,

$$\mu_c = m_c/m_0, \tag{3.27}$$

is called the effective mass ratio. The factor K is a constant, independent of the material,

$$K = \frac{2\pi^2 m_0 k_B}{e\hbar} = 14.7[\mathrm{T/K}]. \tag{3.28}$$

The factor z is rewritten as

$$z = \frac{Kp\mu_c T}{B} = 2\pi^2 p\left(\frac{k_B T}{\hbar\omega_c}\right). \tag{3.29}$$

This expression makes it easy to understand the physical meaning of $R_T(z)$; the factor z is proportional to the ratio of the width of the Fermi edge $k_B T$ to the Landau level spacing $\hbar\omega_c$. For $z > 1$, we have $R_T \approx ze^{-z}$, which exponentially decreases with increasing z. Among the above three reduction factors, only R_T depends on temperature, including an unknown parameter μ_c. Therefore, by fitting the temperature dependence of the oscillation amplitude with R_T, the effective mass m_c can be calculated. The oscillation amplitude is given by the peak height of the Fourier spectrum. Usually, the peak height divided by the temperature is plotted as a function of temperature, which is called the mass plot. The oscillation amplitudes of the higher harmonics ($p \geq 2$) are reduced more, due to the larger effective mass ratio, $p\mu_c$.

Figures 3.2a and b show the SdH oscillations at various temperatures for a quasi-2D organic conductor (BEDT-TTF)$_2$Br(DIA) [6] in the magnetic field perpendicular to the layers ($B \| b^*$) and the Fourier spectra of the oscillations in the range between 13 and 14 T, respectively [7]. This organic conductor has a triclinic structure with a single quasi-2D Fermi surface as depicted in the inset. Figure 3.2c shows the mass plot of the oscillation, where the oscillation amplitude divided by the temperature is plotted against the temperature. The solid curve is the result fitted with $1/\sinh(K\mu_c T/B)$, showing that the effective mass of the

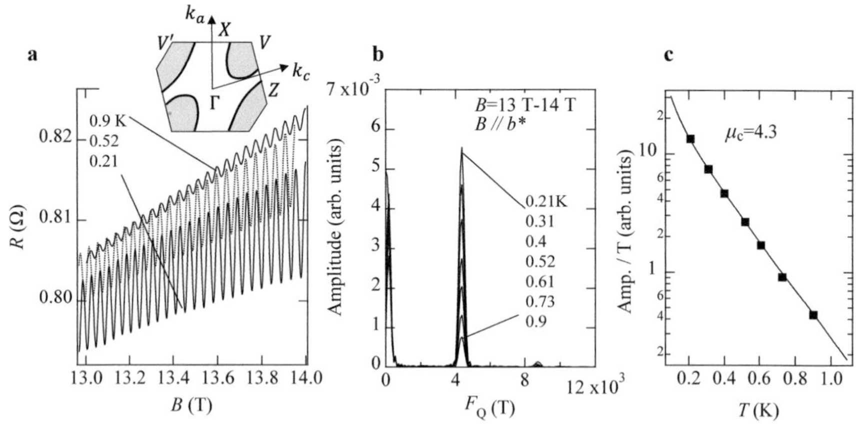

Fig. 3.2 **a** SdH oscillations in the interlayer resistance at perpendicular magnetic field ($B \| b^*$) for the quasi-2D organic conductor (BEDT-TTF)$_2$Br(DIA). The magnetic field dependence of the resistance in a wide range is presented in Fig. 2.17. Inset: calculated Fermi surface. **b** Fourier spectra of the SdH oscillations. **c** Mass plot obtained from **b**. The solid curve shows $1/\sinh(K\mu_c T/B)$ with $\mu_c = 4.3$

electrons is 4.3 m_0. In the fitting, the magnetic field is the average value of the Fourier transform range between B_1 and B_2, $1/B_{av} = (1/B_1 + 1/B_2)/2$.

Since the reduction factor R_T decreases exponentially as a function of z for $z > 1$, a larger magnetic field or a lower temperature is required to observe the quantum oscillations of electrons with heavy effective masses. In a Ce compound CeRu$_2$Si$_2$ [8, 9] and a U compound UPt$_3$ [10], which are classified as heavy electron systems, the electrons with masses 100 times heavier than free electrons ($\mu_c > 100$) have been observed. Their quantum oscillation measurements require low temperatures below 50 mK and high magnetic fields above 15 T, as discussed in Sect. 3.2.

3.2.2 Dingle Reduction Factor

The Dingle reduction factor results from the Landau level broadening due to scattering. It is given by

$$R_D = \exp\left(-\frac{Kp\mu_c T_D}{B}\right), \tag{3.30}$$

which has a maximum value of 1 for $B \to \infty$ or $T_D \to 0$, and exponentially decreases with decreasing B or increasing T_D. This factor is derived in Appendix 3.2. The factor T_D has the dimension of temperature [K] and is called the Dingle temperature. It is defined as

$$T_D = \frac{\hbar}{2\pi k_B \tau}, \tag{3.31}$$

where τ is the scattering time due to impurities or defects, a good indicator of the sample quality. By definition, the lower the value of T_D, the better the sample quality. In conductors where quantum oscillations are observed, T_D is roughly in the range of 0.1–10 K. Although a closed Fermi surface is expected, quantum oscillations are not observed in many conductors. The reason is generally ascribed to the large T_D values (low sample quality). Equation (3.30) is rewritten as

$$R_D = \exp\left(-\frac{p\pi}{\omega_c \tau}\right). \tag{3.32}$$

This expression may make it easier to understand the physical meaning of R_D; the quantum oscillation is observed when the cyclotron motion is completed before the scattering ($\omega_c \tau > 1$). The amplitudes of the higher harmonics with $p \geq 2$ are also greatly reduced by the factor R_D.

Figure 3.3a shows the oscillatory part of the interlayer resistance, $\rho(B)/\rho_0(B) - 1$ at 0.6 K in the quasi-2D organic conductor (BEDT-TTF)$_2$Br(DIA). Since this electronic state is an almost perfect 2D system, it is appropriate to use the 2D formula, $\tilde{D}_{2D}(B)$ for the analyses of the oscillation. From Eq. (3.25), the magnetic field dependence of the fundamental oscillation ($p = 1$) is written as

$$\frac{\tilde{\rho}_{xx}(B)}{\rho_0(B)} = \frac{\rho(B)}{\rho_0(B)} - 1 \propto \tilde{D}_{2D}(B) R_{\mathrm{T}} \propto R_{\mathrm{D}} R_{\mathrm{T}} \cos\left[2\pi\left(\frac{F_Q}{B} - \frac{1}{2}\right)\right], \tag{3.33}$$

which includes the unknown parameters μ_c and T_D. The factor R_{S} does not depend on the magnetic field as will be seen later. If μ_c is obtained from the mass plot, only T_{D} is the unknown parameter and thus T_{D} can be obtained from the magnetic field dependence of the oscillation amplitude A_{osc}. Figure 3.3b shows the so-called Dingle plot, a semi-log plot of A_{osc}/R_T versus 1/*B*. The effective mass ratio $\mu_c = 4.3$ (Fig. 3.2c) is used in the calculation of R_T. From the relation $A_{\mathrm{osc}}/R_T \propto R_D$ in Eq. (3.33), the slope in the semi-log plot gives $K\mu_c T_{\mathrm{D}}$. The solid line in Fig. 3.3b gives $T_D = 0.9$ K, which corresponds to the Landau level broadening, $(\hbar/\tau)/k_B = 7$ K from Eq. (3.31).

The scattering time τ can be approximately calculated from the resistivity $\rho_{xx} = \rho_0 = m_c/N_e e^2 \tau$ in Eq. (3.14). Note that the dominant scattering process to the resistivity is the large-angle scattering. In contrast, τ involved in R_D results from any scattering that changes the phase of the electronic state; even small angle scattering breaks the Landau quantization. Therefore, τ calculated from the resistivity is usually longer than that obtained from the quantum oscillation measurement.

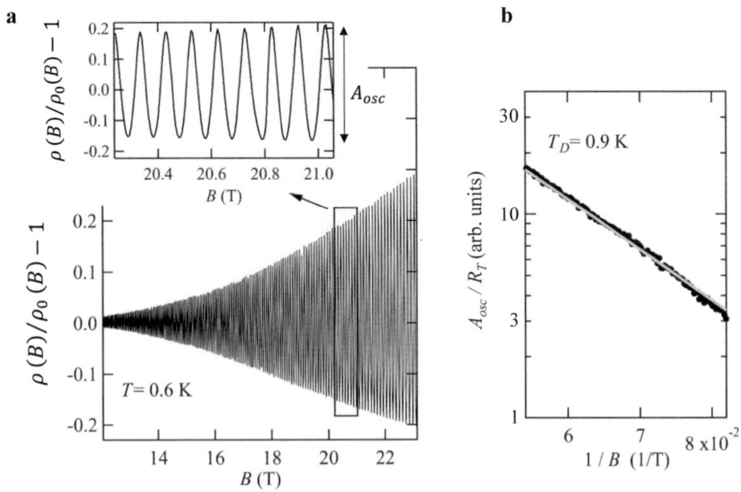

Fig. 3.3 **a** SdH oscillations observed in the interlayer resistance for (BEDT-TTF)$_2$Br(DIA). Inset: close-up of the oscillation near 20 T. **b** Dingle plot, semi-log plot of $A_{\rm osc}/R_T$ versus 1/*B*. The solid line shows the calculated result by $R_D = \exp(-K\mu_c T_D/B)$ with $T_D = 0.9$ K

3.2.3 Spin Splitting Reduction Factor

Until now, we have not considered the spin degree of freedom of electrons. In magnetic fields, each Landau level undergoes spin splitting due to the Zeeman effect. For a 2D electron system in a magnetic field perpendicular to the layer ($B \| z$), the spin-split Landau levels are expressed as

$$E_N = \hbar\omega_c\left(N + \frac{1}{2}\right) - g_c\mu_B s_z B, \tag{3.34}$$

where g_c is the effective g-factor and $s_z(= \pm 1/2)$ is the z-component of the electron spin. The Zeeman splitting, $\Delta E = g_c\mu_B B$ in the 2D electron system or in the electronic state of the extremal cross section of a 3D electron system is depicted in Fig. 3.4. As the magnetic field increases, the Landau levels with the up and down spins alternately cross the Fermi level. This spin splitting leads to a phase shift $\Delta\varphi$ of the quantum oscillation, given by $\Delta\varphi = \mp\pi(\Delta E/\hbar\omega_c)$ for $s_z = \pm 1/2$. In Eq. (3.2a), the oscillatory part for $p = 1$ is written as

$$\cos\left[2\pi\left(\frac{F_Q}{B} - \frac{1}{2}\right) + \varphi_0\right] = \cos(\Phi_{\rm B}). \tag{3.35}$$

Assuming that both spins ($s_z = \pm 1/2$) have the same values of R_T and R_D (μ_c and T_D), the sum of the oscillatory parts arising from both spins is

$$\cos(\Phi_B - \Delta\varphi) + \cos(\Phi_B + \Delta\varphi) = 2\cos(\Phi_B)\cos\left(\frac{\pi\Delta E}{\hbar\omega_c}\right)$$

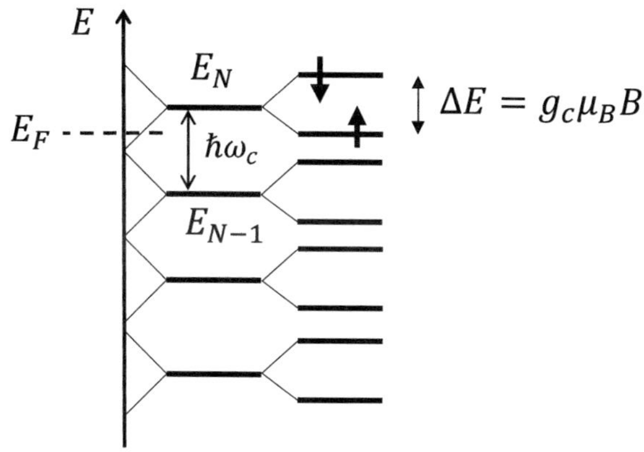

Fig. 3.4 Landau levels and spin splitting due to Zeeman effect. The thick arrows indicate the up ($s_z = 1/2$) and down ($s_z = -1/2$) spins

$$= 2\cos(\Phi_B)\cos\left(\frac{\pi g_c \mu_c}{2}\right). \tag{3.36}$$

The last term $R_S = \cos(\pi g_c \mu_c/2)$ indicates the reduction of the quantum oscillation due to the Zeeman splitting. Since the spin degree of freedom is already taken into account in Eq. (3.2), the factor 2 is removed from R_S. By extending the Zeeman effect to the higher harmonics, we can define the reduction factor as

$$R_S = \cos\left(p\pi S_{\rm spin}\right), \quad S_{\rm spin} = \frac{\Delta E}{\hbar\omega_c} = \frac{g_c\mu_c}{2}. \tag{3.37}$$

The factors R_S and $S_{\rm spin}$ are called the spin splitting reduction factor and the spin splitting factor, respectively. Note that g_c is the effective g-factor, renormalized by many-body effects [11], similar to the effective mass as discussed in Sect. 3.3.

The spin splitting reduction factor causes interesting phenomena in quantum oscillations. When $S_{\rm spin} = 1/2, 3/2, 5/2, \ldots$ we have $R_S = 0$ for $p =$ odd but $R_S = 1$ for $p =$ even. This condition, the disappearance of the fundamental oscillation ($R_S = 0$ for $p = 1$) is called the spin splitting zero condition. In Fig. 3.1b, we observe the relatively large amplitude of the second harmonic ($p = 2$) for the quasi-2D organic conductor (BEDT-TTF)$_2$KHg(SCN)$_4$. This is because the spin splitting zero condition is almost satisfied; the parameters $g_c \approx 2$ and $\mu_c \approx 1.5$ give $S_{\rm spin} \approx 3/2$. The Landau quantization under the spin splitting zero condition is discussed in Appendix 3.2.

As seen in Fig. 2.11c, the oscillatory part of the thermodynamic grand potential $\widetilde{\delta\Omega}(B)$ exhibits pronounced peaks at magnetic fields where the Fermi level coincides with the Landau level, $E_F = E_N = (N + 1/2)\hbar\omega_c$. Taking into account the Zeeman splitting, as the magnetic field increases, the Zeeman splitting becomes sufficiently larger than the thermal broadening due to the finite temperature $k_B T$ and than the Landau level broadening due to the scattering $\hbar/\tau$. Under such conditions, the sharp peaks of the thermodynamic grand potential are split, called spin splitting. Figure 3.5 shows the simulated results of $\widetilde{\delta\Omega}(B)$ given by Eq. (3.2a) with

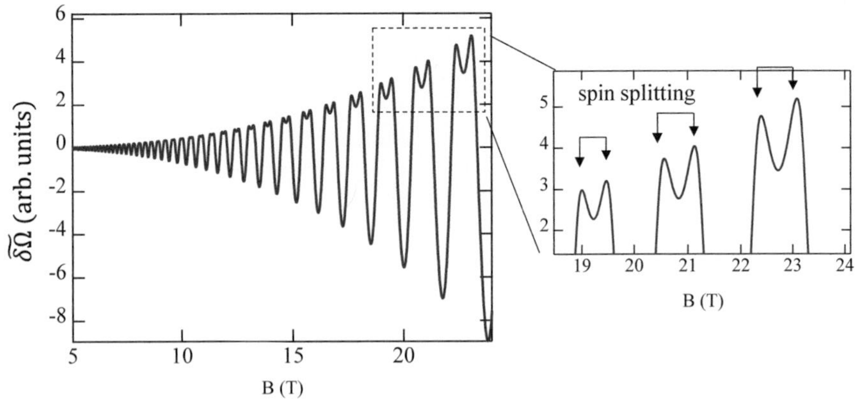

Fig. 3.5 Simulated oscillatory part of thermodynamic grand potential $\delta\widetilde{\Omega}(B)$ with parameters, $F_Q = 250$ T, $\mu_c = 0.6$, $T_D = 0.5$ K, $g_c = 2$, and $T = 0.5$ K. Inset: close-up of oscillation with spin splitting

the parameters listed in the caption. Note that the peak is split at high magnetic fields. The split peaks correspond to the Landau levels of the up and down spins.

3.2.4 L–K Formula for 2D Electron System

Here, let us look at a situation unique to a 2D electron system in the xy plane, as shown in Fig. 3.6a. As seen in Eq. (2.11), the Landau quantization is equivalent to the quantization of the magnetic flux through the closed orbit in real space. When the magnetic field is tilted from the z-axis, the magnetic field through the closed orbit is reduced as $B\cos\theta$ and thus the cyclotron frequency is written as

$$\hbar\omega_c = \frac{\hbar e}{m_c} B\cos\theta. \tag{3.38}$$

Similarly, the terms of the L–K formula are replaced as follows;

$$\cos\left[2\pi p\left(\frac{F_Q}{B} - \frac{1}{2}\right) + \varphi_0\right] \to \cos\left[2\pi p\left(\frac{F_Q}{B\cos\theta} - \frac{1}{2}\right) + \varphi_0\right], \tag{3.39}$$

$$R_T \to \frac{Kp\mu_c T/B\cos\theta}{\sinh(Kp\mu_c T/B\cos\theta)}, \tag{3.40}$$

$$R_D \to \exp\left(-\frac{Kp\mu_c T_D}{B\cos\theta}\right). \tag{3.41}$$

Therefore, we can see that the effect of the tilted magnetic field results in the transformation $F_Q \to F_Q/\cos\theta$ and $\mu_c \to \mu_c/\cos\theta$. The angular dependence of

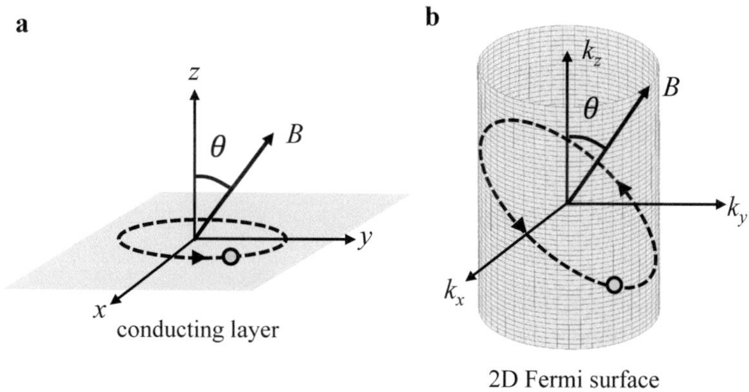

Fig. 3.6 **a** Cyclotron orbit in *xy* plane for 2D electron system, **b** 2D Fermi surface and cyclotron orbit

the frequency for the 2D system is already shown in Fig. 2.14b, $F_Q = F_{Q0}/\cos\theta$. The effective mass obtained by the mass plot also increases as $\mu_c = \mu_{c0}/\cos\theta$. The quantum oscillation amplitude is reduced by the factors R_T and R_D as the magnetic field is tilted from the perpendicular direction. Note that the electron motion is limited in the *xy* plane, but the closed orbit on the cylindrical Fermi surface is always perpendicular to the magnetic field as shown in Fig. 3.6b.

The L–K formula discussed above assumes that the spin–orbit interaction of the conduction electrons is sufficiently weak. This allows the Landau quantization and the Zeeman effect to be treated independently. If the spin–orbit interaction is very strong and comparable to the Fermi energy, the electron spin $\boldsymbol{s}$ is no longer a good quantum number and the angular momentum $\boldsymbol{J} = \boldsymbol{L} + \boldsymbol{s}$, where $\boldsymbol{L}$ is the orbital angular momentum, becomes a good quantum number. In this case the L–K formula is not directly available. The special cases will be discussed in Sect. 3.10.

3.3 Effective Mass

We reconsider the mass of the electrons undergoing cyclotron motion in a magnetic field. In Sect. 3.2, we showed that the effective mass of the conduction electrons can be determined from the temperature dependence of the amplitude of the quantum oscillations. This effective mass is directly related to the cyclotron angular frequency,$\omega_c = eB/m_c$. We will first look at the relationship between the effective mass and the band structure.

3.3.1 Definition of Effective Mass

When the electrons undergo cyclotron motion on the Fermi surface, the energy of the electronic state is written as a function of the area A_F of the closed orbit, $E(A_F)$. The velocity $v_\perp$ in the plane perpendicular to the magnetic field given by Eq. (1.45) is expressed as

$$v_\perp = \frac{1}{\hbar}\frac{\partial E(k_\perp)}{\partial k_\perp} = \frac{1}{\hbar}\left(\frac{\partial E(A_F)}{\partial A_F}\right)\left(\frac{\partial A_F}{\partial k_\perp}\right) = \frac{1}{\hbar}\left(\frac{\partial A_F}{\partial k_\perp}\right)\Big/\left(\frac{\partial A_F}{\partial E}\right), \quad (3.42)$$

where $k_\perp$ is the radius of the cyclotron motion. Assuming a circular cyclotron orbit, the area A_F is given by

$$A_F = \int_0^{k_\perp} \mathrm{d}k_\perp 2\pi k_\perp,$$

and thus $\partial A_F/\partial k_\perp = 2\pi k_\perp$. The velocity in Eq. (3.42) is rewritten as

$$v_\perp = \frac{2\pi k_\perp}{\hbar}\Big/\left(\frac{\partial A_F}{\partial E}\right). \quad (3.43)$$

By comparing Eq. (3.43) with $v_\perp = \hbar k_\perp/m_c$, we can express the effective mass m_c as

$$m_c = \frac{\hbar^2}{2\pi}\left(\frac{\partial A_F}{\partial E}\right). \quad (3.44)$$

Although Eq. (3.44) is derived under the assumption of a circular cyclotron orbit, it is available in the general case. The effective mass is proportional to $\partial A_F/\partial E$, the change rate of the cross section to the energy. When A_F increases with increasing energy E, the Fermi surface is an electron surface with positive mass ($m_c > 0$). Conversely, when A_F decreases, the Fermi surface is a hole surface with negative mass ($m_c < 0$).

We hypothetically consider two energy bands A and B with different dispersions as illustrated in Fig. 3.7a. At $E = E_F$, both bands A and B have the same cross section A_F with the Fermi wave number k_F as depicted in Fig. 3.7b. When E_F increases up to E_{F+}, the cross section of band A increases more than that of band B; the electrons in band A have a larger mass. In this way, the effective masses strongly depend on the band dispersion even for the same cross section. The number of states is proportional to its area, $n(E_F) \propto A_F$, then we obtain the relation,

$$m_c \propto \frac{\partial A_F}{\partial E} \propto \left.\frac{\partial n(E)}{\partial E}\right|_{E=E_F} = D(E_F).$$

The effective mass is proportional to the density of states, as we have seen in Eqs. (1.7)–(1.9) irrespective of the dimensionality of the electronic state.

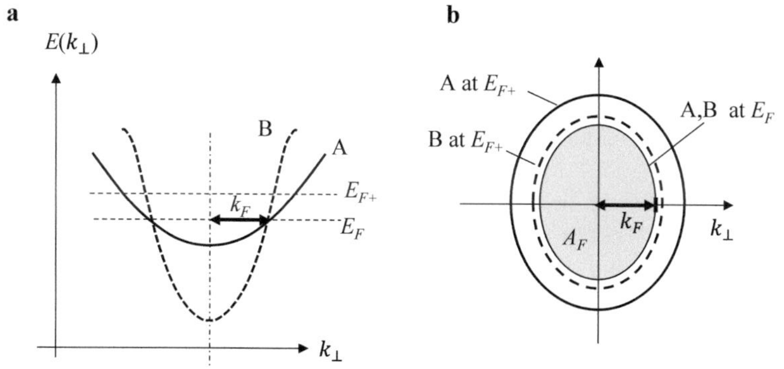

Fig. 3.7 **a** Bands A and B with the same cross section at $E = E_F$ but with different dispersion. **b** Cross sections of Fermi surfaces formed by bands A and B at E_F and E_{F+}

3.3.2 Mass Enhancement

Conventional band calculations do not properly account for the many-body effects such as electron–phonon and electron–electron interactions. The mass obtained by such band calculations according to Eq. (3.44) is called the band mass, written as m_b. In real conductors, the effective mass m_c obtained from quantum oscillation measurements can be heavier than the calculated band mass due to many-body interactions. The relation between m_c and m_b is written as

$$m_c = m_b\left(1 + \lambda_{ep}\right)(1 + \lambda_{ee}), \tag{3.45}$$

where λ_{ep} and λ_{ee} are the mass enhancement factors due to electron–phonon and electron–electron interactions, respectively. The cyclotron motions of the conduction electrons will be slowed down by the Coulomb attraction with the lattice. In addition, the cyclotron motions will also be slowed down by the strong Coulomb repulsive interaction between the electrons, as the electrons tend to move away from each other. These are intuitive interpretations of the mass enhancement due to the many-body effects.

When the electromagnetic wave, whose energy $h\nu$ is equal to the Zeeman splitting $\hbar\omega_c$, is applied to the sample, the electrons are excited from one Landau level to the next higher level. This process is accompanied by the absorption of the electromagnetic wave, called cyclotron resonance. In the classical picture, the cyclotron motions of all the electrons are made in-phase by the oscillating electric field; the relative distance between the electrons remains unchanged. This situation suggests that the cyclotron motion is not affected by the Coulomb repulsion, and therefore, the mass determined by the cyclotron resonance measurement m_{cyc} is not enhanced by the electron–electron interaction but only by the electron–phonon interaction. Therefore, we expect $m_c \geq m_{cyc} \geq m_b$ and the relations will be expressed as

$$m_{\rm cyc} = m_b\left(1 + \lambda_{\rm ep}\right), \quad m_c = m_{\rm cyc}(1 + \lambda_{\rm ee}). \tag{3.46}$$

For the organic conductor κ-(BEDT-TTF)$_2$Cu(NCS)$_2$ shown in Fig. 2.8b, it is reported that $m_{\rm cyc} = 1.18m_0$ is smaller than $m_c = 3.5m_0$ [12]. From the calculated band mass $m_b = 0.89m_0$ [13], the enhancement factors are obtained as $\lambda_{\rm ep} = 0.3$ and $\lambda_{\rm ee} = 2.0$. The electron–electron interaction is dominant as the mass enhancement mechanism in κ-(BEDT-TTF)$_2$Cu(NCS)$_2$. This fact provides important information for understanding the Mott insulator transition and the superconductivity of this material.

Good examples of strong mass enhancement by many-body effects are Ce and *U* compounds, commonly referred to as "heavy fermion systems," where the *4f* and *5f* electronic states, respectively, play an essential role. In these compounds, the hybridization between the itinerant conduction electron and localized *f* electron states leads to a very sharp peak in the density of states near the Fermi level E_F, as shown in Fig. 3.8a. This peak is the direct cause of the heavy electrons. Here E_f is the energy level of a single f orbital without the hybridization, $E_f + U_{\rm ee}$ is the level at which two electrons occupy the same f orbit, and $U_{\rm ee}$ is the Coulomb energy between the f electrons at the same site. In a typical heavy fermion system $CeRu_2Si_2$, the large effective mass ratio of $\mu_c = 120$ is obtained by quantum oscillation measurements of the magnetic susceptibility [8]. The inset of Fig. 3.8b shows the Fourier spectrum of the quantum oscillation, where the arrow represents the oscillation (Ψ) originating from the large Fermi surface presented in Fig. 3.8c. The mass plot of the oscillation is shown in the main panel of Fig. 3.8b. Note that the mass is determined in a very low temperature range below 50 mK due to its very heavy mass.

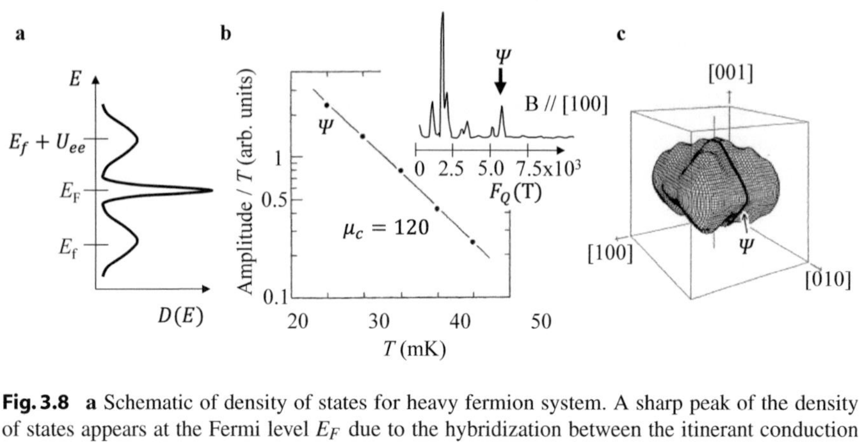

Fig. 3.8 **a** Schematic of density of states for heavy fermion system. A sharp peak of the density of states appears at the Fermi level E_F due to the hybridization between the itinerant conduction electron and localized f electron states. **b** Mass plot of quantum oscillation in $CeRu_2Si_2$ [8]. The solid curve shows the calculated results with the effective mass ratio of $\mu_c = 120$. Insets: Fourier spectrum of the quantum oscillation. **c** Calculated Fermi surface of $CeRu_2Si_2$

Here we simply rewrite Eq. (3.45) as

$$m_c = m_b(1 + \lambda^*). \tag{3.47}$$

The enhancement factor λ^* can be calculated from the effective mass m_c obtained from the quantum oscillation and the calculated band mass m_b. Figure 3.9a presents the superconducting transition temperature T_c versus λ^* for various BEDT-TTF and BETS-based organic superconductors, where BETS stands for *bis(ethylenedithio)tetraselenafulvalene* [13]. The BETS molecule is obtained by replacing the central four S atoms in the BEDT-TTFT molecule in Fig. 2.8b with Se atoms as shown in Fig. 3.9b. For both series, T_c increases with increasing λ^*.

The results are well fitted with

$$T_c \propto \exp\left(-\frac{1}{\alpha\lambda^*}\right), \tag{3.48}$$

where α is a fitting parameter. This equation is similar to the formula obtained by the standard theory of superconductivity, the Bardeen–Cooper–Schrieffer (BCS) theory,

$$k_B T_c = \hbar\omega_D \exp\left(-\frac{1}{D(E_F)\lambda'}\right), \tag{3.49}$$

on the assumption of the Cooper pairs mediated by the electron–phonon interaction. The quantity ω_D is the Debye frequency and λ' is the electron–phonon coupling constant. The superconductivity of these organic superconductors has been discussed in terms of non-BCS mechanisms. Nevertheless, it is interesting

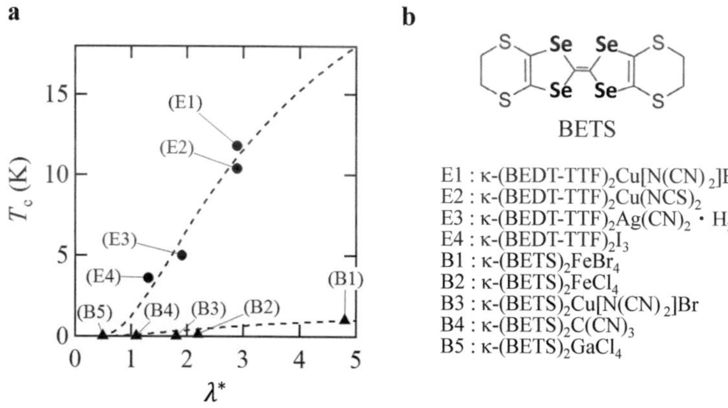

E1 : κ-(BEDT-TTF)$_2$Cu[N(CN)$_2$]Br
E2 : κ-(BEDT-TTF)$_2$Cu(NCS)$_2$
E3 : κ-(BEDT-TTF)$_2$Ag(CN)$_2$ · H_2O
E4 : κ-(BEDT-TTF)$_2$I$_3$
B1 : κ-(BETS)$_2$FeBr$_4$
B2 : κ-(BETS)$_2$FeCl$_4$
B3 : κ-(BETS)$_2$Cu[N(CN)$_2$]Br
B4 : κ-(BETS)$_2$C(CN)$_3$
B5 : κ-(BETS)$_2$GaCl$_4$

Fig. 3.9 **a** T_c versus λ^* plots for various BEDT-TTF (E1–E4) and BETS (B1–B5)-based organic superconductors (reproduced with permission from [13]), **b** BETS molecule

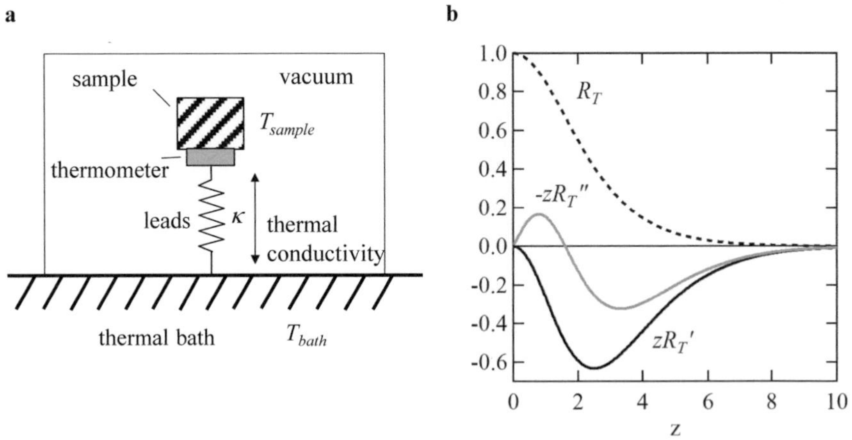

Fig. 3.10 **a** Schematic setup of magnetocaloric effect experiment. The sample and thermometer are placed in vacuum and they are thermally connected to the thermal bath through the leads. **b** $R_T(z)$, $zR_T'(z)$ and $-zR_T''(z)$ as a function of z

that T_c is given by Eq. (3.48) similar to the BCS formula. The correlation between T_c and λ^* provides valuable information for understanding the superconductivity mechanism.

As shown in Fig. 3.9a, the BETS-based superconductors have lower T_c values than the BEDT-TTF-based superconductors. The Se atoms in BETS have more extended π orbitals than the S atoms, so the transfer integral between the neighboring molecules is larger (the bandwidth is wider) in the BETS-based superconductors. Therefore, the BETS-based superconductors have lower densities of states $D(E_F)$, which is closely related to lower T_c values.

Information obtained from quantum oscillation experiments comes only from the extremal cross sections of the Fermi surface, and thus the rest of the Fermi surface is unknown. The effective mass of the conduction electrons can also be estimated from resistivity, heat capacity, and magnetic susceptibility experiments. These experiments provide the mass averaged over the entire Fermi surface and thus it is not necessarily directly comparable to the effective mass obtained from the quantum oscillation experiments.

3.4 Quantum Oscillation in Other Physical Quantities

In Sect. 3.3, we focused on quantum oscillations observed in magnetization and resistance. In principle, quantum oscillations can be observed in all thermodynamic quantities since the Landau quantization causes oscillations in the internal energy and the thermodynamic grand potential of the electron system. Here is an overview of quantum oscillations in some physical quantities other than magnetization and resistance.

3.4.1 Magnetocaloric Effect

Suppose that the sample is mounted in a quasi-adiabatic condition, almost thermally isolated from the thermal bath, as depicted in Fig. 3.10a. If the magnetic entropy S_m of the sample changes as the magnetic field increases, the sample temperature changes accordingly. This sample temperature change is called the magnetocalorimetric effect. The change of the sample entropy ΔS_m results in a heat $\Delta Q = T\Delta S_m$. Since the entropy is a function of T and B, we can write

$$\Delta Q = T\Delta S_m = T\left(\frac{\partial S_m}{\partial T}\right)_B \Delta T + T\left(\frac{\partial S_m}{\partial B}\right)_T \Delta B. \tag{3.50}$$

The heating power is defined as $\Delta P = \Delta Q/\Delta t$ and thus

$$\Delta P = T\frac{\Delta S_m}{\Delta t} = C_s\frac{\partial T}{\partial t} + T\left(\frac{\partial S_m}{\partial B}\right)_T \frac{\partial B}{\partial t}, \tag{3.51}$$

where $\Delta/\Delta t$ is replaced by $\partial/\partial t$ and C_s is the heat capacity of the sample given by $C_s = T(\partial S_m/\partial T)_B$. The heating power ΔP is expressed as the heat transfer from the thermal bath to the sample $\Delta P = \kappa\left(T_{\text{bath}} - T_{\text{sample}}\right) = -\kappa\Delta T$, where κ is the thermal conductivity of the thermometer leads connecting the sample to the thermal bath. Since T_{bath} is constant, replacing $\partial T/\partial t$ by $\partial(\Delta T)/\partial t$ yields

$$\begin{aligned}\Delta T &= T_{\text{sample}} - T_{\text{bath}} \\ &= -\tau_r\frac{\mathrm{d}(\Delta T)}{\mathrm{d}t} - \frac{T_{\text{bath}}}{\kappa}\frac{\mathrm{d}B}{\mathrm{d}t}\left(\frac{\partial S_m}{\partial B}\right)_T,\end{aligned} \tag{3.52}$$

where $\tau_r = C_T/\kappa$ is the thermal relaxation time between the bath and sample, and C_s is replaced by the total heat capacity of the sample and thermometer C_T in the experimental setup in Fig. 3.10a. The quantity $\mathrm{d}B/\mathrm{d}t$ corresponds to the sweep rate of the magnetic field. When the magnetic field sweep is stopped ($\mathrm{d}B/\mathrm{d}t = 0$), we observe the thermal relaxation of the sample, $\Delta T \propto \exp(-t/\tau_r) \to 0$ as $t \to \infty$ from the relation $\Delta T = -\tau_r\mathrm{d}(\Delta T)/\mathrm{d}t$; the sample temperature is relaxed to the thermal bath temperature within an order of τ_r. When the magnetic entropy S_m of the sample changes with increasing magnetic field, an endothermic ($\Delta T < 0$) or exothermic ($\Delta T > 0$) process occurs, depending on the sign of $\partial S_m/\partial B$. For paramagnetic samples, ΔT is sufficiently smaller than T_{bath} in a typical experimental setup.

From the Maxwell relation in thermodynamics,

$$\left(\frac{\partial S_m}{\partial B}\right)_T = \left(\frac{\partial M_\parallel}{\partial T}\right)_B, \tag{3.53}$$

the quantum oscillation of the magnetocaloric effect is written as

$$\Delta\tilde{T} \propto \frac{T_{\text{bath}}}{\kappa}\left(\frac{\partial \tilde{M}_\parallel}{\partial T}\right)_B. \tag{3.54}$$

For a 2D case, the oscillatory part of the magnetization $\tilde{M}_{\parallel}$ is replaced by $\tilde{M}_{2D\parallel}$ in Eq. (3.4b). From the relation $\tilde{M}_{\parallel} \propto R_T(z)(z = Kp\mu_c T/B)$, the temperature dependence of the quantum oscillation is given by

$$\Delta\tilde{T} \propto z\frac{\partial R_T(z)}{\partial z} \equiv zR'_T(z), \tag{3.55}$$

where the thermal conductivity κ is assumed to be constant. Figure 3.10b shows the z dependence of $zR'_T(z)$ as well as $R_T(z)$ and $-zR''_T(z)$. We note $zR'_T = 0$ at $z = 0(T = 0\text{K})$ and zR'_T has a minimum at $z = 2.5$. The absence of the magnetocaloric effect $\left(\Delta\tilde{T} = 0\right)$ at $T = 0\text{K}$ corresponds to the disappearance of the entropy.

Figure 3.11a presents the magnetocaloric effect for $T_{\text{bath}} = 1$ K in a quasi-2D organic conductor β''-(BEDT-TTF)$_2$SF$_5$CH$_2$CF$_2$SO$_3$ [14]. The magnetic field is applied perpendicular to the layers. This conductor also has a layered structure, similar to κ-(BEDT-TTF)$_2$Cu(NCS)$_2$ in Fig. 2.8b. The band calculation shows the presence of 2D electron and hole Fermi surfaces as depicted in the inset of Fig. 3.11b. Due to the equal area of the cross sections, a single frequency quantum oscillation is observed. The quantum oscillations of ΔT for the up and down field sweeps are shown in Fig. 3.11a. The oscillation is reversed between the two sweeps, which is due to the sign change of $\mathrm{d}B/\mathrm{d}t$. Figure 3.11b presents the Fourier spectrum of the oscillation. The harmonics up to 4α are evident in addition to the fundamental oscillation α.

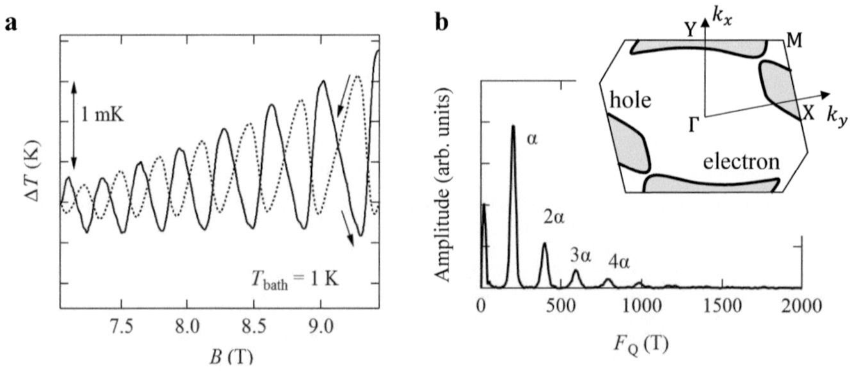

Fig. 3.11 **a** Quantum oscillations of magnetocaloric effect in perpendicular magnetic fields for 2D organic conductor β''-(BEDT-TTF)$_2$SF$_5$CH$_2$CF$_2$SO$_3$ [14]. **b** Fourier spectrum of the oscillation. Inset: calculated 2D Fermi surfaces of this conductor

3.4.2 Heat Capacity

Using the relation $S = -(\partial\Omega/\partial T)_B$ from Eq. (2.44), the magnetic heat capacity is given by

$$C = T\left(\frac{\partial S_m}{\partial T}\right)_B = -T\frac{\partial}{\partial T}\left(\frac{\partial \Omega}{\partial T}\right)_B. \tag{3.56}$$

his gives the quantum oscillation of the heat capacity,

$$\tilde{C} = -T\frac{\partial}{\partial T}\left(\frac{\partial \tilde{\Omega}}{\partial T}\right)_B \propto -z\frac{\partial^2 R_T(z)}{\partial z^2} = -zR_T''(z). \tag{3.57}$$

For a 2D case, the oscillatory part of the grand potential $\tilde{\Omega}$ is replaced with $\tilde{\Omega}_{2D\|}$ in Eq. (3.2b). The z dependence of $-zR_T''(z)$ is already shown in Fig. 3.10b. This function has two nodes at $z = 0$ and $z = 1.61$. Figure 3.12a shows the quantum oscillation of the heat capacity arising from the 2D Fermi surface of the quasi-2D organic conductor κ-(BEDT-TTF)$_2$Cu(NCS)$_2$ [15]. The magnetic field is applied perpendicular to the layers. The oscillation amplitude has a minimum at $B = 16.8$ T for $T = 0.58$ K as indicated by the arrow. This magnetic field corresponds to the node of $-zR_T''(z)$ at $z = K\mu_c T/B = 1.61$ in Fig. 3.10b, from which we obtain the effective mass ratio $\mu_c = 3.2$. Figure 3.12b presents the temperature dependence of the oscillation amplitude at $B = 17$T. The dotted curve indicates the calculated result of $-zR_T''(z)$ for $\mu_c = 3.2$, which is consistent with the result at 0.58 K in Fig. 3.12a.

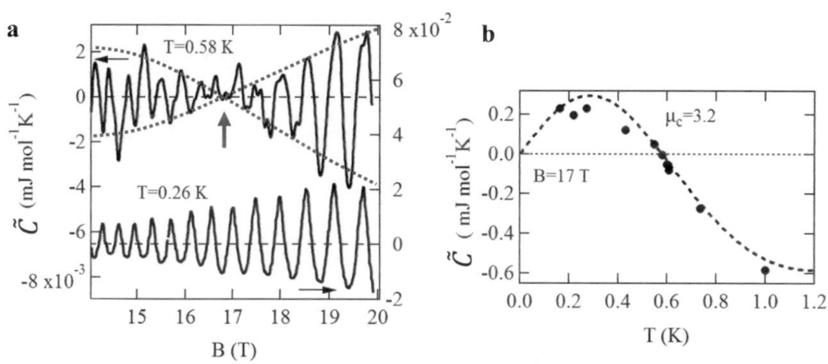

Fig. 3.12 **a** Quantum oscillations of heat capacity in perpendicular magnetic fields for 2D organic conductor κ-(BEDT-TTF)$_2$Cu(NCS)$_2$ [15]. The arrow indicates the minimum of the oscillation amplitude. The dotted curves are guides for the eye. **b** Temperature dependence of the oscillation amplitude in the heat capacity

3.4.3 Elastic Constant and Ultrasonic Attenuation

When ultrasonic waves propagate through a solid, the lattice is periodically distorted. The lattice distortion affects the thermodynamic grand potential of the Landau-quantized electron system in a magnetic field through the electron–phonon interaction. From the opposite perspective, the damping and velocity of the ultrasonic waves are changed by the Landau quantization as a function of the magnetic field; these quantities exhibit quantum oscillations. These quantum oscillations are commonly referred to as the acoustic dHvA effect [16–18]. The elastic constant C_L is related to the sound velocity (v_s) and the density of the sample (ρ_s), $C_L = \rho_s v_s^2$. Its quantum oscillation for a 3D electron system is written as

$$\tilde{C}_L(B) = -\left(\frac{e}{2\pi\hbar}\right)^{3/2}\left(\frac{e\hbar}{m_c}\right)\frac{4F_Q^2 B^{1/2}}{\left(A_F''\right)^{1/2}}|\Lambda_L|^2 \times \sum_{p=1}^{\infty}\frac{(-1)^p}{p^{1/2}}R_T R_D R_S \cos\left[2\pi p\frac{F_Q}{B}+\varphi_0\right], \tag{3.58}$$

where Λ_L is the quantity related to the deformation of the Fermi surface due to the ultrasonic waves. This quantity depends on the polarization and propagation directions of the ultrasonic waves. Unlike other quantities, the quantum oscillation of the elastic constant contains information on the magnitude of the electron–phonon interaction through Λ_L. Figure 3.13a shows the quantum oscillation of the elastic constant for a rare earth compound $SmCu_6$ [19]. $SmCu_6$ with a tetragonal structure shows an antiferromagnetic transition at $T_N = 9$ K. Below T_N, the elastic constant is measured for a longitudinal ultrasonic wave with a frequency of several tens of MHz, whose polarization and propagation directions are along the *c*-axis. The Fourier spectrum of the oscillation is shown in Fig. 3.13b, where several peaks are evident. The results are consistent with the quantum oscillations observed in the magnetization.

The attenuation of the ultrasonic waves per unit length for a 3D electron system is expressed as [20]

$$\frac{\tilde{\Gamma}_q(B)}{\Gamma_0} = \sqrt{2}\left(\frac{e\hbar}{m_c}\right)^{1/2}\frac{(q_s\tau)}{\pi^{3/2}}\left(\frac{B}{A_F'' m_c}\right)^{1/2} \times \sum_{p=1}^{\infty}\frac{(-1)^p}{p^{1/2}}R_T R_D R_S \cos\left[2\pi p\frac{F_Q}{B}+\varphi_0\right]. \tag{3.59}$$

Here $\tilde{\Gamma}_q$ and Γ_0 are the oscillatory and non-oscillatory parts of the attenuation, respectively, and q_s is the wave number of the ultrasonic wave. Acoustic dHvA effects have been observed in various compounds [18, 19, 21].

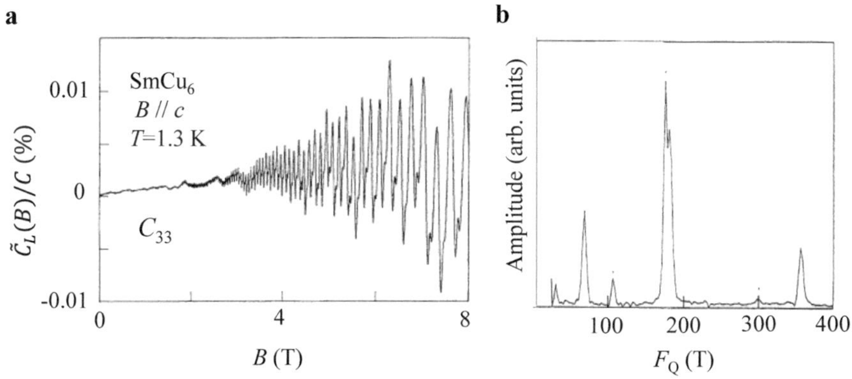

Fig. 3.13 **a** Quantum oscillations of elastic constant for $SmCu_6$ (reproduced with permission from [21]). **b** Fourier spectrum of the oscillation

3.5 Magnetic Breakdown

The discussion of quantum oscillations so far has assumed the existence of closed Fermi surfaces, but this does not mean that quantum oscillations can be observed only for closed Fermi surfaces. In this section, we will see that quantum oscillations can sometimes be observed even for open Fermi surfaces, where the interband tunneling effect plays an essential role.

3.5.1 What Is Magnetic Breakdown?

Figure 3.14a depicts the Fermi surface of a quasi-2D organic conductor θ-(BEDT-TTF)$_2$I$_3$. This conductor has a layered structure similar to κ-(BEDT-TTF)$_2$Cu(NCS)$_2$, as shown in Fig. 2.8b, and the energy bands are well described by the tight binding approximation. There exist a closed 2D Fermi surface (2D FS) and open quasi-1D Fermi surfaces (1D FS). The Fermi velocities $\boldsymbol{v}_F$ are indicated by the thin arrows. In a magnetic field perpendicular to the conduction layer, the Lorentz force $\boldsymbol{F} = -e\boldsymbol{v}_F \times \boldsymbol{B}$ causes the electrons to move in the direction of the thick arrow. In classical mechanics, the electrons remain on each Fermi surface as long as they are not scattered. At the zone boundary, the electron wave number shifts by the reciprocal lattice vector $\pm\boldsymbol{g}$, which is called the Bragg reflection of the electron. The bands from point X to M are depicted in Fig. 3.14b. The orbits on the 1D FS are not closed; the electronic states are not Landau-quantized as long as the electrons remain on the 1D FS.

As the magnetic field increases, so does the Lorentz force $\boldsymbol{F}$, and the electrons try to move in the $\boldsymbol{F}$ (tangential) direction as much as possible. Although there is an energy gap (or energy barrier) E_g between the two bands at the zone boundary, the electrons can quantum-mechanically tunnel through the barrier from the

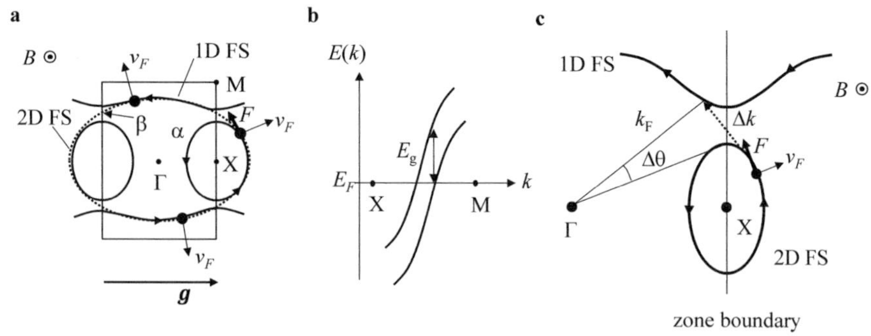

Fig. 3.14 **a** Closed 2D Fermi surface (2D FS) and open quasi-1D Fermi surfaces (1D FS) in quasi-2D organic conductor θ-(BEDT-TTF)$_2$I$_3$. A small closed orbit α (solid curve) near the zone boundary and a large closed orbit β formed by magnetic breakdown (dotted curve) are indicated. **b** Energy bands from point X to M with an energy gap (barrier) E_g. **c** Fermi surfaces near zone boundary

2D to 1D FS. This is called magnetic breakdown. Similarly, the magnetic breakdown can occur from the 1D to 2D FS at the opposite zone boundary. In this way, the electrons can travel the large closed orbit β in Fig. 3.14a. This magnetic breakdown process is a coherent process without energy dissipation. Therefore, if a closed orbit is formed by the successive processes, the electronic state is Landau-quantized, causing quantum oscillations.

3.5.2 Probability of Magnetic Breakdown

It may be easy to understand the probability of the magnetic breakdown in a semi-classical picture. If the gap E_g between the two bands is smaller than the energy uncertainty of the electron ΔE, then the electron would quantum-mechanically tunnel through the gap. The probability of the magnetic breakdown will be given by

$$P = \exp\left(-E_g/\Delta E\right). \tag{3.60}$$

The uncertainty relation between the energy and time, $\Delta E \Delta t \approx \hbar$, suggests that the electron can pass through the gap within Δt. The time to pass through the gap is given by $\Delta t = \Delta\theta/\omega_c$ as depicted in Fig. 3.14c. Here ω_c is the cyclotron angular frequency of the magnetic breakdown orbit β. Taking Δk as the wave number change in the tunneling process, $\Delta\theta \approx \Delta k/k_F \approx E_g/E_F$, we obtain

$$\Delta E \approx \frac{\hbar}{\Delta t} = \frac{\hbar\omega_c}{\Delta\theta} \approx \frac{\hbar\omega_c E_F}{E_g} \rightarrow \frac{E_g}{\Delta E} \approx \frac{E_g^2}{\hbar\omega_c E_F}. \tag{3.61}$$

The probability of the magnetic breakdown Eq. (3.60) is rewritten as

$$P = \exp\left(-\frac{E_g^2}{\hbar\omega_c E_F}\right) = \exp\left(-\frac{B_{MB}}{B}\right), \tag{3.62}$$

where the magnetic breakdown field B_{MB} is defined as

$$B_{MB} = \frac{m_c}{e\hbar}\left(\frac{E_g^2}{E_F}\right). \tag{3.63}$$

The value B_{MB} corresponds to the magnetic field at which the magnetic breakdown becomes pronounced. Although the more general argument shows that the magnetic breakdown field also depends on the relation between the zone boundary and the orientation of the velocity in the plane perpendicular to the magnetic field [22, 23], it does not result in a significant qualitative difference. The probability of no magnetic breakdown, which corresponds to the probability of the Bragg reflection, is given by

$$Q = 1 - P. \tag{3.64}$$

The orbit produced by the magnetic breakdown is called the magnetic breakdown orbit. As is clear from Eq. (3.62), the magnetic breakdown does not occur ($P \approx 0$) in the low magnetic field limit but the magnetic breakdown frequently occurs ($P \rightarrow 1$) in the high magnetic field limit. If the magnetic breakdown occurs successively at the zone boundaries, the closed magnetic breakdown orbit β is formed as shown in Fig. 3.14a.

Figure 3.15a shows the quantum oscillations of the resistance at 50 mK in θ-(BEDT-TTF)$_2$I$_3$, whose Fermi surface is depicted in Fig. 3.14a [24]. The magnetic field is applied perpendicular to the layers. As shown in the inset of Fig. 3.15a, the fast oscillation (β) becomes evident for $B \gtrsim 11$T in addition to the slow oscillation (α). From the Fermi surface structure shown in Fig. 3.14a, the quantum oscillation arising from the α orbit near the zone boundaries should be observed at low magnetic fields. At higher magnetic fields, the magnetic breakdown occurs at the zone boundaries, and thus the quantum oscillation from the large β orbit should be observed. In fact, the Fourier spectrum in Fig. 3.15b shows peaks corresponding to the α and β orbits. In addition to α and β, the $\beta + \alpha$ peak is also observed, the origin of which is discussed in Sect. 3.5.

In θ-(BEDT-TTF)$_2$I$_3$, the BEDT-TTF molecules are dimerized, forming a bonding and an antibonding orbitals, as shown in Fig. 2.12 **b**. The bonding orbital is fully occupied and the antibonding orbital is occupied by a single electron, forming the half-filled conduction band. Since there are two dimerized BEDT-TTF molecules in the unit cell, the number of states in the conduction band is equal to the total number of states in the first Brillouin zone; the area of the β orbit, $A_F^{(\beta)} = 2\pi e F_Q^{(\beta)}/\hbar$ must be equal to that of the first Brillouin zone.

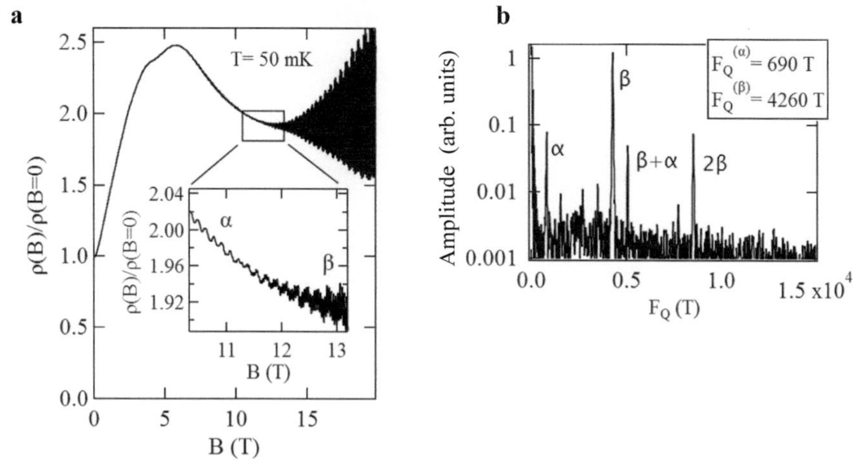

Fig. 3.15 **a** Quantum oscillations of the resistance at 50 mK in θ-(BEDT-TTF)$_2$I$_3$. Inset: close-up of the oscillation near 12 T. **b** Fourier spectrum of the quantum oscillation in **a.** Peaks arising from the α orbit near the zone boundaries and the magnetic breakdown β orbit are evident

Since the oscillation from the β orbit appears abruptly around 11 T in Fig. 3.15a, one might think that there is a phase transition with a change of the Fermi surface, but this interpretation is incorrect. This is a crossover phenomenon from the Bragg reflection-dominant process to the magnetic breakdown-dominant process. The probability of the magnetic breakdown varies continuously with magnetic field, which cannot be an order parameter of a phase transition.

3.5.3 Various Oscillations Due to Magnetic Breakdown

Even for a simple Fermi surface structure, combinations of magnetic breakdown and Bragg reflection can form various closed orbits; quantum oscillations with many frequencies can appear. Figure 3.16 depicts possible closed orbits for the Fermi surfaces shown in Fig. 3.14a due to the combinations of the magnetic breakdown and the Bragg reflection. The probability of each closed orbit can be roughly obtained by counting how many magnetic breakdowns (P) and Bragg reflections (Q) occur at the zone boundaries. In Fig. 3.16a and b, we note that the α orbit requires two Bragg reflections ($\times$) and the β orbit requires four magnetic breakdowns ($\bigcirc$). The probabilities that the α and β orbits are formed, P_α and P_β will be intuitively expressed as

$$P_\alpha \propto Q^2 = \left[1 - \exp\left(-\frac{B_{MB}}{B}\right)\right]^2, \quad P_\beta \propto P^4 = \exp\left(-\frac{4B_{MB}}{B}\right). \tag{3.65}$$

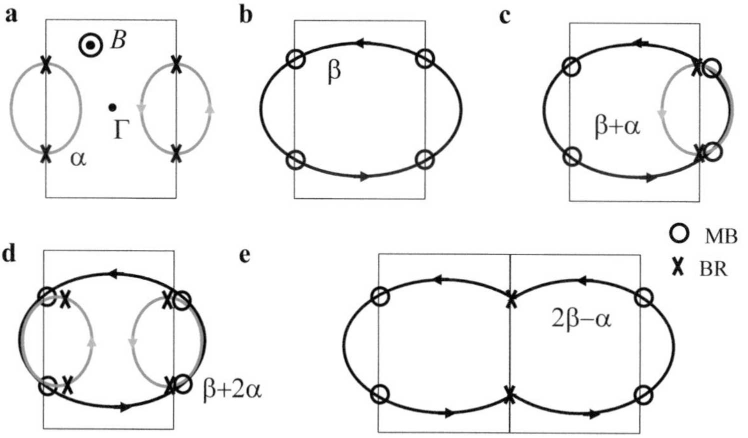

Fig. 3.16 Possible closed orbits due to magnetic breakdown (MB) and Bragg reflection (BR) processes. **a** α orbit, **b** β orbit, **c** $\beta + \alpha$ orbit, **d** $\beta + 2\alpha$ orbit, **e** $2\beta - \alpha$ orbit. The symbols ○ and × indicate MB and BR, respectively

As seen in Fig. 3.16c, the $\beta + \alpha$ orbit requires four magnetic breakdowns and two Bragg reflections. Similarly, it is easy to obtain the probabilities of the $\beta + 2\alpha$ and $2\beta - \alpha$ orbits,

$$P_{\beta+\alpha} \propto P^4 Q^2, \; P_{\beta+2\alpha} \propto P^4 Q^4, \; P_{2\beta-\alpha} \propto P^4 Q^2. \tag{3.66}$$

The probabilities of the closed orbits including both magnetic breakdown and Bragg reflection in the above semiclassical picture are intuitive but too simplified because the phase factors are not incorporated. According to the previous arguments [23, 25], the factor modifying the quantum oscillation amplitude C_{nm}, called "breakdown reduction factor," is given by $C_{nm} \propto (ip_T)^n (q_T)^m$. Here, p_T and q_T are the transition amplitudes of the magnetic breakdown and Bragg reflection, defined as $P = |p_T|^2$ and $Q = |q_T|^2$, respectively, and n and m are the numbers of their processes. The factor i in C_{nm} gives an additional phase of the oscillation. For complicated closed orbits, produced by several non-equivalent orbits, the phase change of p_T or q_T must be independently taken into account in adding up each orbital contribution, and resultantly complicated magnetic field dependence may emerge [23, 26]. The factor C_{nm} is different from what is intuitively expressed by Eq. (3.65). For instance, the breakdown reduction factor of the β orbit is given by $C_{40} \propto |p_T|^4 = P^2$, which is different from $P_\beta \propto P^4$ in Eq. (3.65). The difference can be simply ascribed to the redefinition of the magnetic breakdown field, $B_{MB} \to B_{MB}/2$ in Eq. (3.62) and thus there is no qualitative difference in the magnetic field dependence of the oscillation amplitude between the two cases, C_{40} and P_β.

As mentioned below, some oscillations experimentally observed cannot be explained by the above semiclassical magnetic breakdown picture. The results show that the semiclassical picture is not relevant and that fully quantum mechanical calculations are required to discuss the origin of the observed various frequencies, and to obtain their magnetic field and temperature dependences quantitatively. In addition, as will be discussed in Sect. 3.8, there is another mechanism producing many combination frequencies of α and β, and it leads to different magnetic field dependences. Therefore, it is very difficult to make reliable analyses of the magnetic field dependences of the various oscillations. For simplicity, therefore, we represent the probabilities of the closed orbits by the semiclassical picture in Eq. (3.65).

For the magnetization in Eq. (3.4a), the oscillation from the $\beta + \alpha$ orbit is written as

$$\tilde{M}_{\|} \propto P_{\beta+\alpha} \sum_{p=1}^{\infty} \frac{B^{1/2}}{p^{3/2}} R_T R_D R_S \sin\left[2\pi p\left(\frac{F_Q^{(\beta+\alpha)}}{B} - \frac{1}{2}\right) + \varphi_0\right], \tag{3.67}$$

where the frequency of the closed orbit is given by the sum of the frequencies

$$F_Q^{(\beta+\alpha)} = F_Q^{(\beta)} + F_Q^{(\alpha)}. \tag{3.68}$$

Note that the parameters m_c,T_D, and g_c in all the reduction factors, R_T, R_D, and R_S in Eq. (3.67) are characterized by the $\beta + \alpha$ orbit. For instance, the effective mass of the $\beta + \alpha$ orbit, $m_{\beta+\alpha}$ should be different from the masses m_α and m_β. The area of the $\beta + \alpha$ orbit is written as

$$A_{\beta+\alpha} = A_\beta + A_\alpha, \tag{3.69}$$

and thus the effective mass will also be given by the sum of these two values from Eq. (3.44),

$$m_{\beta+\alpha} = m_\beta + m_\alpha. \tag{3.70}$$

Since Eq. (3.67) is proportional to $P_{\beta+\alpha} \propto P^4 Q^2$, the amplitude of the $\beta + \alpha$ oscillation does not increase monotonically with increasing magnetic field. The amplitude is expected to have a maximum value at an order of B_{MB}, and the $\beta + \alpha$ oscillation disappears in the high magnetic field limit because $Q \to 0$ for $B \to \infty$. There is no significant difference in the magnetic field dependence of the oscillation with $F_Q^{(\beta+\alpha)}$ even in the case for $C_{42} \propto |p_T|^4 |q_T|^2 = P^2 Q$.

How many frequencies are actually observed? Fig. 3.17a and b present the quantum oscillations of the magnetic torque for κ-(BEDT-TTF)$_2$Cu(NCS)$_2$, and their Fourier spectra in the range between 20.1 and 30 T [27]. The magnetic field is applied perpendicular to the layers. The Fermi surface structure of this conductor is equivalent to that in Fig. 3.14a, and thus various closed orbits can be formed by the magnetic breakdown and Bragg reflection, as shown in Fig. 3.16. In fact, many peaks, assigned to α, β, their harmonics and combinations are evident in

Fig. 3.17b. Most of them can be explained by the semiclassical picture but $\beta - \alpha$ and $2\beta - 2\alpha$ cannot. The reason is that the semiclassical picture does not take into account quantum interference effects between the different oscillations. Fully quantum–mechanical calculations of the Landau-quantized states yield the oscillations ($\beta - \alpha$, etc.) which are forbidden in the above semiclassical picture [28, 29]. The results also suggest that the magnetic field dependences of the oscillations should be carefully discussed when the magnetic breakdown occurs.

When many oscillations are observed as in Fig. 3.17b, one might think that it would be difficult to determine which peak is assigned to the β orbit. From the number of carriers in κ-(BEDT-TTF)$_2$Cu(NCS)$_2$, the area of the β orbit A_β must be equal to that of the first Brillouin zone. Therefore, the frequency of the β orbit is easily calculated from the lattice constants, $F_Q^{(\beta)} \approx 4000$ T. On the other hand, since the effective mass is $m_c \propto \partial A_F/\partial E$ as defined in Eq. (3.44), the larger the cross section, the heavier the mass should be. For the $\beta - \alpha$ and $2\beta - 2\alpha$ oscillations, which cannot be explained by the semiclassical picture, the fully quantum mechanical calculations show that their masses become heavier, $m_{\beta-\alpha} > m_\beta$ and $m_{2\beta-2\alpha} > m_{2\beta}$. Thus, the quantum oscillations with these combination frequencies are suppressed more strongly with increasing temperature. In Fig. 3.17b, many peaks observed at 0.44 K ($\beta - \alpha$, $\beta + \alpha$, $\beta + 2\alpha$, $2\beta - 2\alpha$) are not visible at 1.65 K and only α, 2α and β with the relatively small effective masses are evident. This result also allows us to clarify which peak is assigned to the α or β orbit. Other examples of closed orbits due to magnetic breakdown and Bragg reflection processes are given in Appendix 3.3.

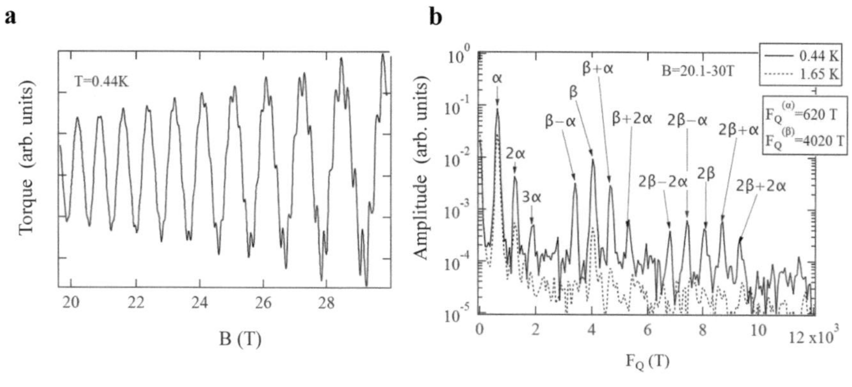

Fig. 3.17 **a** Quantum oscillation of magnetic torque for quasi-2D organic conductor κ-(BEDT-TTF)$_2$Cu(NCS)$_2$ in a high magnetic field range [27]. **b** Fourier spectra of quantum oscillations at 0.44 and 1.65 K. The frequencies arising from the α and β orbits are 620 and 4020 T, respectively. The area of the β orbit A_β with $F_Q^{(\beta)}$ is equal to that of the first Brillouin zone

3.5.4 Maximum of Resistance Due to Magnetic Breakdown

For normal conductors, the resistance increases monotonically with magnetic field. However, as shown in Fig. 3.15a, the resistance for θ-(BEDT-TTF)$_2$I$_3$ has a broad maximum around 6 T. How can we explain the resistance maximum? This conductor has quasi-1D and 2D Fermi surfaces as shown in Fig. 3.18a. In the low magnetic field limit, these two Fermi surfaces contribute to the conduction independently, and thus the resistance along the x-axis is given by $\rho_L = 1/\left(\sigma_{1D} + \sigma_{2D}^{\alpha}\right)$, where σ_{1D} and σ_{2D}^{α} indicate the conductivities from the open orbits of the 1D Fermi surfaces and the α orbit, respectively. If the magnetic breakdown does not occur, we expect $\rho_L(B) \rightarrow 1/\sigma_{1D}$ in the high magnetic field limit because $\sigma_{2D}^{\alpha} \propto 1/\omega_c^2\tau^2 \rightarrow 0$ as given in Eq. (3.16). On the 1D Fermi surface, where the electron motion is restricted along the x-axis, σ_{1D} will be almost independent of the magnetic field. Therefore, the magnetic field dependence of $\rho_L(B)$ is dominated by σ_{2D}^{α}; $\rho_L(B)$ will gradually increase with magnetic field and saturate at high magnetic fields as illustrated in Fig. 3.18b.

Assuming $E_g = 0$ at the zone boundaries, the conduction is given by the β orbit; $\rho_H \approx 1/\sigma_{2D}^{\beta}$, and its magnetic field dependence should be smaller than ρ_L as shown in Fig. 3.18b because of the larger effective mass for the β orbit. At zero magnetic field, the resistance will be almost the same $\rho_L \approx \rho_H$ in both cases $E_g = 0$ and $E_g > 0$. For $E_g > 0$, the magnetic breakdown will occur frequently at the zone boundaries for $B > B_{MB}$. Therefore, the resistance should shift from the ρ_L curve at low magnetic fields to the ρ_H curve at high magnetic fields, as shown by the solid curve in Fig. 3.18b. In this way, the resistance maximum is qualitatively interpreted by the magnetic breakdown effect. Detailed arguments are made in the literature [30]. In fact, at higher magnetic fields after the emergence

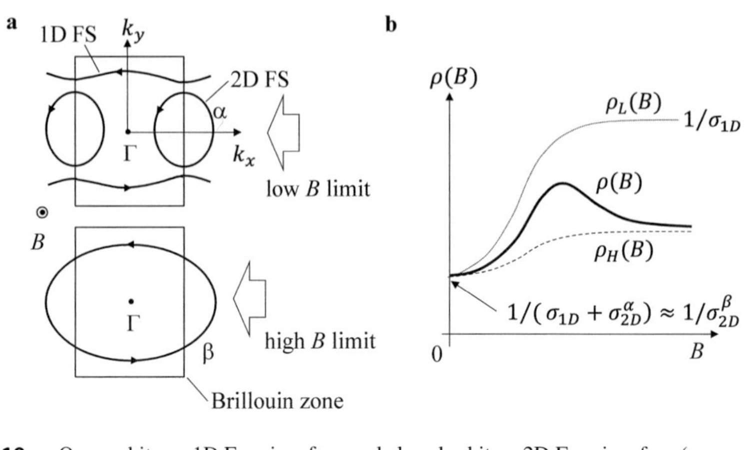

Fig. 3.18 **a** Open orbits on 1D Fermi surface and closed orbit on 2D Fermi surface (upper panel), and magnetic breakdown orbit (lower panel). **b** Schematic magnetic field dependence of resistance in θ-(BEDT-TTF)$_2$I$_3$ (see text in detail)

of the resistance maximum, the quantum oscillation from the β orbit is evident as shown in Fig. 3.15a. Again, the resistive maximum is the crossover behavior due to the magnetic breakdown, not a phase transition. The magnetic field of the resistive maximum depends on the magnetic breakdown field B_{MB}, the scattering time τ, and the effective mass m_c. As τ becomes shorter, the resistance maximum shifts to higher magnetic fields and is smeared out.

3.6 Effect of Internal Magnetic Field

We consider quantum oscillations in the presence of an internal magnetic field. As illustrated in Fig. 3.19a, we suppose that there exist large local spins ($\boldsymbol{S}_{\text{loc}}$) in addition to the conduction electrons with the spins ($\boldsymbol{s}$). When the exchange interaction $E_{\text{ex}} = J_{\text{ex}}\boldsymbol{s} \cdot \boldsymbol{S}_{\text{loc}}$ is present between $\boldsymbol{S}_{\text{loc}}$ and $\boldsymbol{s}$, the internal magnetic field $\boldsymbol{B}_{\text{int}}$ acting on the conduction electron spins is written as

$$\boldsymbol{B}_{\text{int}} = -\frac{J_{\text{ex}}\boldsymbol{S}_{\text{loc}}}{g_c\mu_B}. \tag{3.71}$$

This equation gives the relation $E_{\text{ex}} = -g_c\mu_B\boldsymbol{s}\cdot\boldsymbol{B}_{\text{int}}$. In a sufficiently high external magnetic field $\boldsymbol{B}\|z$ at low temperatures, the local spins are almost completely polarized by $\boldsymbol{B}$ and thus $\boldsymbol{B}_{\text{int}}$ can be taken as a constant.

The Landau levels are Zeeman-split as shown in Fig. 3.19b, where the energy splitting is given by

$$\Delta E = g_c\mu_B(B + B_{\text{int}}). \tag{3.72}$$

As seen in Eqs. (3.35) and (3.36), the Zeeman effect leads to a phase shift of the quantum oscillation. The sum of the oscillatory parts for $s_z = \pm 1/2$ is written

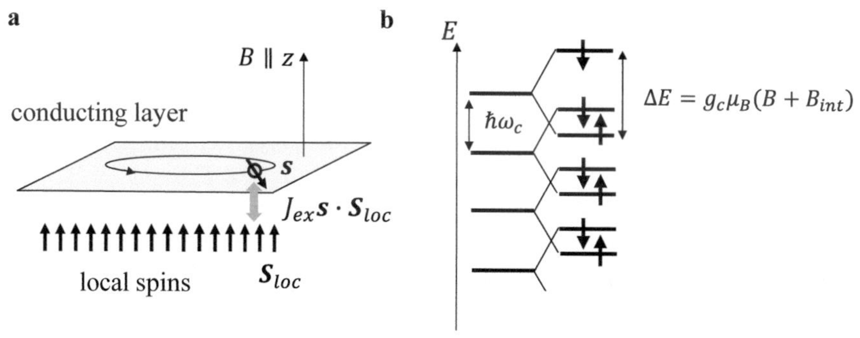

Fig. 3.19 **a** Schematic of conducting layer and local spins. J_{ex} is exchange interaction between local spins $\boldsymbol{S}_{\text{loc}}$ and conduction electron spins $\boldsymbol{s}$. **b** Zeeman splitting of Landau levels due to external and internal magnetic fields

as

$$\cos\left[\Phi_B+\frac{\pi\Delta E}{\hbar\omega_c}\right]+\cos\left[\Phi_B-\frac{\pi\Delta E}{\hbar\omega_c}\right]$$
$$=\cos\left[2\pi\left(\frac{F_{Q+}}{B}-\frac{1}{2}\right)+\varphi_0+\pi S_{\rm spin}\right]$$
$$+\cos\left[2\pi\left(\frac{F_{Q-}}{B}-\frac{1}{2}\right)+\varphi_0-\pi S_{\rm spin}\right], \quad (3.73)$$

where $F_{Q+}=F_Q+\Delta F$ and $F_{Q-}=F_Q-\Delta F$ are the frequencies for the down and up spins, respectively, and ΔF is given by

$$\Delta F=\frac{g_c}{4}\mu_c B_{\rm int}. \quad (3.74)$$

Thus, the presence of the constant internal magnetic field $B_{\rm int}$ results in two frequencies, F_{Q+} and F_{Q-} even for a single extremal cross section.

Figure 3.20a shows the quantum oscillations of the resistance for the quasi-2D organic conductors λ-(BETS)$_2$MCl$_4$ (M = Ga and Fe) [31]. They are isostructural layered conductors with the same 2D Fermi surface. The oscillations are normalized by the non-oscillatory backgrounds $\rho_0(B)$. These conductors consist of conducting layers formed by the BETS molecules (Fig. 3.9b) and insulating layers formed by the MCl_4^{-1} (M = Ga or Fe) ions as schematically shown in the inset of Fig. 3.20a. The Ga^{3+} ions in the insulating layers are nonmagnetic ($S_{\rm loc}$ = 0) while the Fe^{3+} ions have large spins ($S_{\rm loc}$ = 5/2). A rather large exchange interaction between the Fe spins and the conduction electron spins results in an internal magnetic field $B_{\rm int}$ in the BETS layers. Since the radius of the cyclotron orbit is much larger than the distance between the Fe^{3+} ions, the conduction electrons experience a uniform $B_{\rm int}$.

As shown in Fig. 3.20a, the nonmagnetic λ-(BETS)$_2$GaCl$_4$ shows a simple sinusoidal quantum oscillation. In contrast, the magnetic λ-(BETS)$_2$FeCl$_4$ shows oscillations with periodic nodes. The nodes are due to an interference between the two frequencies F_{Q+} and F_{Q-}. Figure 3.20b shows the Fourier transform spectra (FT, solid curves) and the spectra calculated by the maximum entropy method (MEM, dotted curves) for both conductors. The MEM technique can provide much higher resolution spectra. For λ-(BETS)$_2$FeCl$_4$, two peaks are evident in the FT and MEM spectra, giving $2\Delta F$ = 128 T. We obtain $B_{\rm int}$ = 32 T from g_c = 2.0 and μ_c = 4.0 using Eq. (3.74). The dotted curves in Fig. 3.20 a show the calculated results by the L–K formula with appropriate parameters, which reasonably reproduce the experimental data. In λ-(BETS)$_2$FeCl$_4$, superconductivity is surprisingly induced in strong magnetic fields parallel to the conduction layers [32, 33]. This magnetic-field-induced superconductivity is closely related to the large internal magnetic field generated by the Fe spins.

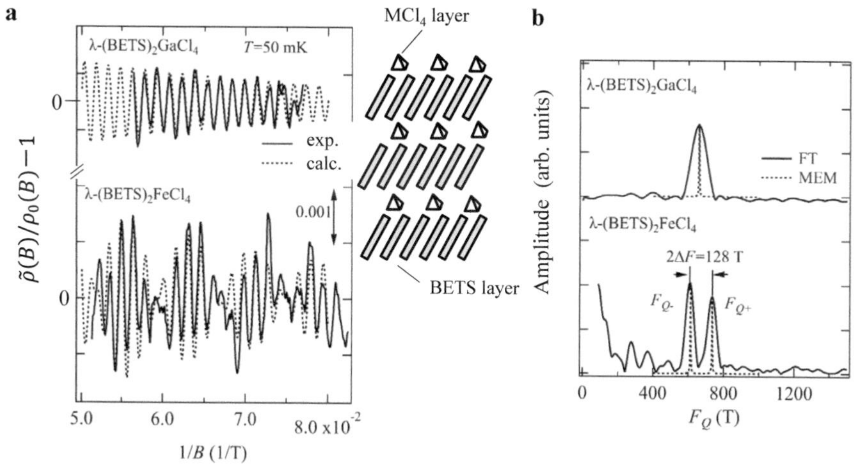

Fig. 3.20 **a** Quantum oscillations of resistance normalized by the non-oscillatory backgrounds $\rho_0(B)$ for quasi-2D organic conductors λ-(BETS)$_2$MCl$_4$ (M = Ga, Fe) [31]. Dotted curves indicate the calculated results by the L–K formula with appropriate parameters. Inset: schematic of layered structure of λ-(BETS)$_2$MCl$_4$. **b** Fourier (solid curves) and MEM spectra (dotted curves) of the oscillations

3.7 Quantum Oscillation of Quasi-2D Fermi Surface

Let us return to the quasi-2D Fermi surface in Fig. 2.19. The energy band is expressed as

$$E(\boldsymbol{k}) = \frac{\hbar^2}{2m_c}\left(k_x^2 + k_y^2\right) - 2t_c\cos(k_z c).$$

The magnetic field is assumed to be parallel to the $k_z(\|z)$-axis. Taking $E_F = \hbar^2 k_F^2/2m_c$, the cross section of the Fermi surface perpendicular to the magnetic field $A_F = \pi k_F^2 = \pi\left(k_x^2 + k_y^2\right)$ is given by

$$A_F = \frac{2m_c\pi}{\hbar^2}(E_F + 2t_c\cos(k_z c)). \tag{3.75}$$

From Eq. (2.60), the frequencies arising from the two extremal cross sections are $F_{Q0} \pm \Delta F_Q$, where

$$F_{Q0} = \frac{m_c}{\hbar e}E_F,\ \Delta F_Q = \frac{2m_c}{\hbar e}t_c. \tag{3.76}$$

If t_c is sufficiently small, the relation $2t_c < \hbar\omega_c$, which is identical to $\Delta F_Q < B$, will be satisfied in the experimental magnetic field range. In this case, the quasi-2D Fermi surface can lie completely between two Landau tubes as shown

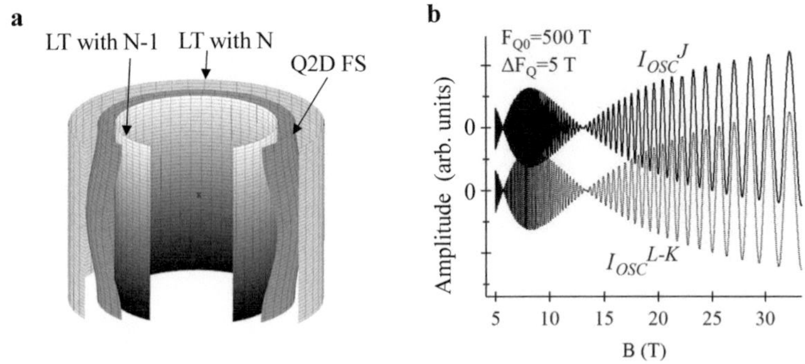

Fig. 3.21 **a** Quasi-2D Fermi surface (Q2D FS) and two Landau tubes (LTs) with indices $N-1$ and N. **b** Calculated quantum oscillations of the thermodynamic grand potential, $I^J_{\rm osc}(B)$ in Eq. (3.77) and $I^{\rm L-K}_{\rm osc}(B)$ in Eq. (3.78)

in Fig. 3.21a. How accurate is the L–K formula in this case? The frequency of the quantum oscillation arising from the cross section on the k_z plane is given by

$$\begin{aligned} F_Q(k_z) = \frac{\hbar}{2\pi e}A_F(k_z) &= \frac{m_c}{\hbar e}(E_F + 2t_c\cos(k_z c)) \\ &= F_{Q0} + \Delta F_Q \cos(k_z c). \end{aligned}$$

On the assumption of the same mass and Dingle temperature for all cross sections of the Fermi surface, the oscillatory part of the thermodynamic grand potential $I_{\rm osc}(B)$ is obtained by integrating Eq. (3.2b) over the whole k-space,

$$\begin{aligned} I^J_{\rm osc}(B) &= \int_{-\pi/c}^{\pi/c} {\rm d}k_z \cos\left[2\pi\left(\frac{F_Q(k_z)}{B} - \frac{1}{2}\right)\right] \\ &= \frac{2\pi}{c}\cos\left[2\pi\left(\frac{F_{Q0}}{B} - \frac{1}{2}\right)\right] J_0\left(\frac{2\pi\,\Delta F_Q}{B}\right), \end{aligned} \tag{3.77}$$

where $J_0(z)$ is the zeroth-order Bessel function defined as $J_0(z) = (1/2\pi)\int_0^{2\pi}\cos(z\sin\theta){\rm d}\theta$. On the other hand, the oscillatory parts arising from the two extremal areas ($F_{Q0} \pm \Delta F_Q$) are given by Eq. (3.2a),

$$\begin{aligned} I^{\rm L-K}_{\rm osc}(B) = &\sqrt{\frac{B}{\left|F''_Q\right|}}\left\{\cos\left[2\pi\left(\frac{F_{Q0}+\Delta F_Q}{B} - \frac{1}{2}\right) - \frac{\pi}{4}\right]\right. \\ &\left. + \cos\left[2\pi\left(\frac{F_{Q0}-\Delta F_Q}{B} - \frac{1}{2}\right) + \frac{\pi}{4}\right]\right\} \\ = &\frac{2}{c}\sqrt{\frac{B}{\Delta F_Q}}\cos\left[2\pi\left(\frac{F_{Q0}}{B} - \frac{1}{2}\right)\right]\cos\left[2\pi\left(\frac{\Delta F_Q}{B}\right) - \frac{\pi}{4}\right]. \end{aligned} \tag{3.78}$$

Here we used the relation, $\left|F''_Q\right| = \Delta F_Q c^2$. For $F_{Q0} = 500$ T and $\Delta F_Q = 5$ T, corresponding to $2t_c/\hbar\omega_c = \Delta F_Q/B = 1/4$ for $B = 20$ T, the oscillations calculated from Eqs. (3.77) and (3.78) are presented in Fig. 3.21b. There is little difference between the two cases. For these parameters, there is a node at about 13.3 T, arising from $2\pi\Delta F_Q/B - \pi/4 = \pi/2$ in Eq. (3.78). No more nodes appear at higher magnetic fields. As can be seen here, the L–K formula is a very good approximation even for highly 2D Fermi surfaces.

3.8 Magnetic Interaction

In Sect. 3.5, we showed that a variety of quantum oscillations can be observed when magnetic breakdown occurs. In this section, we will see that various quantum oscillations can appear owing to a different mechanism than the magnetic breakdown.

3.8.1 Origin of Magnetic Interaction

In the discussion so far, we have assumed that the magnetization M of the sample is sufficiently small and the external magnetic field $\mu_0 H$ is approximated as B inside the sample, $B = \mu_0(H + M) \approx \mu_0 H$. In organic conductors, M is actually small and this assumption is appropriate. However, in some strongly correlated electron systems, for example, M can be non-negligible. The large M makes the observed quantum oscillations even more complicated. Coming back to Eq. (3.4a), we write the oscillatory part of the magnetization for $p = 1$ as

$$\tilde{M} = \sum_i C_{\mathrm{osc}}^{(i)} \sin\left[\frac{2\pi F_Q^{(i)}}{B}\right] = \sum_i C_{\mathrm{osc}}^{(i)} \sin\left[\frac{2\pi F_Q^{(i)}}{\mu_0(H + M)}\right]. \tag{3.79}$$

For simplicity, the phase of the oscillations is neglected. For the general discussion, we assume that the Fermi surface structure has multiple extremal cross sections; the magnetization shows quantum oscillations with multiple frequencies $F_Q^{(i)}$ $(i = 1, 2, 3, \ldots)$. The coefficient $C_{\mathrm{osc}}^{(i)}$, indicating the amplitude of the oscillation with the frequency $F_Q^{(i)}$, contains many parameters. In a conventional experimental setup, the magnetization is measured as a function of H produced by a magnet. The electrons in the sample feel $B = \mu_0(H + M)$ and $M = M_0 + \tilde{M}$, where M_0 and $\tilde{M}$ are the non-oscillatory and oscillatory parts, respectively. From the relation $H \gg M$, the sine term of Eq. (3.79) is rewritten as

$$\sin\left[\frac{2\pi F_Q^{(i)}}{\mu_0(H + M)}\right] \approx \sin\left[\frac{2\pi F_Q^{(i)}}{\mu_0 H}\left(1 - \frac{M}{H}\right)\right]$$

$$
\begin{aligned}
&= \sin\left[\frac{2\pi F_Q^{(i)}}{\mu_0 H}\left(1-\frac{M_0}{H}\right)-\frac{2\pi F_Q^{(i)}}{\mu_0 H}\frac{\tilde{M}}{H}\right] \\
&\approx \sin\left[\frac{2\pi F_Q^{(i)}}{\mu_0 H}-\frac{2\pi F_Q^{(i)}}{\mu_0 H}\frac{\tilde{M}}{H}\right].
\end{aligned} \tag{3.80}
$$

The non-oscillatory part of the magnetization $M_0 = \chi_0 H \ll H$, where χ_0 is the non-oscillatory magnetic susceptibility, gives only a small constant correction to the frequency and thus does not deform the waveform. Since $F_Q^{(i)} \gg \mu_0 H$ in most cases, the last term in the square bracket may satisfy

$$
\frac{2\pi F_Q^{(i)}}{\mu_0 H}\frac{\tilde{M}}{H} \approx 1, \tag{3.81}
$$

when $\tilde{M}$ is sufficiently large. In this case, the sinusoidal waveform, $\sin\left(2\pi F_Q^{(i)}/\mu_0 H\right)$ is strongly deformed by $\tilde{M}$ in Eq. (3.80). This waveform deformation mechanism by $\tilde{M}$ is called magnetic interaction. Since it is difficult to solve $\tilde{M}$ in Eq. (3.79) analytically, the effect of the magnetic interaction has been discussed under some assumptions [34]. In the following, we will see how the waveform is deformed by the magnetic interaction and which oscillation components appear as a result of the magnetic interaction by means of numerical simulations.

3.8.2 Self-Magnetic Interaction

Let us first look at the case of a single frequency. The case in which a single frequency deforms its own waveform is called self-magnetic interaction. In the following simulations of the magnetization, we explicitly include the temperature reduction factor R_T to discuss the temperature dependence,

$$
\tilde{M} = Z_{\mathrm{osc}} R_T \sin\left[\frac{2\pi F_Q^{\alpha}}{\mu_0 (H+M)}\right], \quad M = M_0 + \tilde{M}. \tag{3.82}
$$

For simplicity, we neglect M_0 giving a small correction ($M_0 = 0$). As the coefficient Z_{osc} increases, the sinusoidal waveform becomes more deformed. Figure 3.22a plots the simulated waveforms at $T = 0.5$ K, $\mu_c = 1$, and $F_Q^{\alpha} =$ 1000 T for negligible magnetic interaction (no MI) with $\mu_0 Z_{\mathrm{osc}} = 2\times10^{-5}$[T] and for non-negligible magnetic interaction (MI) with $\mu_0 Z_{\mathrm{osc}} = 2\times10^{-3}$[T]. Note that the waveform for MI is tilted and approaches a sawtooth shape by the magnetic interaction. Because of the waveform deformation, harmonic components (2α, 3α, …) in addition to the fundamental one (α) appear in the Fourier spectrum as shown in Fig. 3.22b. In the Fourier spectrum alone, the harmonics produced by the magnetic interaction are indistinguishable from the $p \geq 2$ terms in the L–K formula given in Eq. (3.4a),

$$\tilde{M}_{\parallel}(B) \propto \frac{1}{p^{3/2}} R_T R_D R_S \sin\left[2\pi p\left(\frac{F_Q^{\alpha}}{\mu_0 H} - \frac{1}{2}\right) + \varphi_0\right].$$

Figure 3.22c shows the mass plots (solid lines) of the fundamental (α) and its harmonics (2α, 3α) produced by the magnetic interaction. The mass plots of α_{LK}, $2\alpha_{\mathrm{LK}}$ and $3\alpha_{\mathrm{LK}}$ expected from the L–K formula are also shown for comparison. We find no significant difference between α and α_{LK}, but 2α and 3α show less temperature dependence than $2\alpha_{\mathrm{LK}}$ and $3\alpha_{\mathrm{LK}}$, respectively.

In actual experiments, the harmonics expected from the L–K formula and those produced by the magnetic interaction can be observed simultaneously. The amplitudes of the harmonics show different temperature and magnetic field dependences between the two cases, depending on the material parameters. Since the main information required is generally the effective mass and the magnetic field angle dependence only of the fundamental oscillation (not the harmonics), the presence of the magnetic interaction is not a serious problem. As the magnetic interaction becomes stronger, the waveform shows a distinct sawtooth shape, resulting in more and larger harmonics [34].

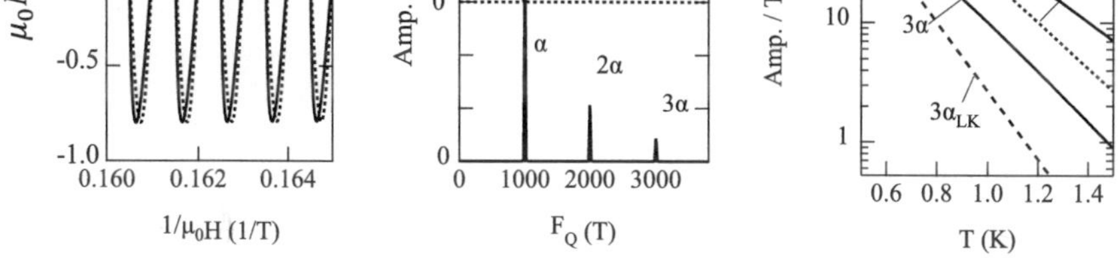

Fig. 3.22 **a** Simulated waveforms of magnetizations at $T = 0.5$ K, $\mu_c = 1$, and $F_Q^{\alpha} = 1000$ T. The dotted and solid curves show the waveforms for negligible magnetic interaction (no MI) with $\mu_0 Z_{\mathrm{osc}} = 2 \times 10^{-5}$[T] and that for non-negligible magnetic interaction (MI) with $\mu_0 Z_{\mathrm{osc}} = 2 \times 10^{-3}$[T]. **b** Fourier spectra of the simulated oscillations. Harmonics appear for MI (solid curve). **c** Mass plots of the oscillations (α, 2α, 3α) for MI. For comparison, the mass plots of α_{LK}, $2\alpha_{\mathrm{LK}}$ and $3\alpha_{\mathrm{LK}}$ expected from the L–K formula are also shown. The curves of α_{LK}, $2\alpha_{\mathrm{LK}}$ and $3\alpha_{\mathrm{LK}}$ are arbitrarily shifted

3.8.3 Mutual Magnetic Interaction

Next, we look at the effect of magnetic interaction in the presence of multiple frequencies. When multiple frequencies affect each other, it is called mutual magnetic interaction. We take the oscillatory parts of the magnetization with two frequencies (α, β) as $\tilde{M}^{(\alpha)}$ and $\tilde{M}^{(\beta)}$, respectively, and thus

$$\begin{aligned}\tilde{M} &= \tilde{M}^{(\alpha)} + \tilde{M}^{(\beta)} \\ &= Z_{osc}^{(\alpha)} R_T \sin\left[\frac{2\pi F_Q^{(\alpha)}}{\mu_0(H+M)}\right] + Z_{osc}^{(\beta)} R_T \sin\left[\frac{2\pi F_Q^{(\beta)}}{\mu_0(H+M)}\right], \\ M &= M_0 + \tilde{M}.\end{aligned} \tag{3.83}$$

We similarly exclude the phase factor and further assume $M_0 = 0$ and $Z_{\text{osc}}^{(\alpha)} = Z_{\text{osc}}^{(\beta)} = Z_{\text{osc}}$ for simplicity. The effective mass ratio in the reduction factor R_T is also assumed to be unity, $\mu_c^{(\alpha)} = \mu_c^{(\beta)} = \mu_c = 1$. Figure 3.23a shows the simulated waveforms at $T = 0.5$ K, $\mu_c = 1$, $F_Q^{\alpha} = 300$ T, and $F_Q^{(\beta)} = 1000$ T for negligible magnetic interaction (no MI) with $\mu_0 Z_{\text{osc}} = 2 \times 10^{-5}$[T] and for non-negligible magnetic interaction (MI) with $\mu_0 Z_{\text{osc}} = 2 \times 10^{-3}$[T]. The waveform for MI is also tilted and approaches a sawtooth shape. This results in harmonic components of α and β (2α, 3α, … 2β, 3β, …) as well as their combinations ($\beta - \alpha$, $\beta + \alpha$, $\beta + 2\alpha$, …) in the Fourier spectrum as seen in Fig. 3.23b. As the magnetic interaction becomes stronger, the waveform shows a distinct complicated sawtooth shape, resulting in more and larger harmonics and combinations [34].

Figure 3.23c presents the mass plots of the fundamental oscillations (α, β), their harmonics and combinations produced by the magnetic interaction. The temperature dependences of the harmonics of the fundamental oscillations (2α, 3α, … 2β, 3β, …) and their combinations ($\beta - \alpha$, $\beta + \alpha$, $\beta + 2\alpha$, $2\beta - \alpha$, …) are all approximately linear in their mass plots. The mass plots of $\alpha_{\text{LK}} (= \beta_{\text{LK}})$ and $2\alpha_{\text{LK}} (= 2\beta_{\text{LK}})$ expected from Eq. (3.4a) are also shown for comparison. For the fundamental oscillations (α, β), there are no significant differences between the two cases. We also note that 2α, $\beta \pm \alpha$, and 2β show smaller temperature dependences than $2\alpha_{\text{LK}} (= 2\beta_{\text{LK}})$. Since the harmonic and combination oscillations decrease faster than the fundamental ones with increasing temperature and decreasing magnetic field, it is generally possible to determine which are the fundamental oscillations from the temperature and magnetic field dependences.

For the Fermi surface in Fig. 3.14a, only β and its harmonic oscillations become dominant at high magnetic fields in the case of the magnetic breakdown. In contrast, various combination frequencies of α and β become prominent at high magnetic fields when the magnetic interaction is sufficiently large. The magnetic field dependence of the combination frequencies could allow us to discuss their origin.

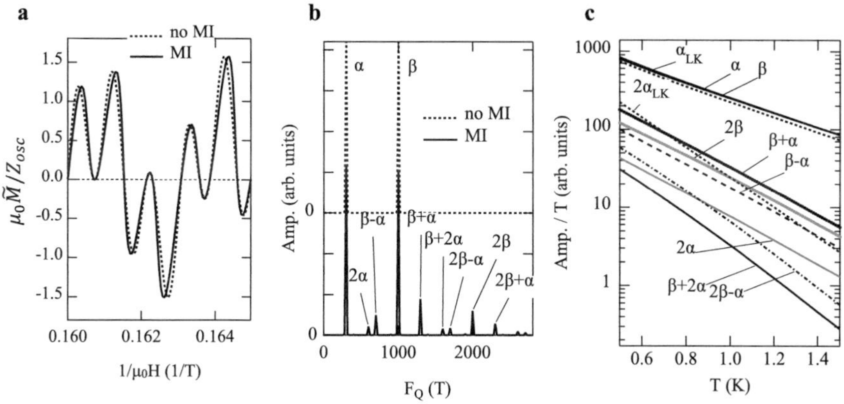

Fig. 3.23 **a** Simulated waveforms of magnetizations at $T = 0.5$ K, $\mu_c = 1, F_Q^{\alpha} = 300$ T, and $F_Q^{(\beta)} = 1000$ T. The dotted and solid curves show the waveforms for negligibly small magnetic interaction (no MI) with $\mu_0 Z_{\text{osc}} = 2 \times 10^{-5}$[T] and that for large magnetic interaction (MI) with $\mu_0 Z_{\text{osc}} = 2 \times 10^{-3}$[T]. **b** Fourier spectra of the simulated oscillations. Various frequencies appear for MI (solid curves). **c** Mass plots of various oscillations. For comparison, the mass plots expected from the L–K formula (α_{LK} and $2\alpha_{\text{LK}}$) are also shown. The curves of α_{LK} and $2\alpha_{\text{LK}}$ are arbitrarily shifted

3.9 Quantum Interference

Up to now it has been assumed that the conduction electrons undergo a cyclotron (closed) motion on a Fermi surface in a magnetic field. In this case, quantum oscillations due to Landau quantization appear in the thermodynamic grand potential as a function of the magnetic field, whose frequency is proportional to the extremal cross section of the Fermi surface. This quantum oscillation appears in all thermodynamic quantities and also in the resistance. In contrast, the resistance in a magnetic field can oscillate even without Landau quantization. Such oscillation arises from quantum interference of the electron wave function, which is induced by magnetic breakdown. It is very similar to the quantum oscillation due to Landau quantization. Let us consider this phenomenon below.

3.9.1 Origin of Quantum Interference

Special geometries of the Fermi surface give rise to a phenomenon called "quantum interference," in which the electron wave function interferes between different paths. Consider a pair of quasi-1D energy bands as shown in Fig. 3.24a. The velocity vectors of the electrons in both bands defined by $\boldsymbol{v}_F = \nabla_k E/\hbar$ point in almost the same direction (k_x-axis). The Fermi surfaces formed by these bands are depicted in Fig. 3.24b, which is the simplest case of the quantum interference. In a magnetic field perpendicular to the $k_x k_y$ plane, the electrons move in the k_y-axis on the Fermi surface by the Lorentz force $\boldsymbol{F} = -e\boldsymbol{v}_F \times \boldsymbol{B}$.

The electron at point A in Fig. 3.24b has two possible paths to point B. On path 1, the electron travels on the same Fermi surface to reach point B through point C. On path 2, the electron reaches point B after two magnetic breakdowns. As discussed in Sect. 3.5, let q_T and p_T be the transition amplitudes of the Bragg reflection and magnetic breakdown, respectively. In general, the wave function changes its phase via quantum tunneling process. We explicitly consider the phase change δ in the magnetic breakdown process, $p_T \to p_T e^{i\delta}$. The phase δ depends on the wave number and the energy gap (barrier). For simplicity, we take q_T and p_T as real numbers, which does not lose generality. The probabilities of the Bragg reflection from point A to point C and the magnetic breakdown from point A to point E are given by $Q = |q_T|^2$ and $P = \left|p_T e^{i\delta}\right|^2 = |p_T|^2$, respectively. The transition amplitude from point A to point B on path 1 is given by $q_T e^{i\varphi_1} q_T$, where φ_1 is the phase change of the wave function on path 1. Similarly, the transition amplitude from point A to point B on path 2 is given by $p_T e^{i\delta} e^{i\varphi_2} p_T e^{i\delta}$, where φ_2 is the phase change on path 2. The same electron at point A passes through both paths and interferes at point B. The transition probability $T_{A\to B}$ from point A to point B is the square of the sum of both transition amplitudes,

$$\begin{aligned} T_{A\to B} &= \left|q_T e^{i\varphi_1} q_T + p_T e^{i\delta} e^{i\varphi_2} p_T e^{i\delta}\right|^2 \\ &= Q^2 + P^2 + 2QP\cos(\varphi_1 - \varphi_2 - 2\delta). \end{aligned} \tag{3.84}$$

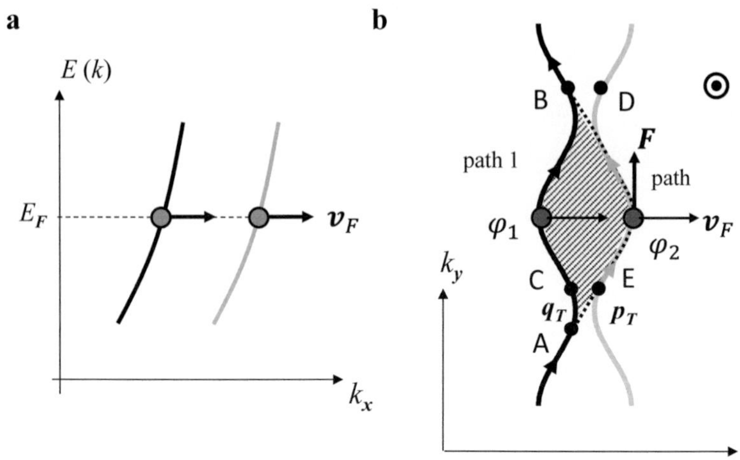

Fig. 3.24 **a** Pair of quasi-1D energy bands. **b** Quasi-1D Fermi surfaces formed by bands in **a**. There are two possible paths (1 and 2) from point A to B

Similarly, the transition probability $T_{A\to D}$ from point A to point D is

$$\begin{aligned} T_{A\to D} &= \left| p_T e^{i\delta} e^{i\varphi_1} q_T + q_T e^{i\varphi_2} p_T e^{i\delta} \right|^2 \\ &= 2QP + 2QP\cos(\varphi_1 - \varphi_2) \end{aligned} \tag{3.85}$$

The sum of Eqs. (3.84) and (3.85) is written as

$$T_A^{\text{sum}} = T_{A\to B} + T_{A\to D} = 1 + 4QP\cos(\delta)\cos(\varphi_1 - \varphi_2 - \delta). \tag{3.86}$$

The second term, which depends on the magnetic field, is the quantum interference of this process. The term $4QP$ has a maximum value of unity for $P = Q = 0.5$. If $B = 0$ or $B \to \infty$, then $P = 0$ or $Q = 0$, respectively, showing no interference.

It is known that the vector potential $\boldsymbol{A}$ changes the phase of the electronic state depending on the path, as shown in Appendix 3.4,

$$\begin{aligned} \varphi_1 - \varphi_2 &= \frac{e}{\hbar}\left(\int_{\text{path 2}} \boldsymbol{A}\cdot \mathrm{d}\boldsymbol{r} - \int_{\text{path 1}} \boldsymbol{A}\cdot \mathrm{d}\boldsymbol{r} \right) = \frac{e}{\hbar}\oint \boldsymbol{A}\cdot \mathrm{d}\boldsymbol{r} \\ &= \frac{e}{\hbar}\int \mathrm{rot}\boldsymbol{A}\cdot \mathrm{d}\boldsymbol{S} = \frac{e}{\hbar}BA_r = \frac{e}{\hbar}\Phi_{1-2}. \end{aligned} \tag{3.87}$$

Here, $\Phi_{1-2} = BA_r$ is the total magnetic flux through the area (A_r) in real space surrounded by paths 1 and 2. We take A_F as the area bounded by paths 1 and 2 in k-space, corresponding to the shaded area in Fig. 3.24b. Using the relation $r = (\hbar/eB)k$ in Eq. (2.6), the area A_r is written as

$$A_r = \left(\frac{\hbar}{eB}\right)^2 A_F. \tag{3.88}$$

The paths 1 and 2 are fixed in k-space, but not in real space; the electrons move faster and the area A_r shrinks as the magnetic field increases. Using Eqs. (3.87) and (3.88), the interference term I_{QI}^{2D} in Eq. (3.86) is given by

$$\begin{aligned} I_{\text{QI}}^{2D} &= 4QP\cos(\delta)\cos(\varphi_1 - \varphi_2 - \delta) \\ &= 4QP\cos(\delta)\cos\left(\frac{2\pi F_Q}{B} - \delta\right), \end{aligned} \tag{3.89}$$

$$\varphi_1 - \varphi_2 = \frac{e}{\hbar}BA_r = \frac{2\pi F_Q}{B}, \quad F_Q = \frac{\hbar}{2\pi e}A_F. \tag{3.90}$$

The interference term shows periodic oscillation with the inverse magnetic field, whose frequency is identical to the quantum oscillation given in Eq. (2.60).

This expression is directly applicable if the energy band in Fig. 3.24a has no energy dispersion in the direction (k_z) perpendicular to the orbital plane. If there exists finite energy dispersion in the k_z direction, the oscillation arising from the extremal area between the two paths can be observed. This is similar to the situation for the quantum oscillations discussed in Sect. 2.4. As given by Eqs. (2.67) and (2.68), the oscillation is written as

$$
\begin{aligned}
I_{\mathrm{QI}} &= \int \mathrm{d}k_z 4QP\cos(\delta)\cos\left(\frac{2\pi F_Q(k_z)}{B} - \delta\right) \\
&= 4\left(\frac{2\pi e}{\hbar}\frac{B}{A_F''}\right)^{\frac{1}{2}} QP\cos(\delta)\cos\left(\frac{2\pi F_Q}{B} - \delta + \varphi_0\right), \qquad (3.91)
\end{aligned}
$$

where F_Q is the frequency corresponding to the extremal area. The phase factor φ_0 is $-\pi/4$ for the maximum cross section and is $\pi/4$ for the minimum cross section.

Since the transition probability T_A^{sum} in Eq. (3.86) has the largest influence on the electrical conductivity in the v_F direction (x-axis), the interference effect I_{QI} should be observed predominantly in the conductivity $\sigma_{xx}(B)$ and thus $\tilde{\sigma}_{xx}(B) \propto I_{\mathrm{QI}}$. The frequency of the quantum interference term I_{QI} is the same as that of the quantum oscillation of the same closed area A_F. However, as seen in Eq. (3.10a), the phase of the quantum oscillation in the conductivity, $\tilde{\sigma}(B) \propto \tilde{D}(B) \propto \cos\left[2\pi\left(F_Q/B - 1/2\right) + \varphi_0\right]$ is different from that of Eq. (3.91). In addition, the magnetic field dependence of the oscillation amplitude is different between the quantum oscillation and quantum interference. Since the magnetic breakdown probability is given by $P = \exp(-B_{MB}/B)$, the amplitude of the quantum interference oscillation will have a maximum value near the magnetic field of $P = Q = 1/\sqrt{2}$. This is in sharp contrast to the case of normal quantum oscillations, in which the oscillation amplitude increases monotonically with the magnetic field.

The quantum oscillation in the conductivity originally results from the oscillation of the thermodynamic grand potential. By contrast, the oscillation due to the quantum interference is the interference of the electronic state between two paths without Landau quantization and thus appears only in the electrical conductivity; the quantum interference cannot be observed in thermodynamic quantities such as heat capacity or magnetization. Therefore, it is possible to experimentally distinguish the oscillation mechanism between quantum oscillation and quantum interference.

3.9.2 Effects of Finite Temperature and Scattering

Figure 3.25a and b schematically depicts the Fermi surfaces where quantum interference and quantum oscillation occur, respectively. In a, the energy bands look like those in Fig. 3.24b, and thus the Fermi surfaces are given by the dotted curves when the Fermi level E_F is slightly raised. In this case, the area A_F bounded by the two paths remains unchanged; the frequency of the quantum interference does not change. In Fig. 3.25b, the area A_F of the closed orbit (electron surface) is enlarged as the Fermi level E_F is raised, as indicated by the dotted curve. The higher E_F causes quantum oscillation with a higher frequency.

The temperature reduction factor R_T can also be defined for the quantum interference. The effective mass defined in Eq. (3.44) is directly related to how much the area enclosed by the orbit changes as the Fermi energy changes,

$$m_c = \left(\frac{\hbar^2}{2\pi}\right) \frac{\partial A_F}{\partial E}\bigg|_{E=E_F} = \left(\frac{\hbar^2}{2\pi}\right)\left[\frac{A_F(E_F+\Delta E) - A_F(E_F)}{\Delta E}\right]. \tag{3.92}$$

For the case of Fig. 3.25a, the effective mass becomes very small, $m_c \propto \partial A_F/\partial E \approx 0$, and thus the reduction factor $R_T \approx 1$, almost independent of temperature and $R_S \approx 1$ for the same reason. This is in sharp contrast to the quantum oscillation in Fig. 3.25b. Other examples of the quantum interference with $m_c \neq 0$ are given in Appendix 3.5.

Next, we see the scattering effect on the quantum interference. As shown in Fig. 3.25a, we assume that the electron moves the length Δk with the time Δt. From $\hbar\dot{k} = F = ev_F B$ and $m_c v_F = \hbar k_F$ for simplicity, we have

$$\Delta t = \frac{\Delta k}{\dot{k}} = \frac{\hbar \Delta k}{ev_F B} = \frac{\Delta k}{k_F}\frac{m_c}{eB} = \frac{2\pi}{\omega_c^{QI}}, \quad \omega_c^{\mathrm{QI}} = 2\pi\frac{k_F}{\Delta k}\left(\frac{eB}{m_c}\right). \tag{3.93}$$

Since the electron undergoes periodic motion on the Fermi surface, we can define the angular frequency ω_c^{QI} in Eq. (3.93). The scattering, which breaks the coherence of the electronic state, reduces the oscillation amplitude. The reduction

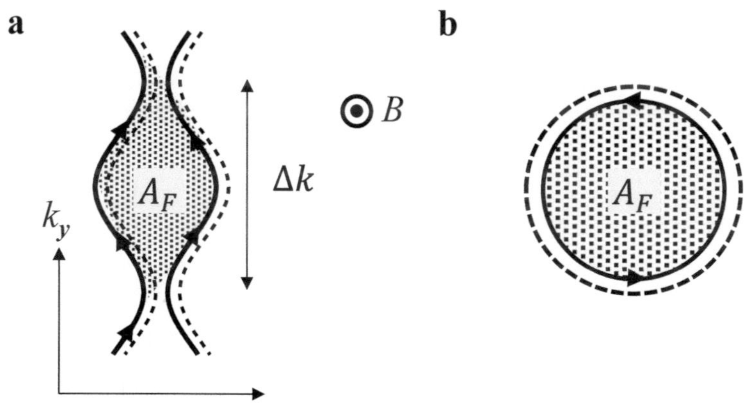

Fig. 3.25 **a** Quasi-1D Fermi surface causing quantum interference. **b** Quasi-2D Fermi surface causing quantum oscillation. The dotted curves in **a** and **b** show the Fermi surfaces with a higher Fermi energy

factor due to the scattering will be given by $R_\tau = \exp(-\Delta t/\tau)$, where τ is the scattering time,

$$R_\tau = \exp\left(-\frac{2\pi}{\omega_c^{\mathrm{QI}}\tau}\right). \tag{3.94}$$

This equation has the same form as the Dingle reduction factor R_D in Eq. (3.32) for the quantum oscillations. The scattering time τ due to the electron-lattice and electron–electron interactions shortens with increasing temperature, and thus the quantum interference will eventually disappear owing to the factor R_τ at higher temperatures.

From the above arguments, the oscillatory part of the quantum interference is given by

$$\frac{\tilde{\sigma}_{xx}(B)}{\sigma_0(B)} \propto I_{\mathrm{QI}} = 4\left(\frac{2\pi e}{\hbar}\frac{B}{A_F''}\right)^{1/2} QPR_\tau \cos(\delta)\cos\left(\frac{2\pi F_Q}{B} - \delta + \varphi_0\right), \tag{3.95}$$

where $\sigma_0(B)$ is the non-oscillatory part depending on the magnetic field. In quasi-1D Fermi surfaces such as Fig. 3.25a, the electrons can move almost only in the x-axis direction, which suggests that the Hall conductivity is sufficiently small, $\sigma_{xx}\sigma_{yy} > \sigma_{xy}^2$. Therefore, the in-plane resistance given by Eq. (3.15) is written as

$$\begin{aligned}\rho_{xx} &= \frac{\sigma_{yy}}{\sigma_{xx}\sigma_{yy} + \sigma_{xy}^2} \approx \frac{1}{\sigma_{xx}} \\ &= \frac{1}{\sigma_0(B) + \tilde{\sigma}_{xx}(B)} \approx \frac{1}{\sigma_0(B)}\left[1 - \frac{\tilde{\sigma}_{xx}(B)}{\sigma_0(B)}\right].\end{aligned} \tag{3.96}$$

The oscillatory part of the resistivity is written as

$$\frac{\tilde{\rho}_{xx}(B)}{\rho_0(B)} \propto -\frac{\tilde{\sigma}_{xx}(B)}{\sigma_0(B)} \propto -B^{1/2} QPR_\tau \cos(\delta)\cos\left(\frac{2\pi F_Q}{B} - \delta + \varphi_0\right), \tag{3.97a}$$

where $\rho_0(B) = 1/\sigma_0(B)$ is the non-oscillatory background. This equation should be compared with Eq. (3.25). For no energy dispersion in the k_z direction, we have

$$\frac{\tilde{\rho}_{xx}(B)}{\rho_0(B)} \propto -QPR_\tau \cos(\delta)\cos\left(\frac{2\pi F_Q}{B} - \delta\right). \tag{3.97b}$$

Figure 3.26a shows the magnetic field dependence of the resistance for a quasi-1D organic conductor $(\mathrm{TMTSF})_2\mathrm{ClO}_4$ [35], whose structure is already depicted in Fig. 1.11. The TMTSF molecules stacked along the a-axis form two pairs of quasi-1D Fermi surfaces as shown in Fig. 1.10f. One FS pair is shown in the inset, which is identical to that in Fig. 3.24b. The stacking structure of the TMTSF molecules leads to the highest conductivity in the a-axis. Although Landau quantization does not occur in magnetic fields, oscillations due to quantum interference can be observed in the resistance along the a-axis, R_{xx} over a wide temperature range. Figure 3.26b plots the oscillatory part with the inverse magnetic field. This oscillation seems quite similar to quantum oscillation due to Landau quantization but not observed in any thermodynamic quantity. The oscillation amplitude decreases with increasing temperature, which is mainly ascribed to the temperature dependence of the scattering time. Comparison between the quantum oscillation and the quantum interference is summarized in Table 3.1.

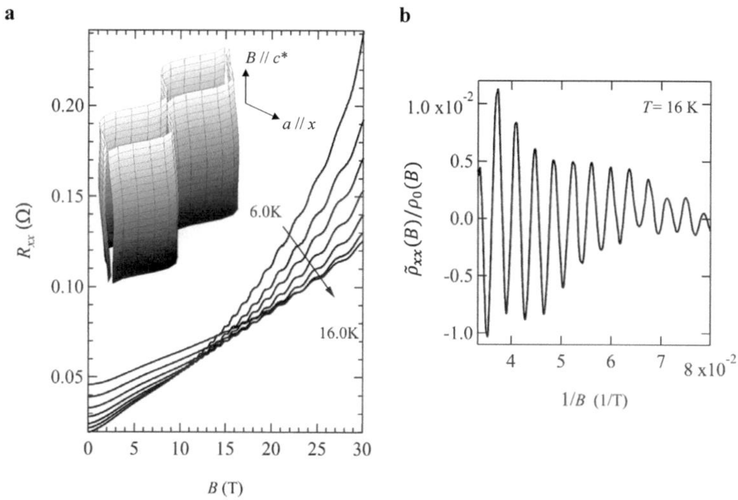

Fig. 3.26 **a** Magnetic field dependence of resistance for quasi-1D organic conductor $(TMTSF)_2ClO_4$ [35]. Inset: schematic of a pair of quasi-1D Fermi surfaces of $(TMTSF)_2ClO_4$, identical to that in Fig. 3.24b. **b** Oscillatory part of resistance due to the quantum interference as a function of the inverse magnetic field

Table 3.1 Comparison between quantum oscillation and quantum interference

	Quantum oscillation	Quantum interference
Origin	Landau quantization	Interference effect of wave function
Observed quantity	All thermodynamic quantities	Only electrical conductivity (resistivity)
Magnetic field dependence	Amplitude monotonically increases with magnetic field	Amplitude has a maximum and then disappears in high magnetic field limit
Temperature dependence	Amplitude exponentially decreases with increasing temperature	Amplitude is less sensitive to temperature in special cases

3.9.3 Aharonov–Bohm Effect

In the quantum interference mentioned above, two paths of the electron on the Fermi surface (k-space) are taken into account. Thus, the area in real space surrounded by the two paths depends on the magnetic field, $A_r \propto 1/B^2$, from Eq. (3.88). In contrast, for a ring-shaped sample with a thin wire, as shown in Fig. 3.27a, the area A_r enclosed by the two paths is fixed, independent of the magnetic field. In this sample, similar quantum interference can occur as long as the electron moves coherently. The electron at point A travels along the two paths simultaneously and then merges at point B, where the two wave functions interfere. This interference is called the Aharonov–Bohm effect. As given in Eq. (3.87), the phase difference between the two paths is written as

$$\varphi_1 - \varphi_2 = \frac{e}{\hbar}\oint \mathbf{A}\cdot \mathrm{d}\boldsymbol{r} = \frac{e}{\hbar}\Phi = 2\pi\frac{\Phi}{\Phi_0}, \tag{3.98}$$

where $\Phi = A_r B$ is the total magnetic flux through the area A_r. The oscillatory part of the electrical conductivity is given by

$$\tilde{\sigma}_{\mathrm{ring}}(B) \propto \cos\left(\frac{2\pi\Phi}{\Phi_0}\right). \tag{3.99}$$

Note that this oscillation is periodic with the magnetic field (not inverse magnetic field). The period of the oscillation corresponds to the magnetic field in which a single flux Φ_0 enters A_r. Figure 3.27b shows the oscillatory part of the resistance for a small Au ring sample [36], where the oscillation with the period of Φ_0 is evident.

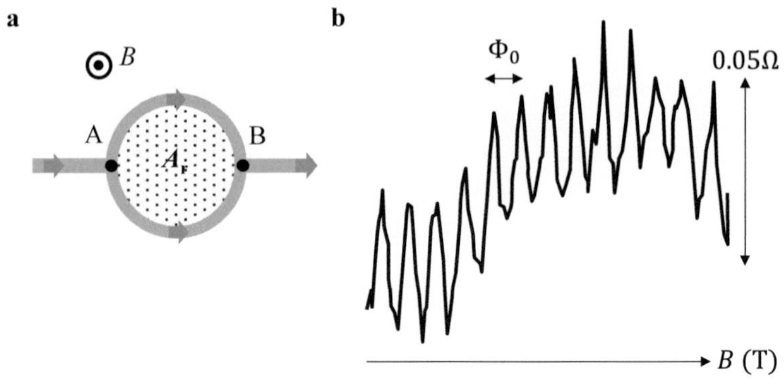

Fig. 3.27 **a** Ring-shaped sample with a thin wire causing quantum interference. **b** Aharonov–Bohm effect for a ring sample with a diameter of 0.8 μm at 10 mK (reproduced with permission from [36])

3.10 Berry Phase

In materials with characteristic energy bands, the electrons acquire a phase as they move along an orbit on the Fermi surface. This phase is called the geometric phase or Berry phase [37]. In this section, we will look at how the Berry phase appears and is experimentally observed.

3.10.1 What Is Berry Phase?

Let $\boldsymbol{k}(t)$ be the wave vector of the time-dependent electronic state. To obtain the eigenstate $\psi_{\boldsymbol{k}}(t)$, we must solve the time-dependent Schrödinger equation,

$$\mathcal{H}(k(t))\psi_k(t) = i\hbar\frac{\partial}{\partial t}\psi_k(t). \tag{3.100}$$

By multiplying by $\psi^*(\boldsymbol{k}(t))$ from the left-hand side and then integrating over space, we have

$$\langle\psi_{\boldsymbol{k}}(t)|\mathcal{H}(\boldsymbol{k}(t))|\psi_{\boldsymbol{k}}(t)\rangle = i\hbar\left\langle\psi_{\boldsymbol{k}}(t)\middle|\frac{\partial}{\partial t}\psi_{\boldsymbol{k}}(t)\right\rangle. \tag{3.101}$$

Let $\phi(\boldsymbol{k})$ be the eigenstate at $t = 0$, $\mathcal{H}(\boldsymbol{k})\phi(\boldsymbol{k}) = E_0\phi(\boldsymbol{k})$. The electron moving adiabatically in k-space does not change its eigenvalue according to the adiabatic theorem, but the phase can change with time. The eigenstate with the phase γ at time t can be written as

$$\psi_{\boldsymbol{k}}(t) = e^{i\gamma(t)}\phi(\boldsymbol{k}). \tag{3.102}$$

Assume that the states $\psi_{\boldsymbol{k}}(t)$ and $\phi(\boldsymbol{k})$ are both normalized. Substituting Eq. (3.102) into the right side of Eq. (3.101), we have

$$\begin{aligned}\left\langle\psi_{\boldsymbol{k}}(t)\middle|\frac{\partial}{\partial t}\psi_{\boldsymbol{k}}(t)\right\rangle &= \left\langle\phi(\boldsymbol{k})\middle|e^{-i\gamma(t)}\frac{\partial}{\partial t}\left[e^{i\gamma(t)}\phi(\boldsymbol{k})\right]\right\rangle \\ &= i\frac{\partial\gamma(t)}{\partial t} + \left\langle\phi(\boldsymbol{k})\middle|\frac{\partial}{\partial t}\phi(\boldsymbol{k})\right\rangle.\end{aligned} \tag{3.103}$$

The left side of Eq. (3.101) is the constant value of E_0 in the adiabatic process and thus

$$i\frac{\partial\gamma(t)}{\partial t} = -\frac{i}{\hbar}E_0 - \left\langle\phi(\boldsymbol{k})\middle|\frac{\partial}{\partial t}\phi(\boldsymbol{k})\right\rangle. \tag{3.104}$$

By integrating Eq. (3.104) over time, the phase change from $t = 0$ to t is given by

$$\gamma(t) = -\frac{E_0}{\hbar}t + i\int_0^t \left\langle \phi(\boldsymbol{k}) \middle| \frac{\partial}{\partial t}\phi(\boldsymbol{k}) \right\rangle \mathrm{d}t. \tag{3.105}$$

The first term is the phase that the electron with kinetic energy E_0 acquires during its motion and is called the dynamical phase. The second term is the phase depending on the path of the electron when $\boldsymbol{k}(t)$ is changed adiabatically. This is called the "geometric phase" or "Berry phase" [37]. By using the relation,

$$\frac{\partial}{\partial t}\phi(\boldsymbol{k}) = \frac{\partial \phi(\boldsymbol{k})}{\partial \boldsymbol{k}} \cdot \frac{\mathrm{d}\boldsymbol{k}}{\mathrm{d}t} = \nabla_k \phi(\boldsymbol{k}) \cdot \frac{\mathrm{d}\boldsymbol{k}}{\mathrm{d}t}, \tag{3.106}$$

we can define the Berry phase as

$$\begin{aligned} \psi_{\mathrm{B}} &= i\int_0^t \left\langle \phi(\boldsymbol{k}) \middle| \frac{\partial}{\partial t}\phi(\boldsymbol{k}) \right\rangle \mathrm{d}t \\ &= i\oint \langle \phi(\boldsymbol{k}) | \nabla_k \phi(\boldsymbol{k}) \rangle \cdot \mathrm{d}\boldsymbol{k} = \oint \boldsymbol{a}_B(\boldsymbol{k}) \cdot \mathrm{d}\boldsymbol{k}. \end{aligned} \tag{3.107}$$

Here, the time integration is replaced by the k-integral on the closed orbit. This is the Berry phase that the electron acquires on its closed orbit. The quantity $\boldsymbol{a}_B$ is called the Berry connection, which is defined as

$$\boldsymbol{a}_B(\boldsymbol{k}) = i\langle \phi(\boldsymbol{k}) | \nabla_k \phi(\boldsymbol{k}) \rangle. \tag{3.108}$$

3.10.2 Berry Phase Around 2D Dirac Point

Let us calculate the Berry phase of the electronic state whose energy band has a Dirac-cone as shown in Fig. 3.28a. The simplest model Hamiltonian, describing the graphene energy band, is already given by Eq. (2.73),

$$\mathcal{H} = v_F\hbar\begin{pmatrix} 0 & k_x + ik_y \\ k_x - ik_y & 0 \end{pmatrix} = v_F\hbar\begin{pmatrix} 0 & ke^{i\varphi} \\ ke^{-i\varphi} & 0 \end{pmatrix}. \tag{3.109}$$

The electrons are confined in the xy plane and $\boldsymbol{k} = \left(k_x, k_y\right) = (k\cos\varphi, k\sin\varphi)$. It is easily verified that the eigenvalues have linear dispersion given by $E_\pm = \pm v_F\hbar k$. The up- and down-spin states are degenerate at zero magnetic field. In a magnetic field, the electronic state is Landau-quantized and each Landau level is further split by the Zeeman effect. In the following, we consider the spinless case for

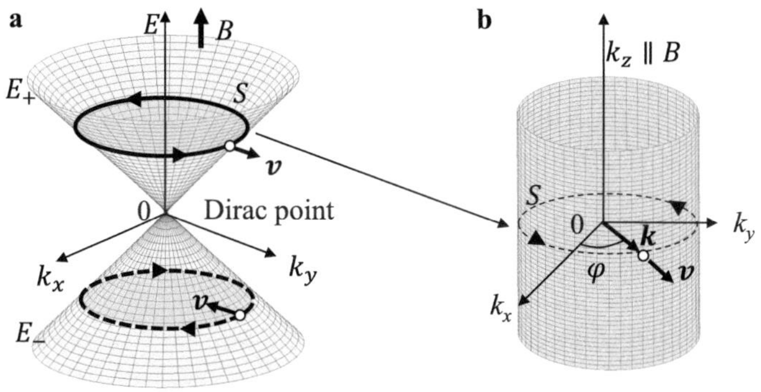

Fig. 3.28 **a** Energy bands with Dirac-cone and cyclotron orbit in a magnetic field along the k_z-axis, **b** 2D Fermi surface and cyclotron orbit for the energy band E_+ ($E_F > 0$)

simplicity. The energy bands form the 2D cylindrical Fermi surface along the k_z-axis, whose cross section for $E_F > 0$ is depicted in Fig. 3.28b. From the relation $\boldsymbol{v} = \nabla_k E(\boldsymbol{k})/\hbar$, the velocity vector $\boldsymbol{v}$ of the band E_+ is parallel to $\boldsymbol{k}$ and that of the band E_- is antiparallel to $\boldsymbol{k}$. The normalized eigenstates for $E_\pm$ are, respectively,

$$\psi_\pm = \frac{1}{\sqrt{2}}\begin{pmatrix} 1 \\ \pm e^{-i\varphi} \end{pmatrix} = \frac{1}{\sqrt{2}}\left[\varphi_A(\boldsymbol{r}) \pm e^{-i\varphi}\varphi_B(\boldsymbol{r})\right],$$
$$\varphi_A(\boldsymbol{r}) = \begin{pmatrix} 1 \\ 0 \end{pmatrix},\ \varphi_B(\boldsymbol{r}) = \begin{pmatrix} 0 \\ 1 \end{pmatrix}. \tag{3.110}$$

In a 2D polar coordinate (k, φ), the gradient is written as

$$\nabla_k = \frac{\partial}{\partial k}\boldsymbol{e}_k + \frac{1}{k}\frac{\partial}{\partial \varphi}\boldsymbol{e}_\varphi, \tag{3.111}$$

where $\boldsymbol{e}_i (i = k, \varphi)$ is the unit vector in the i-axis. Since $\psi_\pm$ is a function only of φ, the Berry connection is easily calculated as

$$\boldsymbol{a}_{B\pm}(k) = i\langle\psi_\pm|\nabla_k\psi_\pm\rangle = \frac{i}{2}\begin{pmatrix} 1 & \pm e^{i\varphi} \end{pmatrix}\nabla_k\begin{pmatrix} 1 \\ \pm e^{-i\varphi} \end{pmatrix}$$
$$= \frac{i}{2}\left[e^{i\varphi}\nabla_k e^{-i\varphi}\right] = \frac{1}{2}\nabla_k\varphi = \frac{1}{2k}\boldsymbol{e}_\varphi. \tag{3.112}$$

In some literature, the Berry connection is defined per unit length ($k = 1$). Both eigenstates give the same Berry connection. Using the relation

$$\mathrm{d}\boldsymbol{k} = \mathrm{d}k\boldsymbol{e}_k + k\mathrm{d}\varphi\boldsymbol{e}_\varphi, \tag{3.113}$$

we obtain the Berry phase

$$\psi_{\mathrm{B}} = \oint \boldsymbol{a}_B(\boldsymbol{k}) \cdot \mathrm{d}\boldsymbol{k} = \oint \frac{1}{2k} \boldsymbol{e}_\varphi \cdot k \mathrm{d}\varphi \boldsymbol{e}_\varphi = \frac{1}{2} \int_0^{2\pi} \mathrm{d}\varphi = \pi, \tag{3.114}$$

assuming a counterclockwise cyclotron motion as shown in Fig. 3.28b. The result shows that the electrons in a magnetic field ($B \| k_z$), which undergo the counterclockwise cyclotron motion around the Dirac point, acquire the Berry phase ($\psi_B = \pi$) as long as they are not scattered. Since the electrons of the band E_-, which forms the hole surface, have opposite velocities, they undergo the clockwise cyclotron motion for $B \| k_z$. The integration range of φ in Eq. (3.114) is $(0, -2\pi)$, and thus the cyclotron motion yields the Berry phase $\psi_B = -\pi$. Although the sign of the Berry phase depends on the eigenstate and on the magnetic field direction, we should note that $\psi_B = \pm\pi$ are physically equivalent.

3.10.3 Berry Curvature

Equation (3.107) is rewritten as

$$\begin{aligned}\psi_B &= \oint \boldsymbol{a}_B(k) \cdot \mathrm{d}\boldsymbol{k} = \int_S \nabla_k \times \boldsymbol{a}_B(k) \cdot \mathrm{d}\boldsymbol{S} \\ &= \int_S \frac{1}{2} \nabla_k \times (\nabla_k \varphi) \cdot \mathrm{d}\boldsymbol{S}.\end{aligned} \tag{3.115}$$

In the second equality, Stokes theorem is used. Note that as long as φ is a differentiable function, $\nabla_k \times (\nabla_k \varphi) = 0$, then ψ_B vanishes. By differentiating $\cos\varphi = k_x/k$ with respect to k_x, we obtain $\partial\varphi/\partial k_x = -\sin\varphi/k$ and thus $\nabla_k \varphi$ diverges at $k = 0$. The Dirac point ($k = 0$) is a singularity in the integral; the Berry phase $\psi_B = \pi$ is ascribed to the integral around this singularity. For a closed orbital plane including no Dirac point (no singularity), we have $\psi_B = 0$. If the off-diagonal terms of the Hamiltonian are not given as in Eq. (3.109), there exists no singularity and thus the Berry phase is zero for any closed orbit on the Fermi surface.

The Berry curvature $\boldsymbol{\Omega}_B(k)$ is defined as

$$\boldsymbol{\Omega}_B(k) = \nabla_k \times \boldsymbol{a}_B(k), \tag{3.116}$$

and thus the Berry phase is written as

$$\psi_B = \int_S \boldsymbol{\Omega}_B(k) \cdot \mathrm{d}\boldsymbol{S}. \tag{3.117}$$

Since $\phi(\boldsymbol{k})$ is the eigenfunction of $\mathcal{H}(\boldsymbol{k})\phi(\boldsymbol{k}) = E_0\phi(\boldsymbol{k})$, $\phi'(\boldsymbol{k}) = \phi(\boldsymbol{k})e^{i\chi(k)}$ with a phase $\chi(k)$ is also the eigenfunction. The Berry connection for $\phi'(\boldsymbol{k})$ is modified as

$$\boldsymbol{a}'_B(\boldsymbol{k}) = i\left\langle \phi'^{(\boldsymbol{k})} | \nabla_k \phi'^{(\boldsymbol{k})} \right\rangle = \boldsymbol{a}_B(\boldsymbol{k}) - \nabla_k \chi(k).$$

By contrast, both $\boldsymbol{\Omega}_B(k)$ and ψ_B are independent of $\chi(k)$, gauge invariant quantities.

3.10.4 Berry Phase in Quantum Oscillation

How does the Berry phase actually affect the quantum oscillation? When the electron acquires the Berry phase ψ_B on a closed orbit, the Bohr quantization condition given by Eq. (2.7) will be modified as,

$$\frac{e}{\hbar}\oint \boldsymbol{p} \cdot \mathrm{d}\boldsymbol{r} = 2\pi N \rightarrow 2\pi N - \psi_B = 2\pi\left(N - \frac{\psi_B}{2\pi}\right). \tag{3.118}$$

In free space, the quantization condition of the electronic state is given by $A_N = 2\pi eB(N + \gamma_0)/\hbar$ in Eq. (2.37), where A_N is the area of the cyclotron orbit at the Nth Landau level in k-space. The factor 1/2, arising from the zero-point energy in Eq. (2.29), is replaced by γ_0 as a general expression. Strictly speaking, $\gamma_0 = 1/2$ is not guaranteed in a non-parabolic band but it gives only a very small correction [38]. Note that $\gamma_0 = 1/2$ is not guaranteed for magnetic breakdown closed orbits either. The phase γ_0 appears in the quantum oscillation of the thermodynamic grand potential, $\cos\left[2\pi\left(F_Q/B - \gamma_0\right)\right]$ in Eqs. (3.2a) and (3.2b). If the electron acquires the Berry phase, the phase of the quantum oscillation is modified as $\gamma_0 \rightarrow \gamma_0 - \psi_B/2\pi$ [39, 40] and thus the L–K formula for $p = 1$ is written as,

$$\tilde{\Omega}(B) \propto \cos\left[2\pi\left(\frac{F_Q}{B} - \gamma_0\right) + \varphi_0 + \psi_B\right], \tag{3.119a}$$

$$\tilde{\Omega}_{2D}(B) \propto \cos\left[2\pi\left(\frac{F_Q}{B} - \gamma_0\right) + \psi_B\right]. \tag{3.119b}$$

Since the Berry phases are added in the same way to the oscillatory parts of all thermodynamic quantities, magnetization, torque, density of states, etc., one can verify the existence of Berry phases by examining the phase of the quantum oscillation of any physical quantity. For simplicity, we take $\gamma_0 = 1/2$ in the following.

Let us see how the Berry phase is actually determined in graphene. The Hamiltonian of the Dirac-cone band in a magnetic field and its eigenvalue are given by Eqs. (2.73) and (2.75), respectively. If the Zeeman effect is taken into account, the Landau level splits into the up- and down-spin levels. However, the Zeeman splitting is much smaller than the Landau level spacing in graphene, $\hbar\omega_c \gg g_c\mu_B B$ and thus the Zeeman splitting can be ignored: the spin splitting is not observed in the quantum oscillation as seen in Fig. 3.5. In a thin film sample such as graphene, the Fermi level (i.e., the number of carriers) can be controlled by applying a voltage to the gate electrode, which is attached to the substrate of the graphene sheet. When the Fermi level E_F is shifted slightly below the Dirac point (inset of Fig. 3.29a), a large quantum oscillation is observed at a low temperature as shown in Fig. 3.29a [41]. The oscillations are periodic with inverse magnetic field. Since the electronic state of graphene can be regarded as a perfect 2D system, the oscillatory part of the density of states is given by

$$\tilde{D}_{2D}(B) \propto \cos\left[2\pi p\left(\frac{F_Q}{B} - \frac{1}{2}\right) + \psi_B\right]. \tag{3.120}$$

As seen in Eq. (3.25), the oscillatory part of the resistivity $\tilde{\rho}_{xx}(B)$ for $p = 1$ is written as

$$\tilde{\rho}_{xx}(B) \propto \tilde{D}_{2D}(B) \propto \cos\left[2\pi\left(\frac{F_Q}{B} - \frac{1}{2}\right) + \psi_B\right] = -\cos\left[2\pi\frac{F_Q}{B} + \psi_B\right]. \tag{3.121}$$

The magnetic fields $B_{\min}$, where the minima of the quantum oscillation of the resistance appear, are given by

$$2\pi\frac{F_Q}{B_{\min}} + \psi_B = 2N_L\pi, \quad N_L = 0, 1, 2, \ldots \tag{3.122}$$

The inverse minimum field $1/B_{\min}$ is plotted against N_L in Fig. 3.29b, which is called a fan plot. In an ideal case, we expect $N_L \to \psi_B/2\pi$ in the limit $1/B \to 0$; $N_L = 1/2$ for $\psi_B = \pi$. In the experiment, the linear extrapolation yields $N_L \approx 0.4$

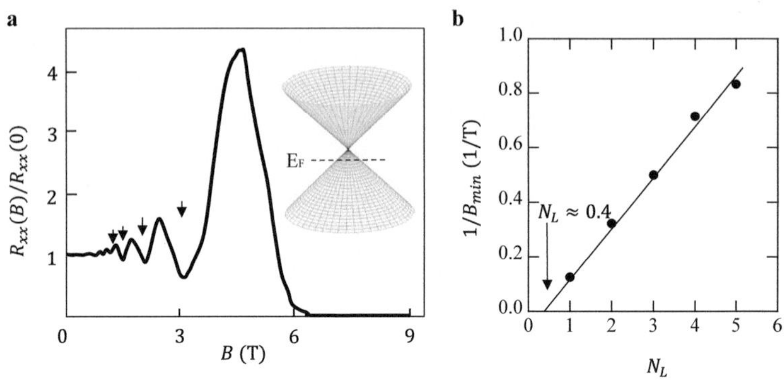

Fig. 3.29 **a** Resistance as a function of magnetic field for graphene with low hole carriers (reproduced with permission from [41]). Inset: Dirac-cone energy band with the Fermi level slightly below the Dirac point. **b** Fan plot of the oscillation shown in **a**. Inverse minimum field $1/B_{\min}$ vs Landau index N_L. The solid line shows that $N_L \approx 0.4$ for $1/B \to 0$.

for $1/B \to 0$, which is a little <1/2. The discrepancy from 1/2 may be due to the experimental error or the nonlinear effect of the bands, which is known to deviate ψ_B from π [42].

3.10.5 Berry Phase Around 3D Dirac Point

For graphene, the electronic state is 2D, and its energy band has a linear dispersion in two directions (k_x, k_y). It is easy to extend the model to a 3D electron system with the Dirac point inside the bulk material. We assume an isotropic 3D material with a linear energy dispersion in all directions $(k_x, k_y, k_z) = (k \sin\theta \cos\varphi, k \sin\theta \sin\varphi, k \cos\theta)$. The model Hamiltonian is written as

$$\mathcal{H} = v_F \hbar \begin{pmatrix} k_z & k_x + ik_y \\ k_x - ik_y & -k_z \end{pmatrix} = v_F \hbar k \begin{pmatrix} \cos\theta & \sin\theta e^{i\varphi} \\ \sin\theta e^{-i\varphi} & -\cos\theta \end{pmatrix}. \tag{3.123}$$

The eigenvalues are easily obtained as

$$E_\pm = \pm v_F \hbar \sqrt{k_x^2 + k_y^2 + k_z^2} = \pm v_F \hbar k. \tag{3.124}$$

The dispersion of the energy band, where the spin states are degenerate, and its spherical Fermi surface are depicted in Fig. 3.30.

The eigenfunctions for $E_\pm$ are respectively,

$$\psi_+ = \begin{pmatrix} \cos(\theta/2) \\ \sin(\theta/2)e^{-i\varphi} \end{pmatrix} = \cos\left(\frac{\theta}{2}\right)\varphi_A(\boldsymbol{r}) + \sin\left(\frac{\theta}{2}\right)e^{-i\varphi}\varphi_B(\boldsymbol{r}), \tag{3.125}$$

$$\psi_- = \begin{pmatrix} \sin(\theta/2) \\ -\cos(\theta/2)e^{-i\varphi} \end{pmatrix} = \sin\left(\frac{\theta}{2}\right)\varphi_A(\boldsymbol{r}) - \cos\left(\frac{\theta}{2}\right)e^{-i\varphi}\varphi_B(\boldsymbol{r}). \tag{3.126}$$

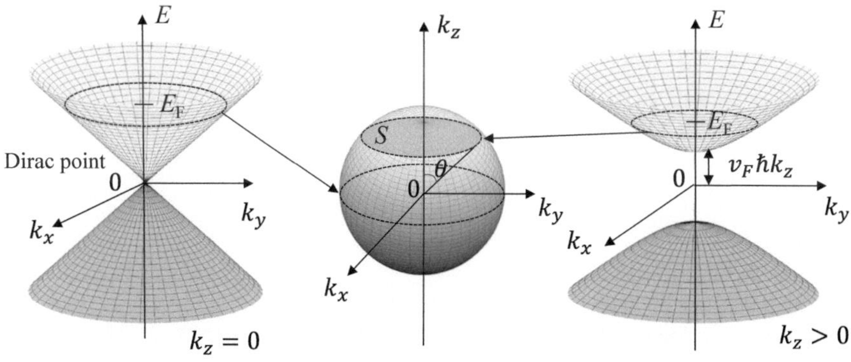

Fig. 3.30 Energy bands of Eq. (3.124) for $k_z = 0$ (left) and $k_z > 0$ (right), and its Fermi surface (center). The area S denotes orbital plane for $k_z > 0$.

Using the unit vectors $\boldsymbol{e}_i(i = k, \theta, \varphi)$ in a 3D coordinate, the gradient is given by

$$\nabla_k = \frac{\partial}{\partial k}\boldsymbol{e}_k + \frac{1}{k}\frac{\partial}{\partial \theta}\boldsymbol{e}_\theta + \frac{1}{k\sin\theta}\frac{\partial}{\partial \varphi}\boldsymbol{e}_\varphi, \tag{3.127}$$

and thus the Berry connection is calculated as,

$$\begin{aligned}
\boldsymbol{a}_{B+}(k) &= i\langle\psi_+|\nabla_k\psi_+\rangle \\
&= i\left[\cos\left(\frac{\theta}{2}\right)\nabla_k\cos\left(\frac{\theta}{2}\right) + \sin\left(\frac{\theta}{2}\right)e^{i\varphi}\nabla_k\sin\left(\frac{\theta}{2}\right)e^{-i\varphi}\right] \\
&= \frac{\sin^2(\theta/2)}{k\sin\theta}\boldsymbol{e}_\varphi, \\
\boldsymbol{a}_{B-}(k) &= i\langle\psi_-|\nabla_k\psi_+\rangle = \frac{\cos^2(\theta/2)}{k\sin\theta}\boldsymbol{e}_\varphi.
\end{aligned} \tag{3.128}$$

The Berry phases obtained by the φ rotation at a fixed θ are calculated as

$$\begin{aligned}
\psi_{\mathrm{B}+} &= \oint \boldsymbol{a}_{B+}(k)\cdot \mathrm{d}\boldsymbol{k} = \int_0^{2\pi}\mathrm{d}\varphi\left[\frac{\sin^2(\theta/2)}{k\sin\theta}\boldsymbol{e}_\varphi\cdot k\sin\theta\boldsymbol{e}_\varphi\right] \\
&= 2\pi\sin^2(\theta/2), \\
\psi_{\mathrm{B}-} &= \oint \boldsymbol{a}_{B+}(k)\cdot \mathrm{d}\boldsymbol{k} = 2\pi\cos^2(\theta/2).
\end{aligned} \tag{3.129}$$

Since the solid angle of the orbital plane S with the angle θ on the spherical Fermi surface in Fig. 3.30 is $4\pi\sin^2(\theta/2)$, the Berry phase ψ_{B+} is 1/2 times the solid angle. In a magnetic field parallel to the k_z-axis, the extremal cross section is the equatorial plane ($k_z = 0$); $\theta = \pi/2$ in Eq. (3.129). Therefore, we obtain the Berry phase $\psi_{B\pm} = \pi$ from the quantum oscillation measurement of this Fermi surface. The sign depends on the direction of the cyclotron motion. The Berry curvatures are calculated as,

$$\begin{aligned}
\boldsymbol{\Omega}_{B+}(k) &= \nabla_k\times\boldsymbol{a}_{B+}(k) \\
&= \frac{\sin\theta}{k}\left[\frac{\partial}{\partial\theta}\left(\sin\theta\boldsymbol{a}_{B+}(k)\cdot\boldsymbol{e}_\varphi\right)\right]\boldsymbol{e}_k = \frac{1}{2k^2}\boldsymbol{e}_k, \\
\boldsymbol{\Omega}_{B-}(k) &= \nabla_k\times\boldsymbol{a}_{B-}(k) = -\frac{1}{2k^2}\boldsymbol{e}_k.
\end{aligned} \tag{3.130}$$

When the Fermi level is close to the Dirac point, the 3D material has few carriers. Such a material is called a Dirac semimetal.

3.10.6 Topological Insulator and 2D Weyl Point

Topological insulators are classified into a special group of materials that are insulating in the bulk but conducting at their surfaces or edges. The bulk states have energy gaps at the Fermi level but the energy bands at the surfaces for 3D topological insulators or at the edges for 2D topological insulators cross the Fermi level, forming conducting states. These peculiar electronic states, where strong spin–orbit interaction of the materials plays an important role, have been of great interest due to their fascinating properties. Substances with the same topology mean that they have the same shape under continuous deformation without being cut or joined. Extending this concept to electronic states, electronic states with different topologies do not mix with each other under continuous deformation. When insulators with different topologies come into contact with each other, their conduction and valence bands cannot be continuously connected, and thus the energy gap closes at the interface, forming a conducting state. Since vacuum is a trivial insulator, topological insulators can have conducting surface states.

The schematic energy bands for topological insulators are depicted in Fig. 3.31a, where the bands cross the Fermi level at the surfaces or edges. These unique conducting states are recognized as robust (symmetry-protected) states against scattering by nonmagnetic impurities or defects. The schematics of the possible edge and surface states are depicted in Fig. 3.31b and c, respectively, where the spin-polarized electrons move in the opposite directions. So far, various classes of topological materials have been found, in which nontrivial states arise owing to the topological energy bands. Here we focus on the electronic states realized on the surface of topological insulators and their spin states.

At the surface of a crystal or at the interface between different materials, the inversion symmetry of the lattice potential is broken, $V(z) \neq V(-z)$, where z is the axis perpendicular to the surface or interface. In such electron systems, the Rashba-type spin–orbit interaction can be induced. The Rashba-type spin–orbit

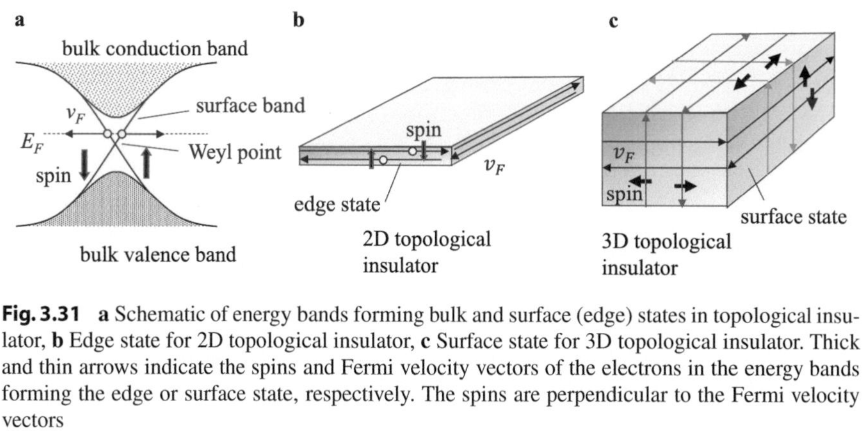

Fig. 3.31 **a** Schematic of energy bands forming bulk and surface (edge) states in topological insulator, **b** Edge state for 2D topological insulator, **c** Surface state for 3D topological insulator. Thick and thin arrows indicate the spins and Fermi velocity vectors of the electrons in the energy bands forming the edge or surface state, respectively. The spins are perpendicular to the Fermi velocity vectors

interaction of a 2D electron system in the xy plane can be written as

$$\mathcal{H}_{SO} = 2\alpha_{SO}\left(k_y s_x - k_x s_y\right), \tag{3.131}$$

where α_{SO} indicates the strength of the spin–orbit interaction. This functional form can be derived in terms of an effective magnetic field as shown in Appendix 3.6. This is recognized as a model Hamiltonian describing the surface state of 3D topological insulators. We define 2×2 Hermitian (Pauli) matrices $\boldsymbol{\sigma}_i$ and a unit matrix $\boldsymbol{I}$ to describe the up- and down-spin states,

$$\begin{aligned} \boldsymbol{\sigma}_x &= \begin{pmatrix} 0 & 1 \\ 1 & 0 \end{pmatrix}, \boldsymbol{\sigma}_y = \begin{pmatrix} 0 & -i \\ i & 0 \end{pmatrix}, \\ \boldsymbol{\sigma}_z &= \begin{pmatrix} 1 & 0 \\ 0 & -1 \end{pmatrix}, \boldsymbol{I} = \begin{pmatrix} 1 & 0 \\ 0 & 1 \end{pmatrix}, \end{aligned} \tag{3.132}$$

where $\boldsymbol{\sigma}_i$ is called the Pauli matrix. The spin operators $\boldsymbol{s}_i$ are expressed as $\boldsymbol{s}_i = (1/2)\boldsymbol{\sigma}_i (i = x, y, z)$; $\boldsymbol{s}_i$ satisfies the spin operator commutation relations $\left[\boldsymbol{s}_x, \boldsymbol{s}_y\right] = \boldsymbol{s}_x\boldsymbol{s}_y - \boldsymbol{s}_y\boldsymbol{s}_x = i\boldsymbol{s}_z, \boldsymbol{s}_i^2 = (1/4)\boldsymbol{I}$, and $\boldsymbol{s}_x^2 + \boldsymbol{s}_y^2 + \boldsymbol{s}_z^2 = s(s+1)\boldsymbol{I} = (3/4)\boldsymbol{I}$. The up- and down-spin states are respectively defined as

$$|\uparrow\rangle = \begin{pmatrix} 1 \\ 0 \end{pmatrix}, |\downarrow\rangle = \begin{pmatrix} 0 \\ 1 \end{pmatrix}. \tag{3.133}$$

It is easy to verify

$$\boldsymbol{s}_z|\uparrow\rangle = \frac{1}{2}|\uparrow\rangle, \boldsymbol{s}_z|\downarrow\rangle = -\frac{1}{2}|\downarrow\rangle, \tag{3.134}$$

and

$$\boldsymbol{s}_+|\downarrow\rangle = \frac{1}{2}|\uparrow\rangle, \boldsymbol{s}_+|\uparrow\rangle = 0, \boldsymbol{s}_-|\uparrow\rangle = \frac{1}{2}|\downarrow\rangle, \boldsymbol{s}_-|\downarrow\rangle = 0, \tag{3.135}$$

where $\boldsymbol{s}_\pm = \boldsymbol{s}_x \pm \boldsymbol{s}_y$ are the spin raising and lowering operators, respectively. We define the $\boldsymbol{k}$ vector of the spin state tilted from the k_x-axis by the angle φ as shown in Fig. 3.32b. Using $k_x = k\cos\varphi$ and $k_y = k\sin\varphi$, the Hamiltonian in Eq. (3.131) is written as

$$\begin{aligned} \mathcal{H}_{SO} &= \alpha_{SO}\left(k_y\boldsymbol{\sigma}_x - k_x\boldsymbol{\sigma}_y\right) \\ &= \alpha_{SO}\begin{pmatrix} 0 & k_y + ik_x \\ k_y - ik_x & 0 \end{pmatrix} = \alpha_{SO}\begin{pmatrix} 0 & ike^{-i\varphi} \\ -ike^{i\varphi} & 0 \end{pmatrix}. \end{aligned} \tag{3.136}$$

We obtain the eigenvalues giving a linear dispersion

$$E_\pm(k) = \pm\alpha_{SO}k,$$

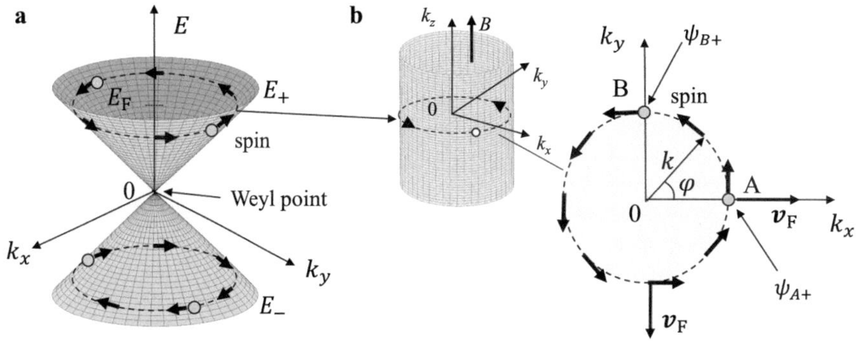

Fig. 3.32 **a** Energy bands for Rashba-type spin–orbit interaction ($\alpha_{SO} > 0$), which correspond to the surface bands in Fig. 3.31**a**. **b** 2D Fermi surface of $E_+(k)$ band and its helical spin state formed on the surfaces for 3D topological insulators

and their eigenfunctions,

$$\psi_{\pm} = \frac{1}{\sqrt{2}}\begin{pmatrix} 1 \\ \mp ie^{i\varphi} \end{pmatrix} = \frac{1}{\sqrt{2}}\Big(|\uparrow\rangle \mp ie^{i\varphi}|\downarrow\rangle\Big). \tag{3.137}$$

The energy bands $E_{\pm}(k)$ are shown in Fig. 3.32a. As seen in graphene, the bands have a Dirac-cone dispersion, but the spin states are very different. In the Dirac-cone states, the up- and down-spin states are degenerate in zero magnetic field. However, the above bands are not degenerate and special spin states are formed as seen below. The apex of this cone is called the Weyl point.

We assume that the Fermi level E_F is in the upper cone ($E_F > 0$) as shown in Fig. 3.32a. The Fermi surface is cylindrical along the k_z-axis with a circular cross section as depicted in Fig. 3.32b. Since $\varphi = 0$ at point A and $\varphi = \pi/2$ at point B, the eigenstates at these points are

$$A : \psi_{A+} = \frac{1}{\sqrt{2}}\begin{pmatrix} 1 \\ -i \end{pmatrix}, \quad B : \psi_{B+} = \frac{1}{\sqrt{2}}\begin{pmatrix} 1 \\ 1 \end{pmatrix}. \tag{3.138}$$

These states are superpositions of the $|\uparrow\rangle$ and $|\downarrow\rangle$ states. The operator $U_j(\alpha)$, which rotates the spin counterclockwise by an angle α around the $j(= x, y, z)$-axis, can be written as

$$U_j(\alpha) = e^{-i(\alpha/2)\sigma_j} = \cos\Big(\frac{\alpha}{2}\Big)\boldsymbol{I} - i\sin\Big(\frac{\alpha}{2}\Big)\boldsymbol{\sigma}_j. \tag{3.139}$$

Using this operator, we rotate a down-spin oriented toward $-k_z$ around the k_x-axis by $\alpha = \pi/2$,

$$U_x(\pi/2)\begin{pmatrix} 0 \\ 1 \end{pmatrix} = \frac{1}{\sqrt{2}}\begin{pmatrix} 1 & -i \\ -i & 1 \end{pmatrix}\begin{pmatrix} 0 \\ 1 \end{pmatrix} = \frac{1}{\sqrt{2}}\begin{pmatrix} -i \\ 1 \end{pmatrix} = \frac{-i}{\sqrt{2}}\begin{pmatrix} 1 \\ i \end{pmatrix} = e^{-\frac{i\pi}{2}}\psi_{A+}. \tag{3.140}$$

This spin state is ψ_{A+} except for the phase factor $e^{-i\pi/2}$. The result shows that the spin at point A is oriented toward k_y. Similarly we rotate a down-spin oriented toward $-k_z$ around the k_y-axis by $\alpha = \pi/2$ and obtain

$$U_y(\pi/2)\begin{pmatrix} 0 \\ 1 \end{pmatrix} = \frac{1}{\sqrt{2}}\begin{pmatrix} 1 & -1 \\ 1 & 1 \end{pmatrix}\begin{pmatrix} 0 \\ 1 \end{pmatrix} = \frac{1}{\sqrt{2}}\begin{pmatrix} -1 \\ 1 \end{pmatrix} = \frac{-1}{\sqrt{2}}\begin{pmatrix} 1 \\ -1 \end{pmatrix} = e^{-i\pi}\psi_{B+}. \tag{3.141}$$

The result shows that the spin at point B is oriented toward $-k_x$. In this way, the electron spins on the Fermi surface are in the helical spin state as shown in Fig. 3.32b. The electrons have the Fermi velocities $\boldsymbol{v}_{\mathrm{F}} = \nabla_{\boldsymbol{k}} E_+(\boldsymbol{k})/\hbar = (\alpha_{SO}/\hbar)\boldsymbol{e}_k$, parallel to $\boldsymbol{k}$ at the surface of the 3D topological insulator and thus the spins are perpendicular to the velocity vector $\boldsymbol{v}$ ($\boldsymbol{s}\perp\boldsymbol{v}$). As depicted in Fig. 3.32c, this surface state can appear on all the surfaces of the 3D topological insulators. A similar calculation for ψ_- can be performed to confirm the helical spin state with the opposite direction. Since the spin direction is conserved for scattering by nonmagnetic impurities or defects, the helical spin state is robust against nonmagnetic scattering.

Next, let us find the Berry phase that the electrons acquire in the orbital motion on this Fermi surface. The Berry connection is calculated as

$$\begin{aligned} \boldsymbol{a}_{B\pm}(k) &= i\langle\psi_\pm|\nabla_k\psi_\pm\rangle = \frac{i}{2}\begin{pmatrix} 1 & \mp ie^{-i\varphi} \end{pmatrix}\nabla_k\begin{pmatrix} 1 \\ \pm ie^{i\varphi} \end{pmatrix} \\ &= -\frac{1}{2}\nabla_k\varphi = -\frac{1}{2k}\boldsymbol{e}_\varphi, \end{aligned} \tag{3.142}$$

which is identical to Eq. (3.112) except for the sign. Therefore, the counterclockwise orbital motion around the Weyl point for the band $E_+(k) = \alpha_{SO}k$ produces the Berry phase $\psi_B = -\pi$. This is very similar to the case of the Dirac point.

As seen in Fig. 3.32b, the cylindrical Fermi surface is formed on the surface of the 3D topological insulator. Therefore, the cyclotron motion on the Fermi surface leads to quantum oscillation in the magnetic field along k_z, where the Berry phase $\psi_B = -\pi$ is added. In the above model, the Weyl point appears at $k = 0$ (Γ point) but it can be anywhere in the Brillouin zone for real materials, depending on the energy bands. In such a case, k in Eq. (3.131) is measured from the Weyl point.

3.10.7 General Model with Spin–Orbit Interaction

So far, only the Rashba-type spin–orbit interaction has been taken into account. Here we consider a more realistic case. The Hamiltonian including the free electron-like kinetic energy $\hbar^2k^2/2m$ and the Zeeman interaction $-g\mu_B s_z B_z$ is written as

$$\mathcal{H} = \frac{\hbar^2k^2}{2m} + 2\alpha_{SO}\left(k_y s_x - k_x s_y\right) - g\mu_B s_z B_z. \tag{3.143}$$

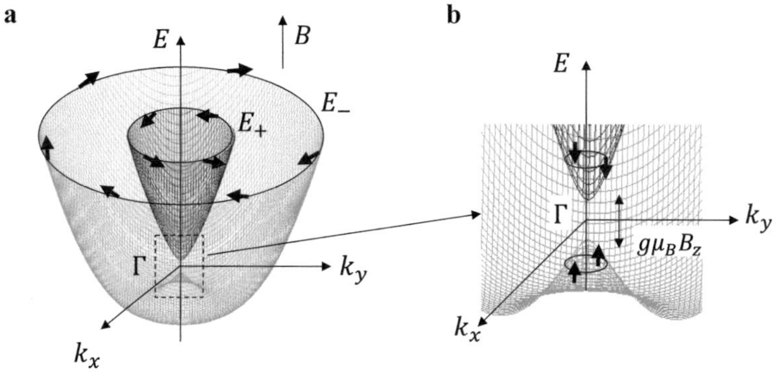

Fig. 3.33 **a** Energy bands for the case including Rashba-type spin–orbit interaction, kinetic energy and Zeeman interaction. Landau quantization is not taken into account. **b** Close-up of the band near Γ point. Thick arrows indicate spins. Band gap opens between two bands, $E_+(k)$ and $E_-(k)$

Using $s_z = (1/2)\sigma_z$, the matrix representation of the Hamiltonian is given by

$$\mathcal{H} = \begin{pmatrix} \frac{\hbar^2 k^2}{2m} - \frac{1}{2} g \mu_B B_z & i\alpha_{\mathrm{SO}} k e^{-i\varphi} \\ -i\alpha_{\mathrm{SO}} k e^{i\varphi} & \frac{\hbar^2 k^2}{2m} + \frac{1}{2} g \mu_B B_z \end{pmatrix}. \tag{3.144}$$

The eigenvalues are easily calculated as

$$E_\pm(k) = \frac{\hbar^2 k^2}{2m} \pm \sqrt{(\alpha_{\mathrm{SO}} k)^2 + (g\mu_B B_z/2)^2}. \tag{3.145}$$

The bands are depicted in Fig. 3.33a. In the close-up near the Γ point, Fig. 3.33b, there exists the Zeeman energy gap of $g\mu_B B_z$ between the $E_+(k)$ and $E_-(k)$ bands.

When the Fermi level is above the gap, two cylindrical Fermi surfaces are formed on the surfaces of the 3D topological insulator: quantum oscillations with two frequencies can be observed. The spin orientations are not completely confined in the $k_x k_y$ plane; the spins are tilted in $-k_z$ for $E_+(k)$ and in k_z for $E_-(k)$ by the Zeeman interaction in magnetic fields. In these energy bands, the Berry phases of the quantum oscillations depend on the Fermi energy E_F. For $E_F \gg g\mu_B B_z$, the spins of both bands $E_\pm(k)$ are almost confined in the $k_x k_y$ plane and then $\psi_B \to \mp\pi$ [43]. However, for $E_F \approx \pm g\mu_B B_z/2$ ($k_F \approx 0$), the spins are polarized in $\mp k_z$ and then $\psi_B \to 0$. The nonlinearity of the band due to the gap deviates the Berry phase from $\psi_B = \pi$.

The L–K formula obtained in Sect. 3.1 is applicable when the spin–orbit interaction– is sufficiently small compared to the Fermi energy. For a Fermi surface with a helical spin structure as shown in Figs. 3.32b or 3.33a, the L–K formula cannot be used directly. Since the spins are fixed on the Fermi surface, each Landau level does not undergo spin splitting. This fact shows that the spin splitting reduction factor should be removed from the L–K formula or $R_S = 1$.

3.10.8 Quantum Oscillation and Berry Phase in Topological Insulator

Here we focus on the phase of the quantum oscillation observed in a topological insulator. It is theoretically and experimentally known that Bi_2Se_3 and Bi_2Te_3 with a tetradimite structure are topological insulators. The electronic state inside the sample is insulating with a gap at the Fermi level. The surface state is a conducting helical spin state, whose energy bands near the Weyl point are well approximated by that given in Fig. 3.32a. If the Fermi level is exactly at the Weyl point, no quantum oscillations are observed. In the real materials, however, the Fermi level is slightly shifted from the Weyl point and thus quantum oscillations can be observed in magnetic fields. Figure 3.34a shows the oscillatory part of the resistance for a thin film of Bi_2Se_3 [44]. The fan plot of the quantum oscillation is presented in Fig. 3.34b. The linear extrapolation for $1/B \to 0$ yields $N_L \approx 0.35$, which is smaller than the ideal value of 1/2. Possible reasons for the discrepancy are experimental error and/or band nonlinearity [45].

For $\omega_c\tau \gg 1$, the conductivity in Eq. (3.16) is written as $\sigma_{xx}(B) = \sigma_0/\left(1+\omega_c^2\tau^2\right) \propto 1/\tau$. From Eq. (3.13), we have a similar relation for the resistivity, $\rho_{xx} = \rho_0 \propto 1/\tau$. Therefore, the oscillatory parts of the conductivity and resistivity are written as $\tilde{\sigma}_{xx}, \tilde{\rho}_{xx} \propto \tilde{D}(B)$. This is true for a single-band electronic state, but not necessarily for multi-band materials. When there are different conduction paths (multiple bands), the measured conductivity is given by the sum of all the conduction paths. For simplicity, we assume an isotropic electron system in the xy plane, $\sigma_{xx} = \sigma_{yy}$. From Eq. (3.15), the total resistivity ρ_{xx}^{total} is given by the sum of the paths ($i = 1, 2, 3, \ldots$),

$$\rho_{xx}^{\text{total}} = \frac{\sum_i \sigma_{xx(i)}}{\left[\sum_i \sigma_{xx(i)}\right]^2 + \left[\sum_i \sigma_{xy(i)}\right]^2}. \tag{3.146}$$

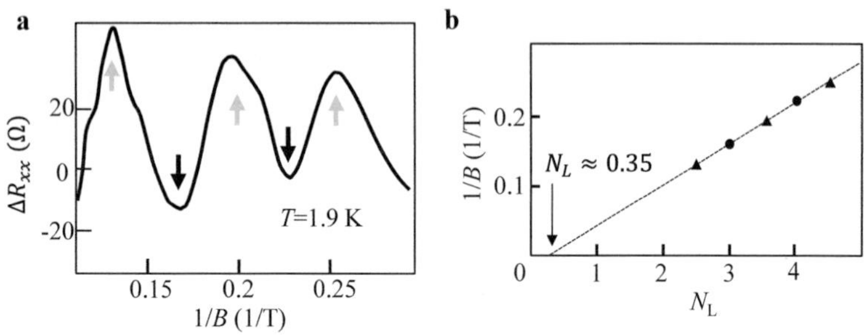

Fig. 3.34 **a** Quantum oscillation of resistance in Bi_2Se_3 thin film (reproduced with permission from [44]). The maxima and minima of the oscillation are indicated by gray and black arrows, respectively. **b** Fan plot of the quantum oscillation. The maximum and minimum fields are indicated by triangles and circles, respectively

If $\sum_i \sigma_{xx(i)} \ll \sum_i \sigma_{xy(i)}$, we have

$$\rho_{xx}^{\text{total}} \approx \sum_i \sigma_{xx(i)} \Big/ \left[\sum_i \sigma_{xy(i)}\right]^2, \tag{3.147}$$

and thus the oscillatory part of the resistivity is

$$\tilde{\rho}_{xx}^{\text{total}} \propto \sum_i \sigma_{xx(i)} \propto \sum_i \tilde{D}_i(B), \tag{3.148}$$

where $\tilde{D}_i(B)$ is the oscillatory part of the density of states for individual band. This result is equivalent to the case for a single band, $\tilde{\rho}_{xx}(B) \propto \tilde{D}(B)$ given by Eq. (3.25). However, since the relation $\sum_i \sigma_{xx(i)} \ll \sum_i \sigma_{xy(i)}$ is not guaranteed in general, ρ_{xx}^{total} becomes a complex functional form of τ_i and thus the resistivity is not simply given by Eq. (3.148). Even for topological insulators, which are supposed to be insulating inside the sample, multiple conduction paths may be formed in the bulk; conduction through impurities and/or by thermal excitation in the bulk may not be negligible. In this case, the linear extrapolation for $1/B \to 0$ in the fan plot does not give the reliable Berry phase.

When observing quantum oscillations in the magnetization, the above problem does not arise. Even in the case of complex Fermi surface structures with various extremal cross sections, the total magnetization can be expressed by the simple sum of the individual contributions. The oscillatory part of the magnetization given by Eq. (3.4b) for $p = 1$, taking into account the Berry phase ψ_B^i, is written as

$$\begin{aligned}\tilde{M}_{2D\parallel}(B) &\propto \sum_i -\sin\left[2\pi\left(\frac{F_Q^i}{B} - \frac{1}{2}\right) + \psi_B^i\right] \\ &= \sum_i \sin\left[2\pi\frac{F_Q^i}{B} + \psi_B^i\right].\end{aligned} \tag{3.149}$$

The magnetic fields B_{max}^i, where the oscillation has maxima, are given by

$$2\pi\frac{F_Q^i}{B_{\text{max}}^i} + \psi_B^i = \left(2N_L^i + \frac{1}{4}\right)\pi,\ N_L^i = 0, 1, 2, \ldots \tag{3.150}$$

The Berry phase of each oscillation can be determined by drawing the fan plot.

3.10.9 3D Weyl Point

In the above, we have seen the model of the 2D electron systems realized on the surface of a crystal or at the interfac between different materials, whose Weyl point is expressed in Fig. 3.32a. Here, we extend the model to a 3D electron system, which has energy dispersion being linear in all directions $\left(k_x, k_y, k_z\right)$. In this case, the Weil point exists inside the sample. When the Fermi level is close to the Weyl point, the electronic system has few carriers and such a material is called a Weyl semimetal.

The model Hamiltonian of the spin states for a 3D Weyl semimetal can be approximated as

$$\begin{aligned}\mathcal{H} &= 2v_F\hbar\left(s_xk_x + s_yk_y + s_zk_z\right)\\ &= v_F\hbar\left(\sigma_xk_x + \sigma_yk_y + \sigma_zk_z\right),\end{aligned} \tag{3.151}$$

where $\left(k_x, k_y, k_z\right) = (k\sin\theta\cos\varphi, k\sin\theta\sin\varphi, k\cos\theta)$ and thus it is rewritten as

$$\mathcal{H} = v_F\hbar\begin{pmatrix} k_z & k_x - ik_y \\ k_x + ik_y & -k_z \end{pmatrix} = v_F\hbar k\begin{pmatrix} \cos\theta & \sin\theta e^{-i\varphi} \\ \sin\theta e^{i\varphi} & -\cos\theta \end{pmatrix}. \tag{3.152}$$

It is easy to check the eigenvalues,

$$E_\pm = \pm\hbar v_F k. \tag{3.153}$$

The energy bands are identical to those of the 3D Dirac material in Eq. (3.124) although the spin states are different. This linear energy dispersion forms a spherical Fermi surface as depicted in Fig. 3.35. The eigenfunctions for $E_\pm$ are respectively,

$$\begin{aligned}\psi_+ &= \begin{pmatrix} \cos(\theta/2) \\ \sin(\theta/2)e^{i\varphi} \end{pmatrix} = \cos\left(\frac{\theta}{2}\right)|\uparrow\rangle + \sin\left(\frac{\theta}{2}\right)e^{i\varphi}|\downarrow\rangle,\\ \psi_- &= \begin{pmatrix} \sin(\theta/2) \\ -\cos(\theta/2)e^{i\varphi} \end{pmatrix} = \sin\left(\frac{\theta}{2}\right)|\uparrow\rangle - \cos\left(\frac{\theta}{2}\right)e^{i\varphi}|\downarrow\rangle.\end{aligned} \tag{3.154}$$

At the north pole ($\theta = 0$) of the spherical Fermi surface, the spin states are $\psi_+ = |\uparrow\rangle$ ($\mathbf{s}\|\mathbf{k}$) and $\psi_- = -e^{i\varphi}|\downarrow\rangle$ ($\mathbf{s}\| - \mathbf{k}$). Using Eq. (3.127), we can calculate the Berry connection,

$$\begin{aligned}\boldsymbol{a}_{B+}(k) &= i\langle\psi_+|\nabla_k\psi_+\rangle = -\frac{\sin^2(\theta/2)}{k\sin\theta}\boldsymbol{e}_\varphi,\\ \boldsymbol{a}_{B-}(k) &= i\langle\psi_-|\nabla_k\psi_+\rangle = -\frac{\cos^2(\theta/2)}{k\sin\theta}\boldsymbol{e}_\varphi.\end{aligned} \tag{3.155}$$

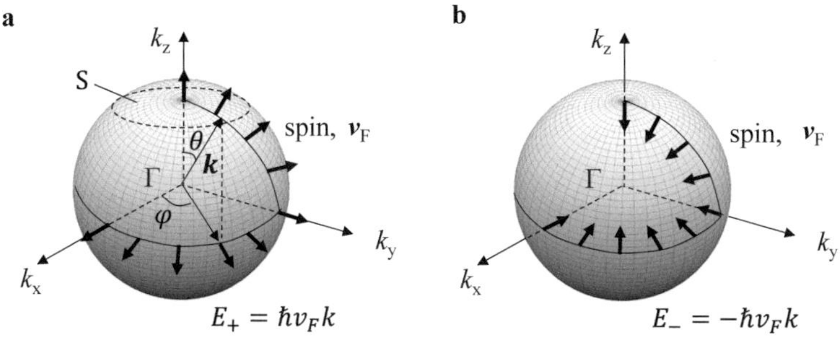

Fig. 3.35 Spherical Fermi surface with **a** $E_F > 0$ and **b** $E_F < 0$ for model Hamiltonian of 3D Weyl semimetal. Thick arrows indicate spin and velocity vectors of electron on surface. All spins are oriented outward from the origin for **a**, forming a magnetic monopole. Antimonopole is formed for **b**

As expected from the Hamiltonian in Eq. (3.151), the spin state has a positive energy when the spins are parallel to the wave number vector $\boldsymbol{s}\|\mathbf{k}$. It corresponds to the state ψ_+. For $E_F > 0$, all the spins are oriented outward from the origin as depicted in Fig. 3.35a, looking like a magnetic monopole. For $E_F < 0$, antimonopole is formed as shown in Fig. 3.35b.

The Berry phases for the eigenfunctions ψ_+ and ψ_- are respectively given by

$$\begin{aligned}\psi_{B+} &= \oint \boldsymbol{a}_{B+}(k)\cdot \mathrm{d}\boldsymbol{k} = -\oint \frac{\sin^2(\theta/2)}{k\sin\theta}\boldsymbol{e}_\varphi \cdot k\sin\theta \mathrm{d}\varphi \boldsymbol{e}_\varphi \\ &= -2\pi\sin^2(\theta/2), \\ \psi_{B-} &= -2\pi\cos^2(\theta/2). \end{aligned} \tag{3.156}$$

Since the solid angle of the orbital plane S with the angle θ on the spherical Fermi surface in Fig. 3.35a is $4\pi\sin^2(\theta/2)$, the Berry phase ψ_{B+} is $-1/2$ times the solid angle. In a magnetic field parallel to the k_z-axis, the extremal cross section is the equatorial plane ($k_z = 0$); $\theta = \pi/2$ in Eq. (3.156). Therefore, the Berry phase $\psi_{B\pm} = \pm\pi$ is obtained in the quantum oscillation measurement, the sign of which depends on the direction of the cyclotron motion. The Berry curvature is given by

$$\boldsymbol{\Omega}_{B\pm}(k) = \nabla_k \times \boldsymbol{a}_{B\pm}(k) = \mp\frac{1}{2k^2}\boldsymbol{e}_k. \tag{3.157}$$

3.11 Summary

In this chapter, we have seen that various information on the Fermi surface can be obtained from the quantum oscillations. We summarize the results in Table 3.2.

Table 3.2 Physical quantities obtained from quantum oscillation measurements and their observation methods

Physical quantity	Method of observation and related expression
Extremal cross section of Fermi surface: A_F	Frequency of quantum oscillation in various thermodynamic quantities or resistance: $F_Q = (\hbar/2\pi e)A_F$
Effective mass: m_c Enhancement factor: λ^*	Temperature dependence of oscillation amplitude Temperature reduction factor: $R_T = \frac{K\mu_c T/B}{\sinh(K\mu_c T/B)}, \quad \mu_c = \frac{m_c}{m_0}, \quad m_c = m_b(1+\lambda^*)$
Scattering timeτ	Magnetic field dependence of oscillation amplitude Dingle reduction factor: $R_D = \exp\left(-\frac{K\mu_c T_D}{B}\right), \quad T_D = \frac{\hbar}{2\pi k_B \tau}$
g factor: g_c	Spin splitting zero: $R_S = 0$ for $p = 1$ Spin splitting reduction factor: $R_S = \cos(\pi S_{\text{spin}}), \quad S_{\text{spin}} = g_c\mu_c/2$
Energy gap: E_g	Magnetic field dependence of oscillation amplitude Magnetic breakdown probability: $P = \exp\left(-\frac{B_{\text{MB}}}{B}\right), B_{\text{MB}} = \frac{m_c}{e\hbar}\left(\frac{E_g^2}{E_F}\right)$ Bragg reflection probability: $Q = 1 - P$
Internal magnetic field: B_{int}	Peak splitting ΔF in Fourier spectrum: $\Delta F = \frac{g_c}{4}\mu_c B_{\text{int}}$

Appendix 3.1

Here we first derive the reduction factor R_T of the quantum oscillation. It is very difficult to calculate R_T directly from Eq. (3.1). Therefore, we obtain R_T according to the literature [1]. The oscillatory part is given by Eq. (3.35),

$$\cos\left[2\pi p\left(\frac{F_Q}{B} - \frac{1}{2}\right) + \varphi_0\right] = \cos(\Phi_B).$$

The following discussion does not depend on the dimensionality of the electronic state. Taking ϕ as an additional phase factor of the quantum oscillation, we write the oscillatory part as $\cos(\Phi_B) \to \cos(\Phi_B+\phi)$ and assume that ϕ is distributed around $\phi = 0$. Let $D(\phi/\lambda)$ be the distribution function of ϕ which is an even function of ϕ and has a peak centered at $\phi = 0$ with its distribution range λ. Then the oscillation I_{osc} will be given by

$$I_{\mathrm{osc}} = \int_{-\infty}^{\infty} \mathrm{d}\phi \cos(\Phi_B + \phi) D(\phi/\lambda) / \int_{-\infty}^{\infty} \mathrm{d}\phi D(\phi/\lambda). \quad (3.158)$$

The denominator is the normalization factor of the oscillation. The larger the value of λ, the more the oscillation amplitude is reduced. Taking $x = \phi/\lambda$, we define

$$g(\lambda) = \int_{-\infty}^{\infty} \mathrm{d}x e^{i\lambda x} D(x). \quad (3.159)$$

Using $g(\lambda)$, the oscillation amplitude is written as

$$I_{\mathrm{osc}} = \mathrm{Re}\left[\frac{e^{i\Phi_B} g(\lambda)}{g(0)}\right] = \cos(\Phi_B)\frac{g(\lambda)}{g(0)}. \quad (3.160)$$

$\mathrm{Re}[Z]$ denotes the real part of Z, $\cos(\Phi_B+\phi) = \mathrm{Re}\left[e^{i\Phi_B} e^{i\phi}\right]$. By the definition, $g(\lambda)$ is the Fourier transform of the distribution function $D(x)$. The above result shows that if there is a factor distributing the phase of the quantum oscillation, the oscillation amplitude $\cos(\Phi_B)$ is reduced by the Fourier transform function of the phase distribution function, $g(\lambda)/g(0) \leq 1$. In the absence of phase distribution, $D(x)$ should be a delta function, $D(x) = \delta(x)$, leading to $I_{\mathrm{osc}} = \cos(\Phi_B)$ in Eq. (3.158).

Equations (2.64) and (2.69) represent the quantum oscillations at $T = 0$ K. Based on the above argument, we introduce the effect of finite temperature into the oscillation amplitude. At finite temperatures ($T > 0$), the electrons are thermally excited to the higher Landau levels and the thermal excitation smears out the sharp feature of the quantum oscillations of the thermodynamic grand potential shown in Fig. 2.11c. We treat the thermal excitation as a phase smearing effect reducing the oscillation. Using the Fermi distribution function $f(E)$, the probability of the thermal excitation is proportional to $f(E)(1-f(E)) \propto -\mathrm{d}f(E)/\mathrm{d}E$. In the low temperature range we are interested in, E_F is taken to be equal to the chemical potential. Therefore, the thermal excitation probability, reducing the quantum oscillation, is expressed as

$$-\frac{\mathrm{d}f(E)}{\mathrm{d}E} \propto \frac{1}{1+\cosh\left\{\frac{(E-E_F)}{k_B T}\right\}}. \quad (3.161)$$

This function has a sharp peak at $E = E_\mathrm{F}$ as depicted in Fig. 2.15b. We can define Eq. (3.161) as the distribution function $D(E)$,

$$D(E) = \frac{1}{1 + \cosh\left\{\frac{(E - E_\mathrm{F})}{k_B T}\right\}}. \tag{3.162}$$

The prefactor in Eq. (3.161) is removed because it is canceled out in $g(\lambda)/g(0)$. Since the size of the Fermi surface (cross section) is a function of the energy E, we expand the frequency of the quantum oscillation $F_Q(E)$ around E_F,

$$F_Q(E) = F_Q(E_F) + \frac{\mathrm{d}F_Q}{\mathrm{d}E}(E - E_F). \tag{3.163}$$

Then the oscillatory part is written as

$$\begin{aligned}\cos\left[2\pi p\left(\frac{F_Q(E)}{B} - \frac{1}{2}\right) + \varphi_0\right] &= \cos\left[2\pi p\left(\frac{F_Q(E_F)}{B} - \frac{1}{2}\right)\right.\\ &\quad \left. + \varphi_0 + \frac{2\pi p}{B}\frac{\mathrm{d}F_Q}{\mathrm{d}E}(E - E_F)\right]\\ &\equiv \cos(\Phi_B + \phi).\end{aligned} \tag{3.164}$$

The phase distribution factor of the quantum oscillation ϕ is defined as

$$\phi = \frac{2\pi p}{B}\left(\frac{\mathrm{d}F_Q}{\mathrm{d}E}\right)(E - E_F) = \frac{2\pi p}{B}\frac{m_c}{e\hbar}(E - E_F), \tag{3.165}$$

by using the relation $F_Q/B = E/\hbar\omega_c$ in Eq. (2.60). We rewrite Eq. (3.165) as

$$\frac{(E - E_F)}{k_B T} = \frac{e\hbar B}{2\pi p m_c}\frac{\phi}{k_B T} = \frac{\phi}{\lambda_T},\ \lambda_T = \frac{2\pi p m_c k_B T}{e\hbar B}. \tag{3.166}$$

Taking $x = \phi/\lambda_T$, the Fourier transform of $D(x) = 1/[1 + \cosh(x)]$ in Eq. (3.162) is given by

$$\frac{1}{1 + \cosh(x)} \to \frac{g(\lambda_T)}{g(0)} = \frac{\pi\lambda_T}{\sinh(\pi\lambda_T)},$$

and thus the oscillation is written as

$$I_{osc} = \cos(\Phi_B)\frac{g(\lambda_T)}{g(0)} = \cos(\Phi_B)\frac{\pi\lambda_T}{\sinh(\pi\lambda_T)}. \tag{3.167}$$

From Eq. (3.167), we obtain the temperature reduction factor in Eq. (3.26),

$$R_T = \frac{z}{\sinh(z)},\ z = \pi\lambda_T.$$

Next we derive the reduction factor R_D. Since the Landau level has a width of $\hbar/\tau$ due to scattering, the quantum oscillation will be reduced by this Landau level broadening. We assume that the broadening of each Landau level with N has a Lorentzian line shape,

$$\frac{1}{(E - E_N)^2 + (\hbar/2\tau)^2}, \tag{3.168}$$

which has a half width at half maximum, $\hbar/2\tau$. This can be taken as the distribution function. Dividing Eq. (3.168) by $(\hbar/2\tau)^2$, we can define the distribution function as

$$D(E) = \frac{1}{\left(\frac{E-E_{\mathrm{F}}}{\hbar/2\tau}\right)^2 + 1}. \tag{3.169}$$

Using $E - E_{\mathrm{F}} = \phi e\hbar B/2\pi p m_c$ from Eq. (3.165), we obtain

$$\frac{E - E_{\mathrm{F}}}{\hbar/2\tau} = \frac{e\hbar B}{2\pi p m_c}\frac{\phi}{\hbar/2\tau} = \frac{\phi}{\lambda_{\mathrm{D}}}, \tag{3.170}$$

where the distribution range λ_{D} is defined as

$$\lambda_{\mathrm{D}} = \frac{\pi p m_c}{Be\tau} = \frac{p\pi}{\omega_c\tau}. \tag{3.171}$$

Taking $x = \phi/\lambda_D$, the Fourier transform of $D(x) = 1/\left(x^2 + 1\right)$ is given by

$$\frac{1}{x^2 + 1} \rightarrow \frac{g(\lambda_T)}{g(0)} = \exp(-\lambda_{\mathrm{D}}),$$

and thus we have

$$I_{osc} = \cos(\Phi_B)\frac{g(\lambda_{\mathrm{D}})}{g(0)} = \cos(\Phi_B)\exp(-\lambda_{\mathrm{D}}). \tag{3.172}$$

Therefore, we obtain the Dingle reduction factor in Eq. (3.32),

$$R_D = \exp(-\lambda_{\mathrm{D}}) = \exp\left(-\frac{p\pi}{\omega_c\tau}\right).$$

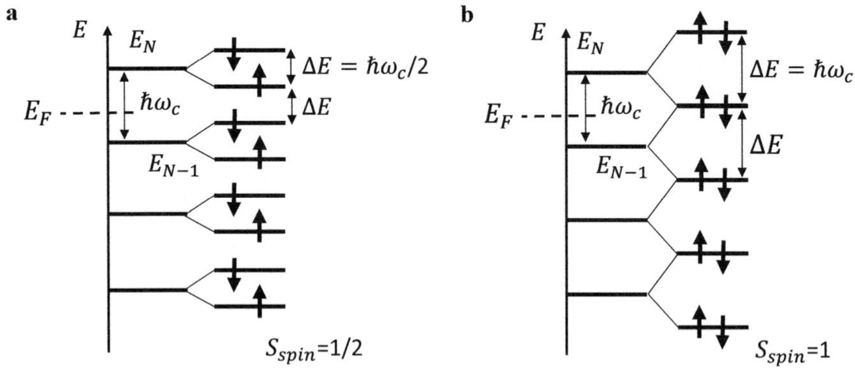

Fig. 3.36 Landau quantization and spin splitting due to Zeeman effect for **a** spin splitting zero condition $S_{\text{spin}} = 1/2$ and **b** for $S_{\text{spin}} = 1$

Appendix 3.2

Let's look at the relation between the Landau quantization and Zeeman splitting. Under a spin splitting zero condition, $S_{\text{spin}} = \Delta E/\hbar\omega_c = 1/2$, the Zeeman-split levels are equally spaced, as shown in Fig. 3.36a. The quantum oscillation arising from the up-spin levels is completely out of phase with that from the down-spin levels. It means that the fundamental oscillation ($p = 1$ in the L–K formula) is canceled out and consequently the second harmonic oscillation ($p = 2$) is enhanced. In contrast, for $S_{\text{spin}} = 1$, as depicted in Fig. 3.36b, the oscillations of the up and down-spin levels are in phase, and thus the fundamental oscillation is observed dominantly.

Figure 3.37 presents the simulated quantum oscillations $\tilde{M}_{2D\parallel}$ in Eq. (3.4b) for $S_{spin} = 1/2$ and $S_{\text{spin}} = 1$. The parameters for this simulation are $T = 0.2$ K, $T_D =$ 0.5K, $F_Q = 100$T, and $\mu_c = 1$. The second harmonic ($p = 2$) and fundamental ($p = 1$) oscillations are evident for $S_{\text{spin}} = 1/2$ and $S_{\text{spin}} = 1$, respectively. Due to the higher harmonic contents, both results show sawtooth-like waveforms.

Appendix 3.3

Here we take a look at possible closed orbits in magnetic fields for the Fermi surface shown in Fig. 1.5e. In Fig. 3.38a, the Fermi surfaces consist of the electron (α) and hole (β) surfaces. The arrows indicate the directions of the electron motion in a perpendicular magnetic field. In the weak magnetic field limit, the Bragg reflection occurs at all the zone boundaries, so the α and β orbits are possible. In the high magnetic field limit, the magnetic breakdown always occurs, so the γ orbit in Fig. 3.38b becomes dominant. In an intermediate magnetic field range, where

Fig. 3.37 Simulated quantum oscillations $\tilde{M}_{2D\parallel}$ in Eq. (3.4b) for $S_{\rm spin} = 1/2$ and $S_{\rm spin} = 1$

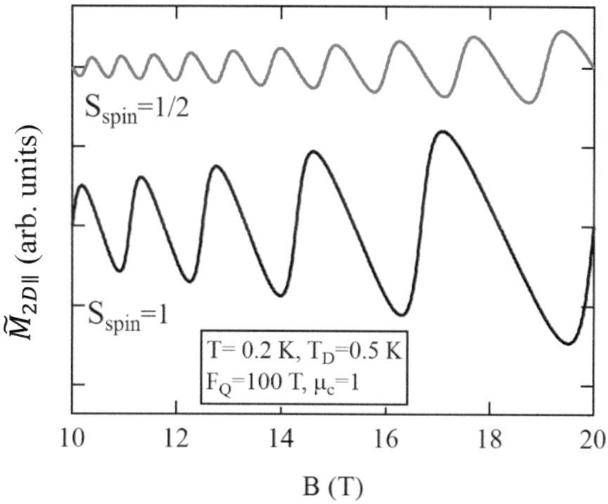

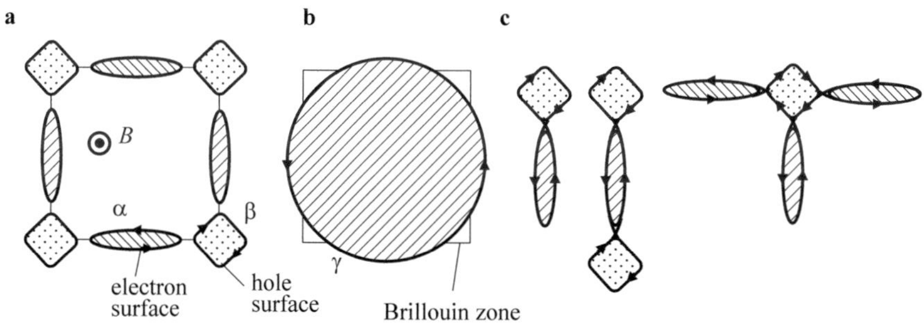

Fig. 3.38 **a** Electron (α) and hole (β) Fermi surfaces, identical to Fig. 1.5e. Arrows indicate the directions of the electron motion in a perpendicular magnetic field. **b** Magnetic breakdown orbit (γ), **c** some other possible orbits

both Bragg reflection and magnetic breakdown occur, various other closed orbits are possible, as shown in Fig. 3.38c.

Appendix 3.4

Using the momentum $\hbar \boldsymbol{k} = \boldsymbol{p} + e\boldsymbol{A} = -i\hbar\nabla + e\boldsymbol{A}$ in a magnetic field $\boldsymbol{B} = \mathrm{rot}\boldsymbol{A}$, the Schrödinger equation is written as

$$\mathcal{H}\Psi = \frac{\hbar^2 \boldsymbol{k}^2}{2m}\Psi = \frac{1}{2m}(-i\hbar\nabla + e\boldsymbol{A})^2\Psi = E\Psi. \tag{3.173}$$

If we take Ψ_0 as the eigenstate of the Schrödinger equation in zero magnetic field, $-(\hbar^2\nabla^2/2m)\Psi_0 = E\Psi_0$, the wave function Ψ is given by

$$\Psi = \Psi_0 e^{i\gamma}, \quad \gamma = -\frac{e}{\hbar}\int_{r_0}^{r} \boldsymbol{A}(\boldsymbol{r}') \cdot \mathrm{d}\boldsymbol{r}', \tag{3.174}$$

where γ is the phase depending on the path of the electron. To prove this equation, we first calculate

$$\begin{aligned}(-i\hbar\nabla + e\boldsymbol{A})\Psi &= (-i\hbar\nabla + e\boldsymbol{A})\Psi_0 e^{i\gamma} \\ &= -i\hbar(\nabla\Psi_0)e^{i\gamma} + \hbar(\nabla\gamma)\Psi_0 e^{i\gamma} + e\boldsymbol{A}\Psi_0 e^{i\gamma}.\end{aligned} \tag{3.175}$$

From (3.174), $\nabla\gamma$ is expressed as

$$\nabla\gamma = -\frac{e}{\hbar}\nabla\int_{r_0}^{r} \boldsymbol{A}(\boldsymbol{r}') \cdot \mathrm{d}\boldsymbol{r}' = -\frac{e}{\hbar}\boldsymbol{A}(\boldsymbol{r}),$$

and thus Eq. (3.175) is written as

$$(-i\hbar\nabla + e\boldsymbol{A})\Psi = -i\hbar(\nabla\Psi_0)e^{i\gamma}.$$

By acting $(-i\hbar\nabla + e\boldsymbol{A})$ again, we have

$$(-i\hbar\nabla + e\boldsymbol{A})^2\Psi = -\hbar^2(\nabla^2\Psi_0)e^{i\gamma}.$$

Therefore, the Schrödinger equation is written as

$$\frac{1}{2m}(-i\hbar\nabla + e\boldsymbol{A})^2\Psi = \left(\frac{-\hbar^2\nabla^2}{2m}\Psi_0\right)e^{i\gamma} = E\Psi_0 e^{i\gamma} = E\Psi. \tag{3.176}$$

The result shows that $\Psi = \Psi_0 e^{i\gamma}$ is the eigenstate of the Schrödinger equation in a magnetic field; the electron moving in the magnetic field obtains the phase γ on its path,

$$\gamma = -\frac{e}{\hbar}\int_{\text{path}} \boldsymbol{A}(\boldsymbol{r}) \cdot \mathrm{d}\boldsymbol{r}.$$

If the electron travels on a closed orbit, the wave function obtains the phase given by Eq. (3.87)

$$\gamma = -\frac{e}{\hbar}\oint e\boldsymbol{A} \cdot \mathrm{d}\boldsymbol{r} = -\frac{e}{\hbar}\int \mathrm{rot}\boldsymbol{A} \cdot \mathrm{d}\boldsymbol{S}$$

$$= -\frac{e}{\hbar} \int \boldsymbol{B} \cdot \mathrm{d}\boldsymbol{S}.$$

When the magnetic flux penetrating the closed orbit is equal to the flux quantum, $\int \boldsymbol{B} \cdot \mathrm{d}\boldsymbol{S} = \Phi_0 = h/e$, we have $\gamma = -2\pi$; the electronic state returns to its initial state.

Appendix 3.5

The quantum interference in Fig. 3.24b is the simplest case. For the Fermi surface structure shown in Fig. 3.14a, various quantum interference paths are possible. In Fig. 3.39, other cases leading to quantum interference are presented. In Fig. 3.39a, the electron at point A has various possible paths to point B. The shortest path 1 includes two magnetic breakdowns with the transition amplitude given by $p_T e^{i\delta} e^{i\varphi_2} p_T e^{i\delta}$. The next shortest path 2 includes two Bragg reflections and magnetic breakdowns with $q_T e^{i\varphi_3} p_T e^{i\delta} e^{i\varphi_2} p_T e^{i\delta} e^{i\varphi_1} q_T$. The phase difference between the two paths is $\varphi_1 - \varphi_3$ and thus the quantum interference term is calculated as

$$I_{\mathrm{QI}}^{2D} = 2QP^2 \cos(\varphi_1 + \varphi_3) = 2QP^2 \cos\left(\frac{2\pi F_Q^{(\beta-\alpha)}}{B}\right),$$
$$F_Q^{(\beta-\alpha)} = F_Q^{(\beta)} - F_Q^{(\alpha)}. \tag{3.177}$$

Another case yielding the interference is depicted in Fig. 3.39b. The quantum interference between paths 1 and 2 occurs, whose transition amplitudes are respectively given by $p_T e^{i\delta} e^{i\varphi_2} p_T e^{i\delta}$ and $q_T e^{i\varphi_3} q_T e^{i\varphi_3} p_T e^{i\delta} e^{i\varphi_2} p_T e^{i\delta} e^{i\varphi_1} q_T e^{i\varphi_1} q_T$. The quantum interference term between the two paths is written as

$$I_{\mathrm{QI}}^{2D} = 2Q^2P^2 \cos(2\varphi_1 + 2\varphi_3) = 2Q^2P^2 \cos\left(\frac{2\pi F_Q^{(2\beta-2\alpha)}}{B}\right),$$

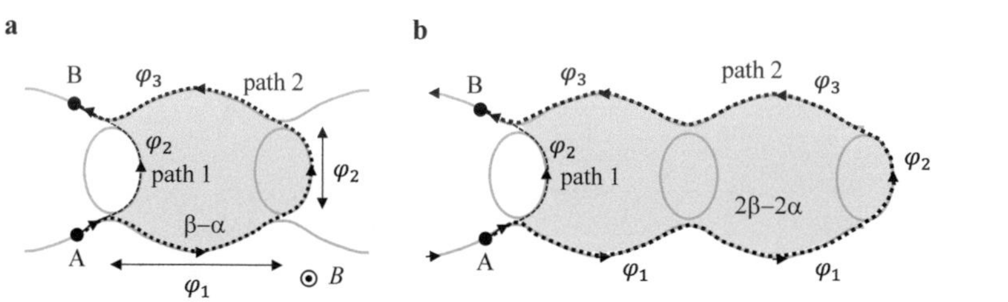

Fig. 3.39 Possible quantum interference paths from point A to point B for the Fermi surface identical to Fig. 3.14a. Paths in **a** and **b** lead to quantum interference with the frequencies $F_Q^{(\beta-\alpha)}$ and $F_Q^{(2\beta-2\alpha)}$, respectively

$$F_Q^{(2\beta-2\alpha)} = 2F_Q^{(\beta)} - 2F_Q^{(\alpha)}. \tag{3.178}$$

In the above two cases, from the relation $m_c \propto \partial A_F/\partial E$ in Eq. (3.44), we expect that the effective mass can be obtained as $m_{\beta-\alpha} = m_\beta - m_\alpha$ for **a** and $m_{2\beta-2\alpha} = 2m_\beta - 2m_\alpha$ for **b**. The oscillation amplitudes in these cases should include the reduction factors R_T and R_S in addition to R_τ in Eq. (3.94), whose parameters are characterized by the area between the two paths. Note that the above frequencies, $F_Q^{(\beta-\alpha)}$ and $F_Q^{(2\beta-2\alpha)}$, do not appear in the quantum oscillation in the semiclassical picture of the magnetic breakdown as discussed in Sect. 3.5. We can draw various other interference paths. In this way, the combinations of the Bragg reflection and the magnetic breakdown can lead to various frequencies in the electrical conductivity.

Appendix 3.6

Here we derive the Rashba-type spin–orbit interaction, Eq. (3.131). As shown in Fig. 3.40, we assume a 2D electron system in the xy plane on the surface of a crystal or at the interface between different materials. Since the inversion symmetry of the lattice potential is broken, $V(z) \neq V(-z)$, the electrons will be exposed to a perpendicular electric field $E \propto -\partial V/\partial z$, $\boldsymbol{E}(0, 0, E)$. The equation of motion of the electron is written as

$$\boldsymbol{F} = -e\boldsymbol{E} = -e\boldsymbol{v} \times \boldsymbol{B}_{\text{eff}} \tag{3.179}$$

where $\boldsymbol{v}$ is the velocity of the electron and $\boldsymbol{B}_{\text{eff}}$ is the effective magnetic field due to $\boldsymbol{E}$. It is perpendicular to both the velocity and electric field,

$$\boldsymbol{B}_{\text{eff}} = \boldsymbol{E} \times \frac{\boldsymbol{v}}{v^2}. \tag{3.180}$$

Assuming the Zeeman interaction with the effective magnetic field, the Hamiltonian is written as

$$\mathcal{H} = -g\mu_B \boldsymbol{s} \cdot \boldsymbol{B}_{\text{eff}} = \frac{-g\mu_B}{v^2} \boldsymbol{s} \cdot (\boldsymbol{E} \times \boldsymbol{v}). \tag{3.181}$$

Using the vector product formula $\boldsymbol{A} \cdot (\boldsymbol{B} \times \boldsymbol{C}) = -\boldsymbol{B} \cdot (\boldsymbol{A} \times \boldsymbol{C})$, we have

$$\mathcal{H} = \frac{-g\mu_B}{v^2} \boldsymbol{s} \cdot (\boldsymbol{E} \times \boldsymbol{v}) = \frac{g\mu_B}{v^2} \boldsymbol{E} \cdot (\boldsymbol{s} \times \boldsymbol{v}) = \frac{g\mu_B}{\hbar v^2} \boldsymbol{E} \cdot [\boldsymbol{s} \times \nabla_{\boldsymbol{k}} E(\boldsymbol{k})] \tag{3.182}$$

Taking $m\boldsymbol{v} = \hbar\boldsymbol{k}$ and $\mu_B = e\hbar/2m$, we obtain the Rashba-type spin–orbit interaction–,

$$\mathcal{H} = \frac{egE}{2k^2} \left(k_y s_x - k_x s_y\right) = \alpha_{SO} \left(k_y s_x - k_x s_y\right).$$

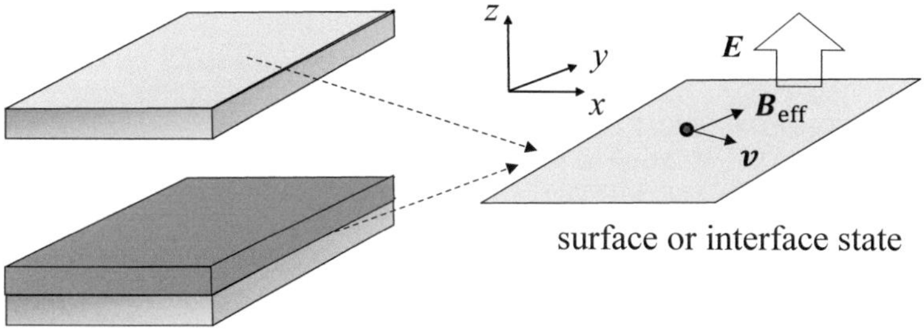

Fig. 3.40 Schematic of 2D electron system at crystal surface or interface between different materials. Electric field $\boldsymbol{E}$ is generated by the potential gradient at crystal surface or interface. Electron motion with velocity $\boldsymbol{v}$ leads to effective magnetic field $\boldsymbol{B}_{\rm eff}$ defined by Eq. (3.180)

The above is an intuitive explanation of the Rashba interaction based on a semi-classical theory. Strictly speaking, this derivation is not correct because Ehrenfest's theorem shows that the confinement of the electrons in the 2D plane vanishes the average electric field in the z-axis ($E = 0$). A more detailed theory shows that the hybridization of the wave functions between the conduction and valence bands gives rise to a finite Rashba term [46].

References

1. Shoenberg, D.: Magnetic Oscillations in Metals, pp. 22–82. Cambridge Univ. Press (1984)
2. Adams, E.N., Holstein, T.D.: Quantum theory of transverse galvano-magnetic phenomena. J. Phys. Chem. Solids **10**, 254–276 (1959). https://doi.org/10.1016/0022-3697(59)90002-2
3. Uji, S., Aoki, H., Brooks, J.S., Perel, A.S., Athas, G.J., Klepper, S.J., Agosta, C.C., Howe, D.A., Tokumoto, M., Kinoshita, N., Tanaka, Y., Anzai, H.: Magnetic breakdown in the organic conductor $(BEDT\text{-}TTF)_2KHg(SCN)_4$. Solid State Commun. **88**, 683–686 (1993). https://doi.org/10.1016/0038-1098(93)90624-V
4. Goldman, V.J., Santos, M., Shayegan, M., Cunningham, J.E.: Evidence for two-dimentional quantum Wigner crystal. Phys. Rev. Lett. **65**, 2189 (1990). https://doi.org/10.1103/PhysRevLett.65.2189
5. Manoharan, H.C., Shayegan, M.: Wigner crystal versus Hall insulator. Phys. Rev. B **50**, 17662 (1994). https://doi.org/10.1103/PhysRevB.50.17662
6. Yamamoto, H.M., Yamaura, J., Kato, R.: Multicomponent molecular conductors with supramolecular assembly: Iodine-containing neutral molecules as building blocks. J. Am. Chem. Soc. **120**, 5905–5913 (1998). https://doi.org/10.1021/ja980024u
7. Uji, S., Terashima, T., Yasuzuka, S., Yamaura, J., Yamamoto, H.M., Kato, R.: Fermi surface and angular-dependent magnetoresistance in the organic conductor $(BEDT-TTF)_2Br(DIA)$. Phys. Rev. B **68**, 064420 (2003). https://doi.org/10.1103/PhysRevB.68.064420
8. Aoki, H., Uji, S., Albessard, A.K., Ōnuki, Y.: Observation of heavy electrons in $CeRu_2Si_2$ via the dHvA effect. J. Phys. Soc. Jpn. **61**, 3457 (1992). https://doi.org/10.1143/JPSJ.61.3457
9. Aoki, H., Uji, S., Albessard, A.K., Ōnuki, Y.: Transition of f electron nature from itinerant to localized: Metamagnetic transition in $CeRu_2Si_2$ studied via the de Haas–van Alphen effect. Phys. Rev. Lett. **71**, 2110 (1993). https://doi.org/10.1103/PhysRevLett.71.2110
10. Taillefer, L., Lonzarich, G.G.: Heavy-fermion quasiparticles in UPt_3. Phys. Rev. Lett. **60**, 1570 (1988). https://doi.org/10.1103/PhysRevLett.60.1570

11. Kaplan, J.I., Glasser, M.L.: *G* factors for an interacting electron gas. Phys. Rev. **186**, 958 (1969). https://doi.org/10.1103/PhysRev.186.958
12. Hill, S., Singleton, J., Pratt, F.L., Doporto, M., Hayes, W., Janssen, T.J.B.M., Perenboom, J.A.A.J., Kurmoo, M., Day, P.: Cyclotron resonance studies of electron dynamics in BEDT-TTF salts. Synth. Met. **55–57**, 2566 (1993). https://doi.org/10.1016/0379-6779(93)90459-A
13. Kawamoto, T., Mori, T.: Many-body effect on the superconducting transition temperature in layered organic superconductors. Phys. Rev. B **74**, 212502 (2006). https://doi.org/10.1103/PhysRevB.74.212502
14. Sugiura, S., Isono, T., Terashima, T., Yasuzuka, S., Schlueter, J.A., Uji, S.: Fulde–Ferrell–Larkin–Ovchinnikov and vortex phases in a layered organic superconductor. NPJ Quant. Mater. **4**, 7 (2019). https://doi.org/10.1038/s41535-019-0147-2
15. Bondarenko, V.A., Uji, S., Terashima, T., Terakura, C., Tanaka, S., Maki, S., Yamada, J., Nakatsuji, S.: First observations of the heat capacity quantum oscillations in the organic superconductor $(BEDT\text{-}TTF)_2Cu(NCS)_2$. Synth. Met. **120**, 1039 (2001)
16. Fukase, T.: Giant quantum oscillations in the magneto-acoustic attenuation and the spin splitting of arsenic. J. Phys. Soc. Jpn. **26**, 964 (1969). https://doi.org/10.1143/JPSJ.26.964
17. Kataoka, M., Goto, T.: Theory of the Acoustic de Haas-van Alphen effect. J. Phys. Soc. Jpn. **62**, 4352 (1993). https://doi.org/10.1143/JPSJ.62.4352
18. Matsui, H., Yamaguchi, M., Endo, S., Inuzuka, T., Uozaki, H., Toyota, N.: Acoustic de Haas-van Alphen effect in organic superconductor α-$(BEDT\text{-}TTF)_2NH_4Hg(SCN)_4$. J. Phys. Soc. Jpn. **67**, 2586 (1998). https://doi.org/10.1143/JPSJ.67.2586
19. Endoh, D., Goto, T., Suzuki, T., Fujimura, T., Onuki, Y., Komatsubara, T.: Structural phase transition and acoustic de Haas-van Alphen effect of $SmCu_6$. J. Phys. Soc. Jpn. **12**, 4489 (1987). https://doi.org/10.1143/JPSJ.56.4489
20. Shoenberg, D.: Magnetic Oscillations in Metals, pp. 160–173. Cambridge Univ. Press (1984)
21. Settai, R., Goto, T., Ōnuki, Y.: Acoustic de Haas-van Alphen effect of YCu_2 and $CeCu_2$. J. Phys. Soc. Jpn. **61**, 609 (1992). https://doi.org/10.1143/JPSJ.61.609
22. Blount, E.I.: Bloch electrons in a magnetic field. Phys. Rev. **126**, 1636 (1962). https://doi.org/10.1103/PhysRev.126.1636
23. Shoenberg, D.: Magnetic Oscillations in Metals, p. 331–368. Cambridge Univ. Press (1984)
24. Tamura, T., Kuroda, H., Uji, S., Aoki, H., Tokumoto, M., Swanson, A.G., Brooks, J.S., Agosta, C.C., Hannahs, S.T.: Analysis of de Haas-van Alphen oscillations and band structure of an organic superconductor, θ-$(BEDT\text{-}TTF)_2I_3$. J. Phys. Soc. Jpn. **63**, 615 (1994). https://doi.org/10.1143/jpsj.63.615
25. Chambers, R.G.: The wave function of a Bloch electron in a magnetic field. Proc. Phys. Soc. **89**, 695–710 (1966). https://doi.org/10.1088/0370-1328/89/3/325
26. Reifenberger, R., Schwarzkopf, D.A.: Magnetic breakdown and the spin-split conduction band in HgSe and $Hg_{1\text{-}x}Mn_xSe$. Phys. Rev. Lett. **50**, 907 (1983). https://doi.org/10.1103/PhysRevLett.50.907
27. Uji, S., Chaparala, M., Hill, S., Sandhu, P.S., Qualls, J., Seger, L., Brooks, J.S.: Effective mass and combination frequencies of de Haas-van Alphen oscillations in κ-$(BEDT\text{-}TTF)_2Cu(NCS)_2$. Synth. Met. **85**, 1573–1574 (1997). https://doi.org/10.1016/S0379-6779(97)80354-2
28. Kishigi, K., Nakano, M., Machida, K., Hori, Y.: DHvA effect with quantum interference oscillation due to magnetic breakdown. J. Phys. Soc. Jpn. **64**, 3043 (1995). https://doi.org/10.1143/jpsj.64.3043
29. Nakano, M.: Unexpected de Haas-van Alphen oscillation in 2D multiband systems due to chemical potential oscillation and its relevance to magnetic breakdown systems. J. Phys. Soc. Jpn. **66**, 19 (1997). https://doi.org/10.1143/jpsj.66.19
30. Falicov, L.M., Sievert, P.R.: Theory of the galvanomagnetic effects in metals with magnetic breakdown: Semiclassical approach. Phys. Rev. **138**, A88 (1965). https://doi.org/10.1103/PhysRev.138.A88

31. Uji, S., Terakura, C., Terashima, T., Yakabe, T., Terai, Y., Tokumoto, M., Kobayashi, A., Sakai, F., Tanaka, H., Kobayashi, H.: Fermi surface and internal magnetic field of the organic conductors λ-$(BETS)_2Fe_xGa_{1-x}Cl_4$. Phys. Rev. B **65**, 113101 (2002). https://doi.org/10.1103/PhysRevB.65.113101
32. Uji, S., Shinagawa, H., Terashima, T., Yakabe, T., Terai, Y., Tokumoto, M., Kobayashi, A., Tanaka, H., Kobayashi, H.: Magnetic-field-induced superconductivity in a two-dimensional organic conductor. Nature **410**, 908–910 (2001). https://doi.org/10.1038/35073531
33. Uji, S., Brooks, J.S.: Magnetic-Field-Induced Superconductivity in Organic Conductors. J. Phys. Soc. Jpn. **75**, 051014 (2006). https://doi.org/10.1143/JPSJ.75.051014
34. Shoenberg, D.: Magnetic Oscillations in Metals, pp. 254–330. Cambridge Univ. Press (1984)
35. Uji, S., Brooks, J.S., Takasaki, S., Yamada, J., Anzai, H.: Origin of rapid oscillation in the metallic phase for the organic conductor $(TMTSF)_2ClO_4$. Solid State Commun. **103**, 387–392 (1997). https://doi.org/10.1016/S0038-1098(97)00209-3
36. Webb, R.A., Washburn, S., Umbach, C.P., Laibowitz, R.B.: Observation of h/e Aharonov–Bohm oscillations in normal-metal rings. Phys. Rev. Lett. **54**, 2696 (1985). https://doi.org/10.1103/PhysRevLett.54.2696
37. Berry, M.V.: Quantal phase factors accompanying adiabatic changes. Proc. R. Soc. Lond. A **392**, 45–47 (1984). https://doi.org/10.1098/rspa.1984.0023
38. Shoenberg, D.: Magnetic Oscillations in Metals, pp. 483–492. Cambridge Univ. Press (1984)
39. Roth, L.M.: Semiclassical theory of magnetic energy levels and magnetic susceptibility of Bloch electrons. Phys. Rev. **145**, 434 (1966). https://doi.org/10.1103/PhysRev.145.434
40. Mikitik, G.P., Sharlai, Y.V.: Manifestation of Berry's phase in metal physics. Phys. Rev. Lett. **82**, 2147 (1999). https://doi.org/10.1103/PhysRevLett.82.2147
41. Zhang, Y., Tan, Y.-W., Stormer, H.L., Kim, P.: Experimental observation of the quantum Hall effect and Berry's phase in graphene. Nature **438**, 201–204 (2005). https://doi.org/10.1038/nature04235
42. Wright, A.R., McKenzie, R.H.: Quantum oscillations and Berry's phase in topological insulator surface states with broken particle-hole symmetry. Phys. Rev. B **87**, 085411 (2013). https://doi.org/10.1103/PhysRevB.87.085411
43. Xiao, D., Chang, M.-C., Niu, Q.: Berry phase effects on electronic properties. Rev. Mod. Phys. **82**, 1959 (2010)
44. He, L., Xiu, F., Yu, X., Teague, M., Jiang, W., Fan, Y., Kou, X., Lang, M., Wang, Y., Huang, G., Yeh, N.-C., Wang, K.L.: Surface-dominated conduction in a 6 nm thick Bi_2Se_3 thin film. Nano Lett. **12**, 1486–1490 (2012). https://doi.org/10.1021/nl204234j
45. Taskin, A.A., Ando, Y.: Berry phase of nonideal Dirac Fermions in topological insulators. Phys. Rev. **84**, 035301 (2011). https://doi.org/10.1103/PhysRevB.84.035301
46. Winkler, R.: Spin-Orbit Coupling Effect in Two-Dimensional Electron and Hole Systems. Springer-Verlag, Berlin (2003)

Angle-Dependent Magnetoresistance Oscillation (AMRO)

4

Abstract

Quantum oscillations are measured by sweeping the magnetic field at a fixed angle. Quantum oscillations appear in all thermodynamic quantities and in electrical resistance. In this chapter, we show that the resistance for low-dimensional electron systems exhibits a peculiar oscillatory behavior as a function of the magnetic field angle, which is called the angle-dependent magnetoresistance oscillation (AMRO). The AMRO is ascribed to the periodic orbital motion of the conduction electrons on the Fermi surface, which is not related to the Landau quantization, and arises even from open orbits on quasi-1D Fermi surface as well as from closed orbits on quasi-2D Fermi surface. From the AMRO we can determine not only the size but also the shape of the Fermi surface. The coherence of the interlayer transport can also be discussed from the measurements.

4.1 What Is AMRO?

The magnetic field angle dependence of the interlayer resistance of the quasi-2D organic conductor θ-(BEDT-TTF)2I3 shows characteristic periodic peaks as shown in Fig. 4.1a at low temperatures and high magnetic fields. The quasi-2D Fermi surface of this conductor is already depicted in Fig. 3.14a. Here the magnetic field is rotated in the ac^* plane, where the c^*- and a-axes are perpendicular and parallel to the conduction layers, respectively, and the magnetic field angle θ is defined in the inset of Fig. 4.1a. As shown in Fig. 4.1b, the peaks are periodic as a function of tanθ. This oscillation is called angle-dependent magnetoresistance oscillation (AMRO) [1–3]. Quantum oscillation measurements can accurately determine the extremal cross sections of the Fermi surfaces, but not the shape of the cross sections. By contrast, AMRO gives the information of not only the size but also the

S. Uji, *Fermiology*, Lecture Notes in Physics 1042,
https://doi.org/10.1007/978-981-95-0820-4_4

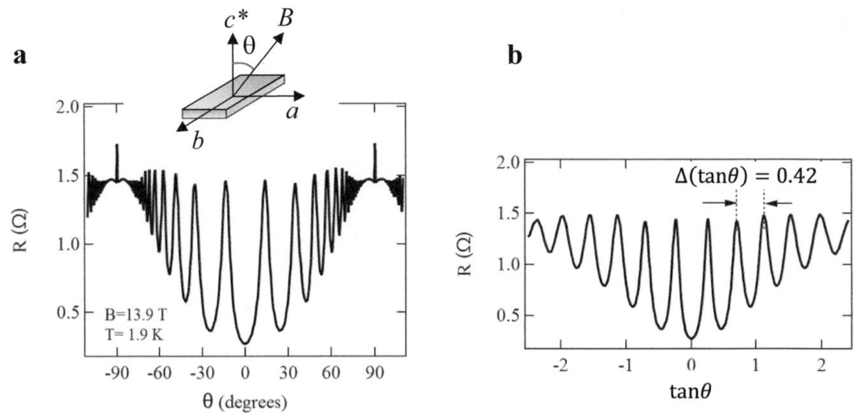

Fig. 4.1 **a** Interlayer resistance as a function of magnetic field angle θ for the quasi-2D organic conductor θ-(BEDT-TTF)$_2$I$_3$. Inset: schematic of sample and definition of the magnetic field angle. **b** Interlayer resistance oscillation with the period $\Delta(\tan\theta) = 0.42$

shape of the Fermi surface cross section. Therefore, AMRO measurements are recognized as another powerful experimental method to determine the structure of the Fermi surface.

4.2 Yamaji Angle

4.2.1 Cyclotron Orbit on Quasi-2D Fermi Surface

The AMRO, as shown in Fig. 4.1a, has been observed in the interlayer resistance of some quasi-2D conductors. To understand its origin, we assume a quasi-2D conductor with a layered structure as depicted in Fig. 4.2a. The energy band with the lattice constant c in the z-axis is written as

$$E = \frac{\hbar^2}{2m_c}\left(k_x^2 + k_y^2\right) - 2t_c \cos(k_z c). \tag{4.1}$$

This energy band uses a free electron approximation in the xy plane and a tight band approximation in the z-axis with the transfer integral t_c. The Brillouin zone in the k_z-axis is defined in the range $-\pi/c < k_z \leq \pi/c$. For a highly 2D case $0 < t_c \ll E_F$, the Fermi surface of the energy band has a corrugated cylindrical shape as shown in Fig. 4.2b. The Fermi surface has a maximum cross section at $k_z = 0$ and minimum cross sections at $k_z = \pm\pi/c$. In a magnetic field parallel to the k_z-axis, the electrons undergo the cyclotron motion in the plane perpendicular to the magnetic field due to the Lorentz force, $\hbar\dot{\boldsymbol{k}} = \boldsymbol{F} = -e\boldsymbol{v} \times \boldsymbol{B}$, as indicated by the dotted curves in Fig. 4.2b. In real space, the velocity of the electrons in the

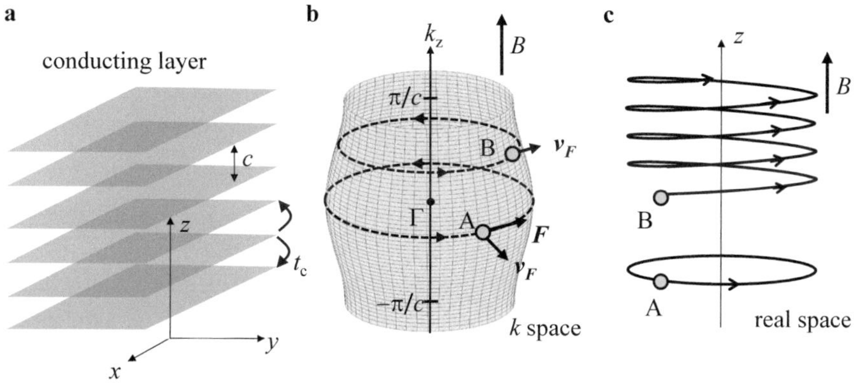

Fig. 4.2 **a** Quasi-2D conductor with a layered structure, **b** Schematic of quasi-2D Fermi surface. Dotted lines indicate cyclotron orbits in a magnetic field parallel to k_z. **c** Orbital motion of the electrons in real space

z-axis is given by

$$v_z = \frac{\partial E}{\hbar \partial k_z} = \frac{2t_c c}{\hbar} \sin(k_z c). \tag{4.2}$$

The electrons at point A ($k_z = 0$) have zero velocity, $v_z = 0$, and thus the cyclotron motion remains in a plane, as shown in Fig. 4.2c. At point B, the electrons have a finite positive value of v_z and thus undergo a helical motion in the z-axis. In this way, the cyclotron motion of the electrons in real space depends on the position of the Fermi surface.

For convenience, we define the Fermi wave number k_F in the $k_x k_y$ plane for $t_c = 0$ as

$$E_F = \frac{\hbar^2}{2m_c} k_F^2, \tag{4.3}$$

where E_F is the Fermi energy. When the magnetic field is tilted from the z-axis by θ as shown in Fig. 4.3, k_z on the cyclotron orbit is approximated as

$$k_z(\varphi) = k_{z0} - k_F \tan\theta \cos\varphi, \tag{4.4}$$

where k_{z0} is the k_z value of the center coordinate of the cyclotron orbit and φ is the angle from the k_x-axis. We project the cyclotron orbit perpendicular to the magnetic field onto the $k_z = k_{z0}$ plane, and define the radius of the orbit as $k_\perp(\varphi)$ as shown in Fig. 4.4a. The area A_F of the cyclotron motion in Fig. 4.3 is calculated from

$$A_F \cos\theta = \int_0^{2\pi} \mathrm{d}\varphi \int_0^{k_\perp} \mathrm{d}k_\perp\, k_\perp(\varphi) = \frac{1}{2} \int_0^{2\pi} \mathrm{d}\varphi k_\perp^2(\varphi). \tag{4.5}$$

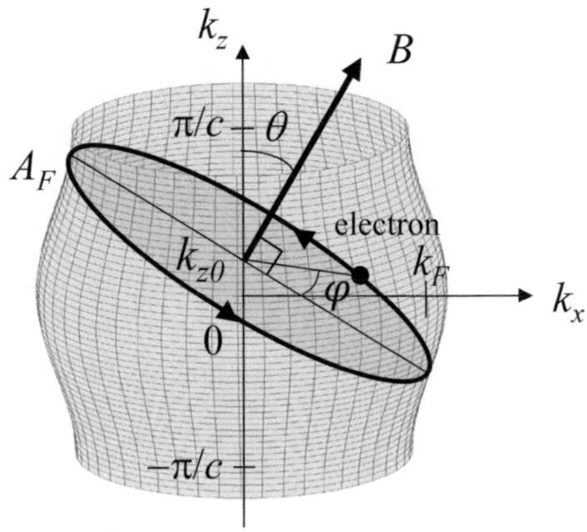

Fig. 4.3 Cyclotron orbit on quasi-2D Fermi surface in magnetic field tilted from the k_z-axis

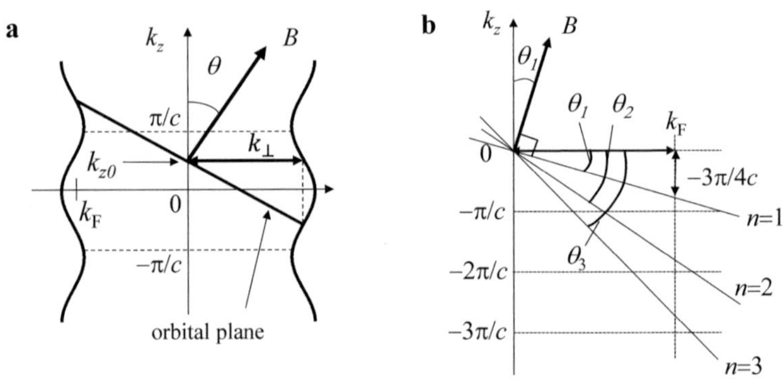

Fig. 4.4 **a** Cyclotron orbit on quasi-2D Fermi surface and radius of cyclotron orbit $k_\perp(\varphi)$ projected onto the $k_z = k_{z0}$ plane in a tilted magnetic field, **b** Orbital planes at Yamaji angles $\theta_N (n = 1, 2, 3)$

From Eqs. (4.1) and (4.3), we have

$$k_\perp^2(\varphi) = k_F^2 + \left(\frac{4m_c t_c}{\hbar^2}\right)\cos[k_z(\varphi)c]. \tag{4.6}$$

Substituting Eq. (4.6) into Eq.(4.5), we have

$$\begin{aligned} A_F \cos\theta &= \frac{1}{2}\int_0^{2\pi} \mathrm{d}\varphi \left[k_F^2 + \frac{4m_c t_c}{\hbar^2}\cos\{(k_{z0} - k_F \tan\theta\cos\varphi)c\}\right] \\ &= \pi k_F^2 + \frac{2m_c t_c}{\hbar^2}\cos(k_{z0}c)\int_0^{2\pi}\cos(k_F c\tan\theta\cos\varphi)\mathrm{d}\varphi \end{aligned}$$

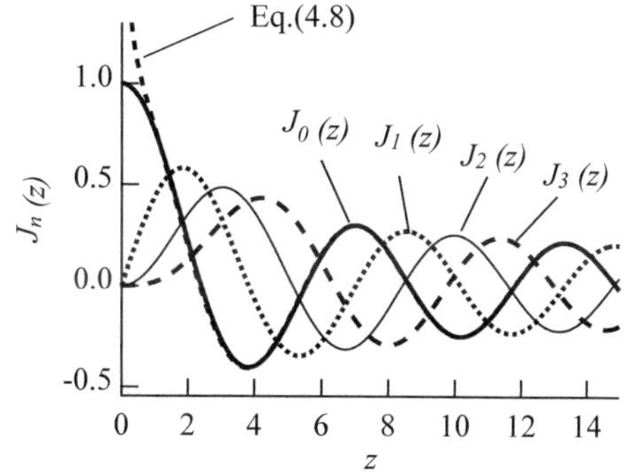

Fig. 4.5 nth order Bessel functions $J_n(z)$. Dotted curve shows the approximate expression of $J_0(z)$, Eq. (4.8). *Note* $J_0(0) = 1$ and $J_{n\geq 1}(0) = 0$

$$= \pi k_F^2 + \frac{4\pi m_c t_c}{\hbar^2} \cos(k_{z0}c) J_0(k_F c \tan\theta), \tag{4.7}$$

where J_0 is the zeroth-order Bessel function, defined as

$$J_0(z) = \frac{1}{2\pi} \int_0^{2\pi} \cos(z \sin\theta) d\theta = \frac{1}{2\pi} \int_0^{2\pi} \cos(z \cos\theta) d\theta.$$

The *nth* order Bessel functions $J_n(z)$ are depicted in Fig. 4.5. They are even functions, $J_n(z) = J_n(-z)$, and $J_0(z)$ is approximated as

$$J_0(z) \approx \left(\frac{2}{\pi |z|}\right)^{1/2} \cos\left(|z| - \frac{\pi}{4}\right) \text{ for } |z| > 1. \tag{4.8}$$

Therefore, we have nodes, $J_0(z) = 0$ for $|z| - \pi/4 = (n - 1/2)\pi$, $(n = 1, 2, 3, \ldots)$. The result means that when the following relation is satisfied,

$$k_F c \tan\theta_n = \pm\pi \left(n - \frac{1}{4}\right), \quad n = 1, 2, 3, \ldots; \tag{4.9}$$

the area A_F is independent of k_{z0},

$$A_F \cos\theta_n = \pi k_F^2. \tag{4.10}$$

At θ_n, all the cross sections perpendicular to the magnetic field have the same area. This characteristic angle θ_n is called "Yamaji" angle, which was first derived by Yamaji [1]. Yamaji angle θ_n is written as $\mathrm{k}_F \tan\theta_n = \pm 3\pi/4c, \pm 7\pi/4c, \pm 11\pi/4c, \ldots$ which reflects the geometric periodic structure of the quasi-2D Fermi surface as shown in Fig. 4.4b. At $\theta = 0°$, the maximum and minimum cross sections of the quasi-2D Fermi surface are respectively given by

$$A_F^{\max} = \pi k_F^2 + \frac{4\pi m_c t_c}{\hbar^2}, \; A_F^{\min} = \pi k_F^2 - \frac{4\pi m_c t_c}{\hbar^2}. \tag{4.11}$$

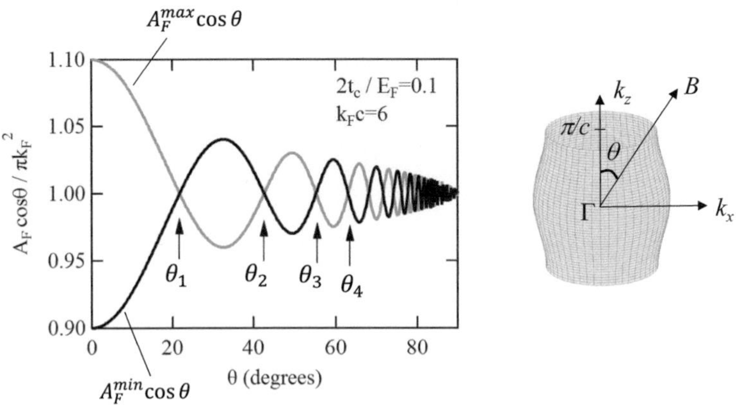

Fig. 4.6 Angular dependence of the two extremal cross sections (left) for quasi-2D Fermi surface (right)

Figure 4.6 shows the two extremal cross sections as a function of the magnetic field angle. The arrows indicate Yamaji angles, at which the two extremal cross sections coincide.

4.2.2 Physical Meaning of Yamaji Angle

The velocity in the z-axis averaged over the cyclotron motion is calculated by

$$\begin{aligned}\langle v_z \rangle &= \int_0^{2\pi} d\varphi v_z(\varphi) / \int_0^{2\pi} d\varphi \\ &= \frac{2t_c c}{\hbar} \frac{1}{2\pi} \int_0^{2\pi} d\varphi \sin(k_{z0}c - k_F c \tan\theta \cos\varphi),\end{aligned}$$

using Eqs. (4.2) and (4.4). From the relation

$$\begin{aligned}\sin(k_{z0}c - k_F c \tan\theta \cos\varphi) &= \sin(k_{z0}c) \cos(k_F c \tan\theta \cos\varphi) \\ &\quad - \cos(k_{z0}c) \sin(k_F c \tan\theta \cos\varphi),\end{aligned}$$

we have

$$\langle v_z \rangle = \frac{2t_c c}{\hbar} \sin(k_{z0}c) J_0(k_F c \tan\theta). \tag{4.12}$$

The relation $\langle v_z \rangle \propto J_0(k_F c \tan\theta)$ leads to nodes $\langle v_z \rangle = 0$ at Yamaji angles. The orbit on the Fermi surface at θ_2 is depicted in Fig. 4.7a. In real space, the electrons

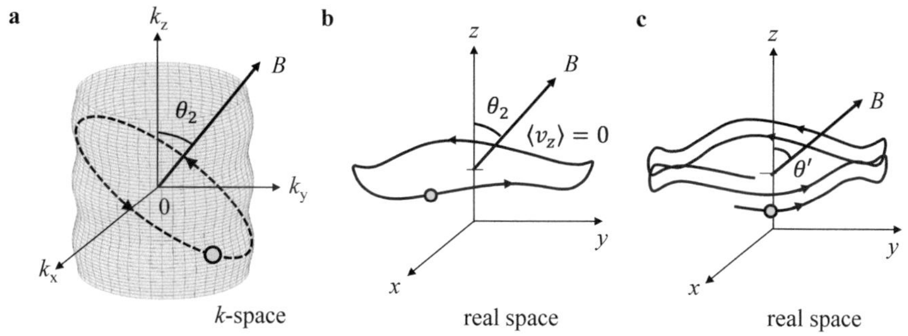

Fig. 4.7 Electron orbit on the quasi-2D Fermi surface at Yamaji angle θ_2 in **a** k-space and **b** real space. **c** Electron orbit for $\theta' > \theta_2$ in real space

remain in the nearly flat plane as presented in Fig. 4.7b. At an angle $\theta' > \theta_2$, the electrons have a drift motion in the z-axis as shown in Fig. 4.7c. Since $\langle v_z \rangle = 0$ at Yamaji angles, the conductivity in the z-axis vanishes, leading to the resistance peak. Therefore, the resistance will have periodic peaks with $\tan\theta$ as shown in Fig. 4.1b. This is the qualitative interpretation of the AMRO. From Eq. (4.9), the period of the AMRO is given by

$$\Delta(\tan\theta) = \frac{\pi}{k_F c}, \tag{4.13}$$

and thus we can estimate k_F from the AMRO period $\Delta(\tan\theta)$.

As we have seen in Chap. 3, the quantum oscillations arise from the Landau quantization of the electronic states at the extremal cross sections of the Fermi surface perpendicular to the magnetic field. If the Fermi surface has the shape as shown in Fig. 4.7a, quantum oscillations with two frequencies arising from the maximum and minimum cross sections are observed for $\theta = 0°$. However, at Yamaji angles, quantum oscillation with a single frequency is observed since the two extremal cross sections coincide, as shown in Fig. 4.6. At Yamaji angles, all electrons of the Fermi surface contribute to the quantum oscillation; the amplitude of the quantum oscillation becomes large.

4.3 Formulation of AMRO

4.3.1 Conductivity Tensor

In this section, we describe the AMRO quantitatively. The conductivity tensor $\boldsymbol{\sigma}$ is defined as follows:

$$\boldsymbol{j} = \boldsymbol{\sigma}\boldsymbol{E} \rightarrow \begin{pmatrix} j_x \\ j_y \\ j_z \end{pmatrix} = \begin{pmatrix} \sigma_{xx} & \sigma_{xy} & \sigma_{xz} \\ -\sigma_{xy} & \sigma_{yy} & \sigma_{yz} \\ -\sigma_{xz} & -\sigma_{yz} & \sigma_{zz} \end{pmatrix} \begin{pmatrix} E_x \\ E_y \\ E_z \end{pmatrix}. \tag{4.14}$$

Each component of the conductivity tensor is expressed by the Chambers formula, including the spin degree of freedom,

$$\sigma_{ij} = 2e^2 \sum_{k(0)} \left(-\frac{\mathrm{d}f(E)}{\mathrm{d}E} \right) v_i(k(0)) \times \int_{-\infty}^{0} \mathrm{d}t \; v_j(k(t)) \exp(t/\tau), \quad i, j = x, y, z, \tag{4.15}$$

where $\sum_{\boldsymbol{k}(0)} v_i(\boldsymbol{k}(0))$ means sum over the i-axis velocity v_i of all states with the wave number $\boldsymbol{k}(0)$ at $t = 0$, and $\int_{-\infty}^{0} \mathrm{d}t \; v_j(k(t))$ means the time integral of the velocity $v_j(\boldsymbol{k}(t))$ for $t < 0$ until it reaches the wave number vector $\boldsymbol{k}(t = 0)$. Due to the $\exp(t/\tau)$ term, the conductivity at $t = 0$ is affected by the past velocity in the range of the scattering time τ. The Chambers formula is derived in Appendix 4.1. Since $k_B T \ll E_F$ under normal experimental conditions, the derivative of the Fermi distribution function can be taken as $-\mathrm{d}f/\mathrm{d}E = \delta(E_F - E)$. Due to the delta function $\delta(E_F - E)$, the electron motion only on the Fermi surface is taken into account in this calculation. The scattering time τ is assumed to be the same everywhere on the Fermi surface. This assumption may not be a good approximation for anisotropic Fermi surfaces. For example, it has been found that in the quasi-2D organic conductor θ-(BEDT-TTF)$_2$IBr$_2$, the mobility $e\tau/m_c$ is about twice as anisotropic in the conduction plane [4]. However, this anisotropy is not so large, and as we will see below, the calculated results can satisfactorily explain the experimental results even for a constant τ. From $v_i = (1/\hbar)\partial E/\partial k_i$ we obtain the velocities in real space,

$$(v_x, v_y, v_z) = \left(\frac{\hbar k_x}{m_c}, \frac{\hbar k_y}{m_c}, \frac{2t_c c}{\hbar} \sin(k_z c) \right) \tag{4.16}$$

for the energy band of the quasi-2D Fermi surface given by Eq. (4.1).

First, we calculate the z-axis conductivity σ_{zz}, giving AMRO. As shown in Fig. 4.3, the magnetic field $\boldsymbol{B}$ in the xz ($k_x k_z$) plane is specified as $(B \sin\theta, 0, B\cos\theta)$. Substituting Eq. (4.16) into the equation of motion $\hbar\dot{\boldsymbol{k}} = -e\boldsymbol{v} \times \boldsymbol{B}$, we have

$$\left(\dot{k}_x(t), \dot{k}_y(t), \dot{k}_z(t)\right) = \left(-k_y\omega_c, k_x\omega_c - \frac{2m_c t_c c}{\hbar^2}\omega_c \sin(k_z c)\tan\theta, k_y\omega_c \tan\theta\right), \tag{4.17}$$

where the cyclotron frequency is defined as

$$\omega_c = \frac{eB\cos\theta}{m_c} = \frac{eB_z}{m_c}.$$

Since Eq. (4.17) is described only by k_i or $\dot{k}_i$ (not v_i), the electron motion in k-space can be discussed. For the highly 2D case ($t_c/E_F \ll 1$), we have $2m_c t_c c/\hbar^2 = t_c c k_F^2/(\hbar^2 k_F^2/m_c) = t_c c k_F^2/E_F \ll k_F$. By approximating as $\dot{k}_y(t) \approx k_x \omega_c$, we obtain the equations of motion; $\ddot{k}_x = -\omega_c^2 k_x$ and $\ddot{k}_y = -\omega_c^2 k_y$ from Eq. (4.17). The solutions are

$$k_x = k_\perp \cos\varphi,\ k_y = k_\perp \sin\varphi,\ \varphi = \omega_c t + \varphi_0, \tag{4.18}$$

where $k_\perp$ is the radius of the cyclotron orbit. For the electrons on the Fermi surface, we have $k_\perp = k_F$. By integrating $\dot{k}_z(t) = k_y(t)\omega_c \tan\theta$ in Eq. (4.17) over time, we obtain

$$k_z(t) = (k_{z0} + k_\perp \cos\varphi_0 \tan\theta) - k_\perp \cos(\omega_c t + \varphi_0)\tan\theta. \tag{4.19}$$

We take initial conditions of the wave numbers; $k_{x0} = k_\perp \cos\varphi_0$, $k_{y0} = k_\perp \sin\varphi_0$ and k_{z0} at $t = 0$. In real space, the velocity is given by

$$\begin{aligned} v_z(t) &= \left(\frac{2t_c c}{\hbar}\right)\sin(k_z c) \\ &= \left(\frac{2t_c c}{\hbar}\right)\sin[(k_{z0}c + k_\perp c\cos\varphi_0\tan\theta) - k_\perp c\cos(\omega_c t + \varphi_0)\tan\theta]. \end{aligned} \tag{4.20}$$

Substituting $v_z(t)$ and $v_z(t = 0)$ into Eq. (4.15), we obtain the AMRO formula [5],

$$\sigma_{zz}(B) = \sigma_{zz}(0)\sum_{n=-\infty}^{\infty}\frac{J_n^2(k_F c\tan\theta)}{1 + (n\omega_c\tau)^2}, \tag{4.21}$$

where the coefficient $\sigma_{zz}(0)$ is

$$\begin{aligned} \sigma_{zz}(0) &= \frac{2e^2 m_c t_c^2 c\tau}{\pi\hbar^4} = 2D(E_F)\left(\frac{e t_c c}{\hbar}\right)^2 \\ &= \left(\frac{v_z^0}{v_F}\right)^2\frac{N_e e^2\tau}{m_c},\ v_z^0 = \frac{2t_c c}{\hbar}. \end{aligned} \tag{4.22}$$

The AMRO formula Eq. (4.21) is derived in Appendix 4.2. As in Eq. (1.8), $N_e = n(E_F)/c = k_F^2/2\pi c$ is the electron number per unit volume. In a perpendicular magnetic field ($\theta = 0$), we have $\sigma_{zz}(B) = \sigma_{zz}(0)$: the conductivity is independent of the magnetic field. In a sufficiently high magnetic field ($\omega_c\tau \gg 1$), Eq. (4.21) is approximated as

$$\sigma_{zz}(B) \approx \sigma_{zz}(0)J_0^2(k_F c\tan\theta).$$

Therefore, $\sigma_{zz}(B)$ vanishes at Yamaji angles; the resistance shows periodic peaks with $\tan\theta$. In the above calculation, note that the effect of the only closed orbits as shown in Fig. 4.3 is taken into account. For $\theta \approx 90°$, Eq. (4.21) is not correct as discussed in Sect. 4.6.

Since the AMRO formula Eq. (4.21) is derived on the assumption of the energy band given by Eq. (4.1), one might think that this is a special case of the free electron approximation in the xy plane. However, we should note that the same result can be obtained even for a more general case, $E = f(k_x, k_y) - 2t_c\cos(k_z c)$. For a quasi-2D electron system, it is reasonable to assume that the energy dispersion in the k_z-axis is given by the tight binding approximation. In this way, we can conclude that Eq. (4.21) is a useful expression for quasi-2D conductors.

Next, we calculate the other components of the conductivity tensor. From Eq. (4.18),

$$\begin{aligned} v_x(t) &= \frac{\hbar k_x}{m_c} = \left(\frac{\hbar k_\perp}{m_c}\right)\cos(\omega_c t + \varphi_0), \\ v_y(t) &= \frac{\hbar k_y}{m_c} = \left(\frac{\hbar k_\perp}{m_c}\right)\sin(\omega_c t + \varphi_0). \end{aligned} \tag{4.23}$$

Substituting these into Eq. (4.15), we obtain

$$\begin{aligned} \sigma_{xx}(B) = \sigma_{yy}(B) &= \sigma_0 \frac{1}{1 + (\omega_c\tau)^2}, \\ \sigma_0 = D(E_F)E_F\frac{e^2\tau}{m_c} &= \frac{N_e e^2\tau}{m_c}. \end{aligned} \tag{4.24}$$

The off-diagonal terms are calculated as

$$\begin{aligned} &\sigma_{xy}(B) = -\sigma_{yx}(B) = -\sigma_0 \frac{\omega_c\tau}{1 + (\omega_c\tau)^2}, \\ &\sigma_{xz}(B) = \sigma_{yz}(B) = 0. \end{aligned} \tag{4.25}$$

Equation (4.25) is derived in Appendix 4.3. The conductivity tensor is summarized as

$$\sigma = \begin{pmatrix} \sigma_0 \frac{1}{1+(\omega_c\tau)^2} & -\sigma_0 \frac{\omega_c\tau}{1+(\omega_c\tau)^2} & 0 \\ \sigma_0 \frac{\omega_c\tau}{1+(\omega_c\tau)^2} & \sigma_0 \frac{1}{1+(\omega_c\tau)^2} & 0 \\ 0 & 0 & \sigma_{zz}(0) \sum_{n=-\infty}^{\infty} \frac{J_n^2(k_F c \tan\theta)}{1+(n\omega_c\tau)^2} \end{pmatrix}. \tag{4.26}$$

The results of $\sigma_{xx}(B)$ and $\sigma_{xy}(B)$ are consistent with Eq. (3.16). All components depend only on the z-component of the magnetic field B_z since $\omega_c = eB_z/m_c$. The resistivity tensor is given by

$$\rho = \begin{pmatrix} \rho_{xx} & \rho_{xy} & 0 \\ -\rho_{xy} & \rho_{xx} & 0 \\ 0 & 0 & \rho_{zz} \end{pmatrix}$$

$$= \begin{pmatrix} \frac{\sigma_{xx}}{\sigma_{xx}^2+\sigma_{xy}^2} & \frac{-\sigma_{xy}}{\sigma_{xx}^2+\sigma_{xy}^2} & 0 \\ \frac{\sigma_{xy}}{\sigma_{xx}^2+\sigma_{xy}^2} & \frac{\sigma_{xx}}{\sigma_{xx}^2+\sigma_{xy}^2} & 0 \\ 0 & 0 & \frac{1}{\sigma_{zz}} \end{pmatrix} = \begin{pmatrix} \frac{1}{\sigma_0} & \frac{\omega_c\tau}{\sigma_0} & 0 \\ -\frac{\omega_c\tau}{\sigma_0} & \frac{1}{\sigma_0} & 0 \\ 0 & 0 & \frac{1}{\sigma_{zz}} \end{pmatrix}. \tag{4.27}$$

Note that the simple relation, $\rho_{zz}(B) = 1/\sigma_{zz}(B)$. The in-plane resistivities $\rho_{ii}(i = x, y)$ are independent of the magnetic field. As discussed in Sect. 3.1, it is due to the assumption that there is only one type of carrier.

The AMRO emerges only in ρ_{zz} but not in the other components. However, the AMRO is often observed experimentally in $\rho_{ii}(i = x, y)$ for quasi-2D conductors. A possible reason is that the current I is not exactly parallel to the c-axis due to some inhomogeneities of the sample or inadequate configuration of the electrical contacts.

Figure 4.8 presents the calculated results of $\rho_{zz}(B) = 1/\sigma_{zz}$ given by Eq. (4.21) at different $\omega_c\tau$ values for $k_F c = 6$. The AMRO clearly appears at $\omega_c\tau = 5$. As $\omega_c\tau$ decreases, the AMRO amplitude steeply decreases, indicating that the high-quality samples (long τ) and/or high magnetic fields are required for the AMRO observation. The origin of the AMRO is quite different from the quantum oscillations due to Landau quantization discussed in Chap. 3; the AMRO arises from the effect of semiclassical orbital motion on the quasi-2D Fermi surface, where the Landau quantization is not taken into account.

As seen in Sect. 3.1.4, the quantum oscillation in the resistance arises from the density of states at the Fermi level through the scattering time τ. Since the frequency and amplitude of the quantum oscillation in general depend on the magnetic field angle, the quantum oscillation is superimposed on the AMRO at sufficiently low temperatures and high magnetic fields. At first sight, the result is very complicated, but it is experimentally easy to distinguish between the AMRO and the quantum oscillation. The angles of the quantum oscillation peaks vary with the magnetic field, but the AMRO peak angles are independent of it. By observing

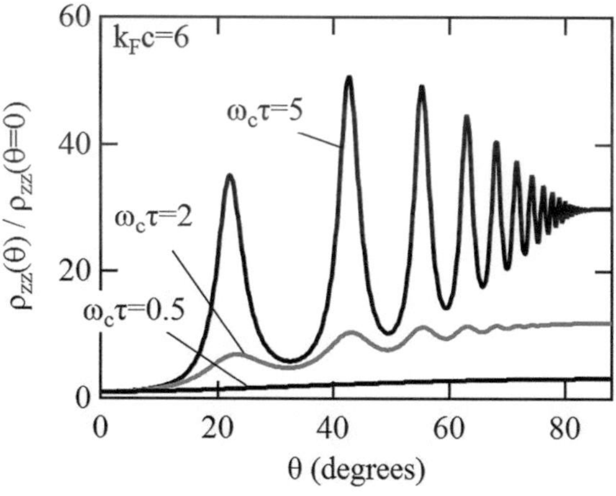

Fig. 4.8 Calculated AMROs, $\rho_{zz}(\theta) = 1/\sigma_{zz}(\theta)$ at different $\omega_c\tau$ values by Eq. (4.21)

the resistance as a function of the magnetic field angle at different magnetic fields, one can distinguish which of the two is the origin of the observed oscillation.

4.3.2 Temperature Dependence of AMRO

Next, we discuss the temperature dependence of the AMRO. Figure 4.9 shows the AMRO measured at various temperatures in the quasi-2D organic conductor θ-(BEDT-TTF)$_2$I$_3$ when the magnetic field is rotated in the ac^* plane. It can be seen that the resistance peak angles are independent of temperature. As the temperature increases, the AMRO amplitude decreases significantly. This is because τ becomes shorter at higher temperatures due to the electron–phonon scattering. At high temperatures, we note that the resistance increases as the magnetic field is tilted from the perpendicular direction, which looks like a sine curve. The qualitative understanding of this behavior is that the electric current and magnetic field are orthogonal at $\theta = 90°$ and the Lorentz force acts strongly on the current, causing the resistance to increase. The sharp peaks are visible at $\theta = \pm 90°$ as indicated by the arrows. This mechanism is not explained by the AMRO formula and will be explained in Sect. 4.6.

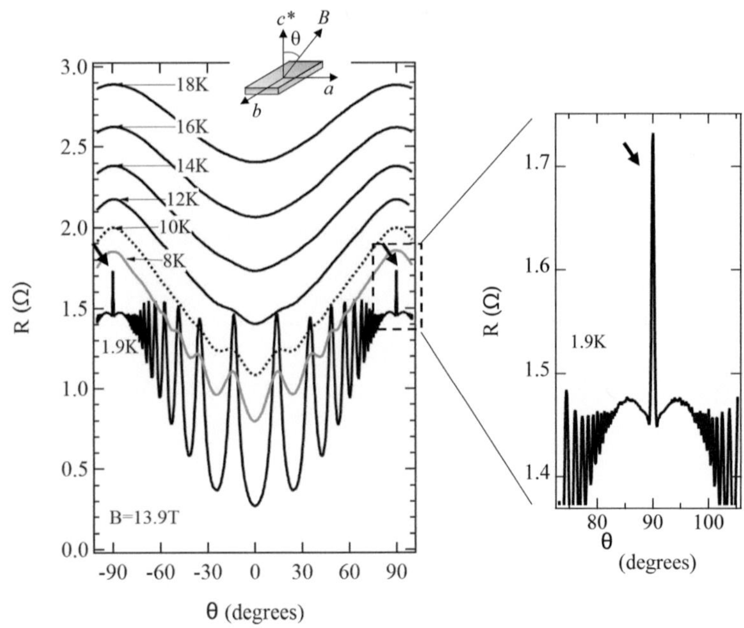

Fig. 4.9 Interlayer resistance as a function of magnetic field direction for θ-(BEDT-TTF)$_2$I$_3$. The amplitude of the AMRO steeply decreases with increasing temperature. Inset: close-up of the AMRO at $\theta \approx 90°$. Arrows indicate sharp peaks at $\theta = \pm 90°$

4.4 Observation of Cross Section of Fermi Surface

So far, we have looked at the AMRO results for a quasi-2D Fermi surface when the magnetic field is rotated from perpendicular to parallel to the plane. If the cross section of the quasi-2D Fermi surface is isotropic (circular), we obtain the same AMRO regardless of the direction in which the magnetic field is tilted from perpendicular. In real materials, however, the cross section of the Fermi surface is anisotropic (not circular). The advantage of the AMRO measurements is that we can even determine the shape of the Fermi surface cross section, as we will see below.

Figure 4.10a shows AMRO results for various rotation planes in θ-(BEDT-TTF)$_2$I$_3$. The rotation plane is defined as the angle φ from the *a*-axis. The interlayer resistance is rather small for the $\phi = 0°$ ($ac*$ plane) rotation, but large for the $\phi = 90°$ ($bc*$ plane) rotation. The AMRO is clearly observed for all rotation planes. To see the AMRO more clearly, the second derivative curves $-\mathrm{d}^2R/\mathrm{d}\theta^2$ are plotted in Fig. 4.10b, where the peaks correspond to the positive curvature of the raw data. The AMRO period $\Delta(\tan\theta)$ is apparently different between the $\phi = 0°$ and $\phi = 90°$ rotations.

Figure 4.11a shows the polar plot of the wave number defined by $k_{\|} = \pi/[c\Delta(\tan\theta)]$, where the $\Delta(\tan\theta)$ is the AMRO period and $c = 1.69$ nm is the distance between the BEDT-TTF layers. As can be seen below, $k_{\|}$ is not identical to k_F. The solid curve shows the cross section of the Fermi surface obtained from $k_{\|}$ [6]. The dotted line is an auxiliary curve for obtaining the Fermi surface. To understand the meaning of the auxiliary curve, suppose there is a quasi-2D

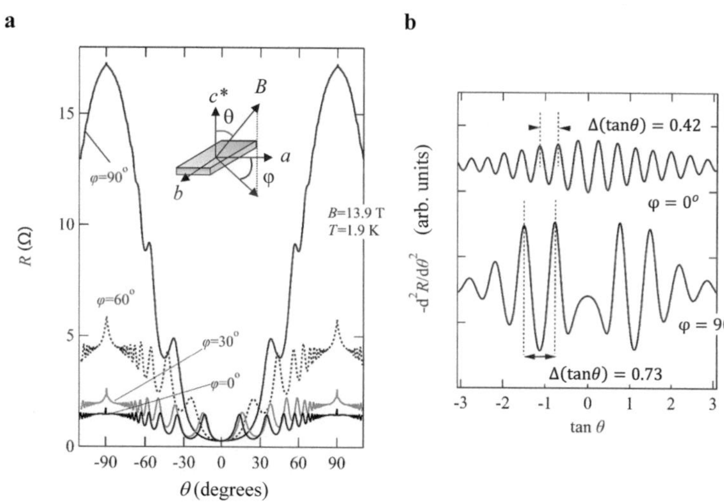

Fig. 4.10 **a** AMRO results for various rotation planes in θ-(BEDT-TTF)$_2$I$_3$. The rotation plane is defined as the angle φ from the *a*-axis. **b** Second derivative curves as a function of tanθ. The AMRO period $\Delta(\tan\theta)$ is different for different rotation planes

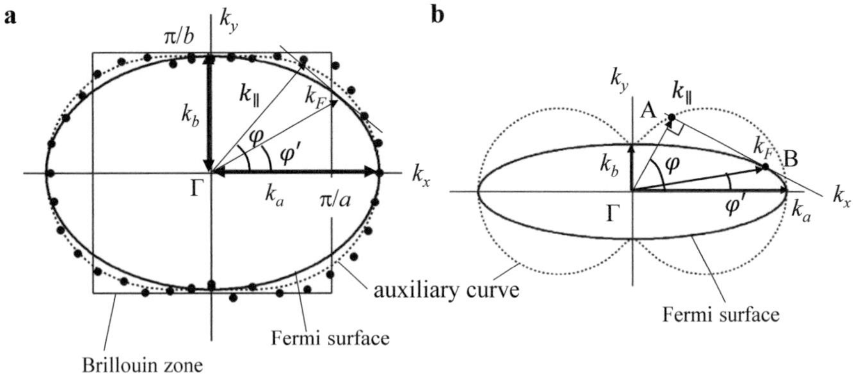

Fig. 4.11 **a** Cross section of the Fermi surface (solid curve) and its auxiliary curve (dotted curve) obtained from AMRO measurements in θ-(BEDT-TTF)$_2$I$_3$, **b** Example of an elongated elliptical cross section (solid curve) and its auxiliary curve (dotted curve). The wave number $k_{\parallel}$ calculated from the AMRO lies on the dotted curve. The Fermi surface is an ellipse inscribed in the auxiliary curve

Fermi surface with an elongated elliptical cross section shown by the solid curve in Fig. 4.11b. If we take the major and minor axes of the elliptical cross section as k_a and k_b, respectively, the Fermi surface cross section (solid curve) can be expressed as

$$1 = \frac{k_x^2}{k_a^2} + \frac{k_y^2}{k_b^2}, \quad k_x = k_F \cos\varphi', \quad k_y = k_F \sin\varphi'. \tag{4.28}$$

The wave number $k_{\parallel}$, determined from the AMRO period, is obtained from the projection of the cross section onto the rotation plane of the magnetic field. As shown in Fig. 4.11b, when the magnetic field is tilted from the perpendicular to the $\Gamma - A$ direction in the conduction plane, $k_{\parallel}$ gives the Fermi wave number k_F at point B on the elliptical Fermi surface. The auxiliary curve giving $k_{\parallel}$ is expressed as [7]

$$k_{\parallel}^2(\varphi) = k_a^2 \cos^2\varphi + k_b^2 \sin^2\varphi. \tag{4.29}$$

Note that $k_{\parallel} = k_a$ for $\varphi = 0°$ and $k_{\parallel} = k_b$ for $\varphi = 90°$. The conductor θ-(BEDT-TTF)$_2$I$_3$ has a small closed quasi-2D Fermi surface and an open quasi-1D Fermi surface, as shown in Fig. 3.14a. From Fig. 4.11a, we see that the Fermi surface obtained from the AMRO has a large closed orbit beyond the first Brillouin zone; the magnetic breakdown orbit (β orbit) in Fig. 3.14a is the origin of the AMRO. This is because the magnetic breakdown probability is large enough at such high magnetic fields. In the experiment, the AMRO from the α orbit cannot be observed in the small angle range even when the magnetic field is reduced. When the magnetic field direction approaches $\theta = \pm 90°$, the AMRO from the α

orbit is observed [8]. The result suggests that the magnetic breakdown probability steeply decreases at high angles where the tunneling distance Δk in Fig. 3.14c becomes large.

The AMRO formula Eq. (4.21) is correct when there exists a single quasi-2D Fermi surface. For θ-(BEDT-TTF)$_2$I$_3$, no magnetic breakdown occurs in the low magnetic field limit. Both the quasi-1D and 2D Fermi surfaces contribute to the interlayer conduction independently, and thus the sum of them determines the total conductivity. In the high magnetic field limit, the magnetic breakdown probability is close to unity, and thus it is reasonable to assume the presence of only the β orbit. In an intermediate magnetic field range, however, the magnetic breakdown probability changes with the magnetic field angle. Strictly speaking, the magnetic breakdown probability also depends on the magnetic field direction. In addition, the difference in the scattering time and effective mass between the quasi-1D and 2D Fermi surfaces must be taken into account in the conductivity simulation. Because of these situations, it is difficult to simulate the overall angular dependence of the resistance quantitatively. The very different results between the $\phi = 0°$ (ac^* plane) and $\phi = 90°$ (bc^* plane) rotations in Fig. 4.10a will be ascribed to the complicated situations described above. Despite these situations, it is evident that the AMRO period provides important information about the Fermi surface structure.

The AMRO has been observed in many organic conductors, but also in other materials with quasi-2D Fermi surfaces [9–14]. The Fe-based superconductor KFe_2As_2 has a layered structure, composed of Fe–As and K layers as shown in the inset of Fig. 4.12a. The interlayer resistance as a function of the magnetic field direction at a low temperature in a high magnetic field is presented in Fig. 4.12a [9]. To see the AMRO more clearly, the second derivative curve $-\mathrm{d}^2R/\mathrm{d}\theta^2$ is plotted in Fig. 4.12b. Two different AMRO periods, *A* and *B,* are observed, originating from the quasi-2D Fermi surfaces at the Γ and M points of the Brillouin zone, respectively. AMRO measurements have been made in other materials, Sr_2RuO_4 [10], graphite intercalation compounds [11], GaAs multilayers[12], and a high-T_c superconductor $Tl_2Ba_2CuO_{6+x}$ [13, 14], and the presence of quasi-2D Fermi surfaces has been discussed.

In the above discussion, the energy dispersion in the perpendicular direction is assumed to have a simple form, $E(k_z) = -2t_c\cos(k_z c)$. It means that only the transfer integrals with the nearest neighbor sites are taken into account. In real materials, the transfer integrals with the next nearest neighbor sites, the higher order terms, may not be negligible. Such higher order terms lead to additional corrugation of the cylindrical Fermi surface and thus lower the symmetry of the Fermi surface structure. The effects on the AMRO have been calculated in some lower symmetry cases [15].

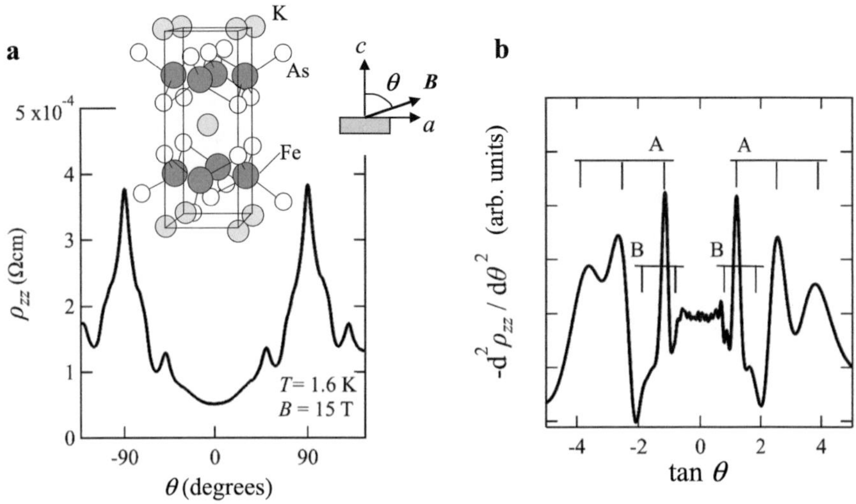

Fig. 4.12 **a** AMRO in Fe-based superconductor KFe_2As_2 (reproduced with permission from [9]). The inset shows the crystal structure of KFe_2As_2. **b** Second derivative curve as a function of $\tan\theta$. Two different periods (A and B) are obtained, which arise from two quasi-2D Fermi surfaces

4.5 Coherence of Interlayer Transport

The AMRO can be described by a semiclassical treatment of the orbital motion on the quasi-2D Fermi surface using the Chambers formula without Landau quantization as shown in Fig. 4.13a. The AMRO observed over a wide range of the magnetic field direction means that the cyclotron motion can span many layers without being scattered in real space ($\omega_c\tau > 1$); the electrons can move coherently between multiple layers.

On the other hand, the interlayer conduction can also be described as a quantum tunneling process from a quantum mechanical point of view (Fig. 4.13b) [16, 17]. We assume that the electrons are frequently scattered in each conducting layer and that coherent quantum tunneling is possible only between the adjacent layers, but not over many layers. This situation occurs when the transfer integral t_c between the layers is smaller than the Landau level broadening $\hbar/\tau$ due to scattering in the layer ($t_c < \hbar/\tau$). In this case, the Bloch state extending many layers cannot be well defined, suggesting that the 2D Fermi surface is defined independently in each layer [16]. This interlayer conduction, where each quantum tunneling is uncorrelated with the next tunneling process, is called weakly incoherent transport. In this regime, the theoretical studies clarify that the tunneling probability changes with the magnetic field angle and the interlayer conductivity yields exactly the same AMRO given by Eq. (4.21). This amazing result shows that the AMRO can also be interpreted by the single quantum tunneling process; the AMRO appears even for a double layer system. In this way, when the AMRO given by Eq. (4.21) is observed for a layered material, the coherent interlayer transport picture over many

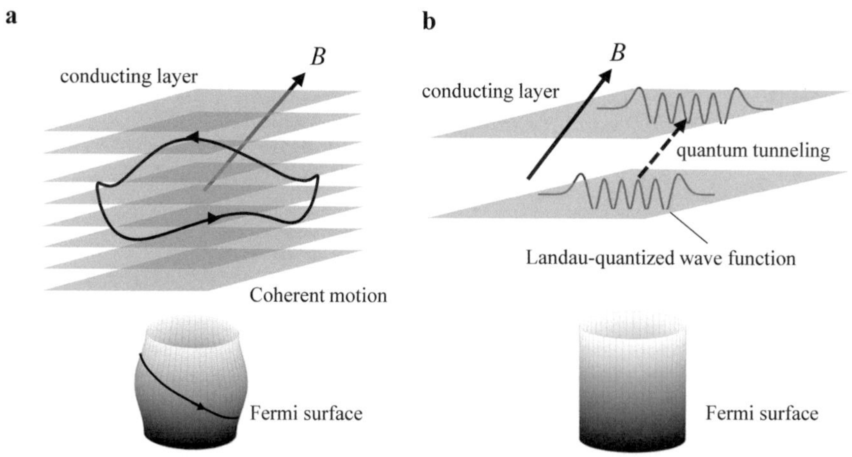

Fig. 4.13 **a** Cyclotron orbit causing AMRO in real space (top) and k-space (bottom). This is the case for coherent interlayer conduction, where large cyclotron orbits around quasi-2D Fermi surface are defined. **b** Case for weakly incoherent transport, where tunneling process between the neighboring layers in real space (top) is dominant. When electrons are frequently scattered within a layer, a 2D Fermi surface is defined independently within each layer (bottom)

layers is appropriate for $t_c > \hbar/\tau$, and the weakly incoherent transport picture for $t_c < \hbar/\tau$. Since exactly the same AMRO can be observed in both cases, we obtain the same information on the shape of the Fermi surface cross section.

What if the scattering times is much shorter, $t_c \ll \hbar/\tau$? In this case, it is reported that the interlayer resistance shows a significant change in the angular dependence. Figure 4.14a presents the angular dependence of the interlayer resistance of the quasi-2D organic conductor (BEDT-TTF)$_2$NH$_4$Hg(SCN)$_4$. The Fermi surface of this conductor consists of quasi-1D and quasi-2D parts, as shown in Fig. 4.14b. The magnetic field is tilted from the b^*-axis to the $\Gamma - V$ axis in the Brillouin zone. For the sample with long τ, the interlayer resistance has minima at $\theta = 0°$ and 180°, and the AMRO arising from the 2D Fermi surface is observed. This is the conventional AMRO explained by Eq. (4.21). For the sample with short τ ($t_c \ll \hbar/\tau$), the interlayer resistance increases and shows no AMRO. It is noticeable that the resistance has a broad minimum at $\theta = 90°$. For θ-(BEDT-TTF)$_2$I$_3$, as shown in Fig. 4.9, the AMRO amplitude is suppressed with increasing temperature, which is ascribed to the shortening of τ. At 18 K, the anisotropy of the resistance follows a sinusoidal curve with a broad maximum at $\theta = 90°$, consistent with Eq. (4.21) for short τ. The maximum at $\theta = 90°$ is interpreted by the large Lorentz force acting on the current because the current is perpendicular to the magnetic field. In contrast, the resistance for short τ in Fig. 4.14a has the opposite anisotropy, the broad minimum at $\theta = 90°$, which cannot be explained by the Chambers formula. A similar opposite anisotropy is reported in the interlayer

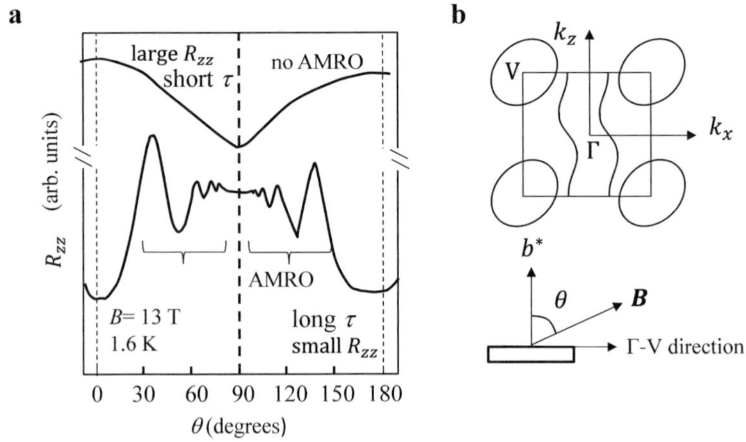

Fig. 4.14 **a** Interlayer resistance of quasi-2D organic conductor (BEDT-TTF)$_2$NH$_4$Hg(SCN)$_4$ as a function of magnetic field angle for two cases (short and long scattering times). No AMRO is observed for short scattering time. **b** Schematic Fermi surface and the definition of the magnetic field angle in the rotation plane

resistance for GaAs/AlGaAs semiconductor superlattices [17]. The microscopic mechanism of this opposite anisotropy for $t_c \ll \hbar/\tau$ is an open question.

4.6 Effect of Closed Orbit on Side of Quasi-2D Fermi Surface

We have seen that the AMRO measurements provide information of the cross section on the quasi-2D Fermi surface in Sect. 4.4. From Eqs. (4.21) and (4.22), the interlayer resistivity is given by $\rho_{zz} \propto 1/\sigma_{zz} \propto 1/t_c^2$. As the two-dimensionality increases ($t_c \to 0$), the resistance steeply increases. The finite value of t_c gives a corrugation to the cylindrical Fermi surface, which is a prerequisite for the observation of the AMRO. However, t_c cannot be obtained directly from the AMRO period. How large is the interlayer transfer integral t_c in real quasi-2D conductors? In the following we will see that the ratio t_c/E_F can be obtained from the interlayer resistance in the magnetic field nearly parallel to the conduction layer.

4.6.1 Electron Orbits in Parallel Magnetic Fields

The analytical expression of the AMRO in Eq. (4.21) does not give correct results when the magnetic field is nearly parallel to the conduction layer. In this case, the interlayer conduction must be evaluated by a different method. Let us return to the Fermi surface in Fig. 4.15a and look at the orbital motion of the electrons on the Fermi surface in a magnetic field parallel to the conduction layer. Note that

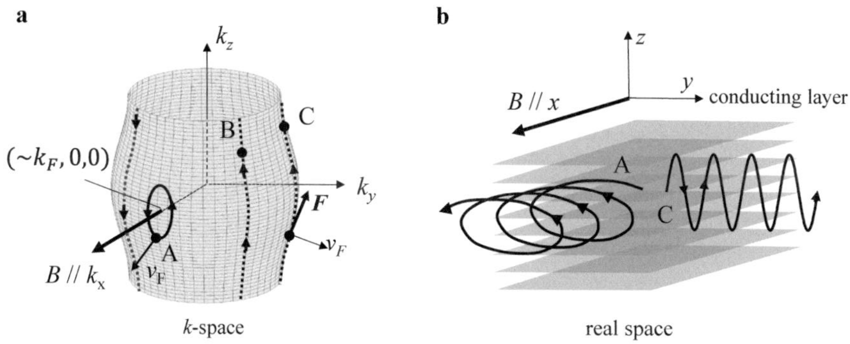

Fig. 4.15 **a** Orbital motion on the quasi-2D Fermi surface in parallel magnetic field ($B \| k_x$). The electron at point A has a closed orbit while the electrons at points B and C have open orbits. **b** Schematic of the orbits of the electrons at points A and C in real space; the electron at point A undergoes a helical motion in the magnetic field direction while the electron at point C undergoes a meandering motion in the y-axis

the orbital plane on the Fermi surface is perpendicular to the magnetic field. From $\hbar\dot{\boldsymbol{k}} = \boldsymbol{F} = -e\boldsymbol{v} \times \boldsymbol{B}$ in the magnetic field $(B_x, 0, 0)$, we have

$$\left(\dot{k}_x(t), \dot{k}_y(t), \dot{k}_z(t)\right) = \left(0, -\frac{ev_z B_x}{\hbar}, \frac{ev_y B_x}{\hbar}\right). \tag{4.30}$$

Using the relations in Eq. (4.16), we obtain

$$\begin{aligned} \ddot{k}_y &= -\frac{eB_x}{\hbar}\dot{v}_z = -\frac{eB_x}{\hbar}\frac{2t_c c^2}{\hbar}\cos(k_z c)\dot{k}_z = -\frac{(eB_x)^2}{\hbar^2}\frac{2t_c c^2}{m_c}\cos(k_z c)k_y, \\ \ddot{k}_z &= \frac{eB_x}{\hbar}\dot{v}_y = -\frac{(eB_x)^2}{\hbar^2}\frac{2t_c c}{m_c}\sin(k_z c). \end{aligned} \tag{4.31}$$

For the orbit around the k_x-axis (point A) on a quasi-2D Fermi surface ($k_z c \ll 1$) as shown in Fig. 4.15a, we can approximate $\cos(k_z c) \approx 1$ and $\sin(k_z c) \approx k_z c$. Then we have

$$\ddot{k}_y = -\Omega_c^2 k_y,\ \ddot{k}_z = -\Omega_c^2 k_z. \tag{4.32}$$

The result shows a cyclotron motion on the side of the Fermi surface, whose cyclotron frequency is given by

$$\Omega_c = \frac{eB_x c}{\hbar}\sqrt{\frac{2t_c}{m_c}}. \tag{4.33}$$

Neglecting the integration constant, the orbital motion of the electron is written as

$$k_y = K_y \cos(\Omega_c t), \quad k_z = K_z \sin(\Omega_c t). \tag{4.34}$$

As depicted in Fig. 4.15a, this is a small closed orbit with the central coordinate ($\sim k_F, 0, 0$), whose orbital plane is perpendicular to the magnetic field. The amplitudes of the cyclotron motion have the relation,

$$K_z = \frac{1}{k_F c}\sqrt{\frac{E_F}{t_c}} K_y. \tag{4.35}$$

Assuming $E_F \gg t_c$ and $k_F c \approx 1$, we obtain $K_z > K_y$. The electron has the velocity $v_x \approx v_F = \hbar k_F / m_c$ in the x-axis and undergoes a resultant helical motion toward the x-axis as depicted in Fig. 4.15b.

The electron at point C ($k_x = 0$) in Fig. 4.15a has the velocity, $v_x = 0$ and $v_y \approx v_F$. This orbit extends upward on the side of the Fermi surface. Using $k_z = k_{z0} + e v_y B_x t / \hbar$ from Eq. (4.30), the velocity in the z-axis is

$$v_z(t) = \left(\frac{2t_c c}{\hbar}\right) \sin(k_z c) = \left(\frac{2t_c c}{\hbar}\right) \sin\left[k_{z0} c + \frac{v_y c e B_x t}{\hbar}\right]. \tag{4.36}$$

The oscillation of $v_z(t)$ with time represents a meandering motion toward the y-axis in real space as depicted in Fig. 4.15b. The electron at point B undergoes a meandering motion in a direction between the x- and y-axes in real space. In this way, the closed orbits on the Fermi surface lead to helical motions in real space, and the open orbits lead to meandering motions as depicted in Fig. 4.15. The orbital motion of the electrons in real space depends on where the electrons are on the Fermi surface.

In the above discussion, we have assumed a layered conductor with the energy band given by Eq. (4.1), using the free electron approximation in the xy plane. Regardless of the approximation method of the energy band, as long as the Fermi surface has a corrugated cylindrical shape as shown in Fig. 4.15, closed orbits appear on the side of the Fermi surface in the magnetic field parallel to the layers.

4.6.2 Coherence Peak

Let us see how the orbital motion on the Fermi surface contributes to the interlayer conduction. In a magnetic field ($B_x, 0, 0$), using Eq. (4.34), the z-axis velocity of the electron at point A in Fig. 4.15a is written as

$$v_z(t) = \left(\frac{2t_c c}{\hbar}\right) \sin(k_z c) = \left(\frac{2t_c c}{\hbar}\right) \sin[K_z c \sin(\Omega_c t)]. \tag{4.37}$$

The velocity averaged over the orbit vanishes,

$$\langle v_z(t) \rangle = \frac{\Omega_c}{2\pi} \int_0^{2\pi/\Omega_c} v_z(t)\mathrm{d}t = 0. \tag{4.38}$$

Therefore, the electrons near the k_x-axis do not contribute to the interlayer conduction. It suggests that the interlayer resistance has a relatively large value when the magnetic field is exactly parallel to the layer. Now we tilt the magnetic field from the x-axis to the z-axis. At angles $\theta < \theta_c$, any cross section perpendicular to the magnetic field will no longer have small closed sections on the side of the Fermi surface as depicted in Fig. 4.16a; no small closed orbits are formed. This means that more electrons contribute to the interlayer transport; the interlayer resistance decreases as the magnetic field is tilted. Indeed, numerical calculations by the Chambers formula show that the interlayer resistance as a function of θ has a peak at $\theta = \pm 90°$[18]. From the energy dispersion in Eq. (4.1), we have

$$\left| \frac{\partial k_z}{\partial k_x} \right| = \frac{\hbar^2 k_x}{2m_c t_c c|\sin(k_z c)|}. \tag{4.39}$$

This function has an inflection point at $k_Z = \pi/2c$ as shown in Fig. 4.16b. Substituting $k_Z = \pi/2c$ in Eq. (4.39), we can define the critical angle θ_c where the small closed orbit on the side of the Fermi surface disappears,

$$\frac{\pi}{2} - \theta_c \equiv \Delta\theta_{\text{peak}} = \left| \frac{\partial k_x}{\partial k_z} \right|_{k_Z=\pi/2c} = \frac{2m_c t_c c}{\hbar^2 k_F} = \frac{t_c}{E_F} k_F c. \tag{4.40}$$

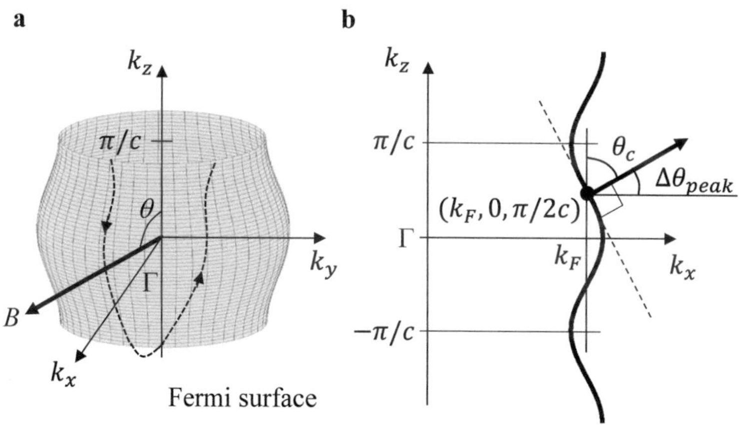

Fig. 4.16 **a** Orbital motion on the quasi-2D Fermi surface in tilted magnetic field. Only large closed orbits indicated by dotted curve around the Fermi surface (no small closed orbits on the side of the Fermi surface) are formed in a tilted magnetic field for $\theta < \theta_c$. **b** Critical angle θ_c where the closed orbit disappears. Small closed orbits on the side of the Fermi surface appear in the limited angle range, $\pi/2 - \Delta\theta_{\text{peak}} < \theta < \pi/2 + \Delta\theta_{\text{peak}}$. Dotted line represents the tangent at $(k_F, 0, \pi/2c)$

Since the angle range where the peak is observed is $\pi/2 - \Delta\theta_{\text{peak}} < \theta < \pi/2 + \Delta\theta_{\text{peak}}$ ($\theta_c < \theta < \pi - \theta_c$), $2\Delta\theta_{\text{peak}}$ can be defined as the peak width.

The peak width is determined only by how the Fermi surface is corrugated along the k_z and thus is independent of magnetic field strength and temperature. The larger the Fermi surface corrugation, the wider the peak width. The interlayer resistances as a function of the magnetic field angle for quasi-2D organic conductors are shown in Fig. 4.17. Figure 4.17a shows the resistance peak at $\theta = 90°$ for (BEDT-TTF)$_2$Br(DIA), whose Fermi surface is shown in the inset of Fig. 3.2a. The peak becomes evident as the magnetic field increases. In the second derivative curves $-\mathrm{d}^2R/\mathrm{d}\theta^2$ in Fig. 4.17b, the peak width $2\Delta\theta_{\text{peak}} \approx 2°$ is well defined, which is independent of the magnetic field strength. Figure 4.17c shows the interlayer resistance for θ-(BEDT-TTF)$_2$I$_3$, where the peak width $2\Delta\theta_{\text{peak}} \approx 1°$ is independent of temperature. From the AMRO period $\Delta(\tan\theta) = 0.42$ in the $ac^*(xz)$ rotation shown in Fig. 4.10, we obtain $k_F c = \pi/\Delta(\tan\theta) \approx 7.5$, where c is the layer distance. It gives $t_c/E_F \approx 0.001$ from Eq. (4.40); the corrugation of the quasi-2D Fermi surface in the k_z is very small for θ-(BEDT-TTF)$_2$I$_3$.

Let's think a little more about the physical meaning of this peak in the interlayer resistance. The closed orbit causing the peak is a helical motion crossing the many layers in real space, as shown in Fig. 4.15b. The observation of the peak shows that the electrons can move over many layers without being scattered, i.e., the coherent motion in the interlayer direction. Therefore, the resistance peak in the parallel magnetic field due to this mechanism is called the "coherence peak." We should note that the effect of the closed orbit on the side of the quasi-2D Fermi surface is not taken into account in Eq. (4.21). Equation (4.21) is applicable in the angle range, $\theta_c > \theta$ or $\theta > \pi - \theta_c$, where only the large closed orbits around the quasi-2D Fermi surface are formed as shown in Fig. 4.7a.

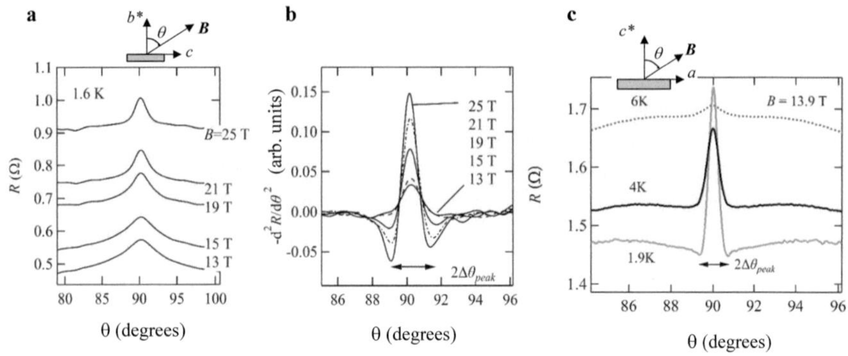

Fig. 4.17 **a** Angular dependence of interlayer resistance and **b** its second derivative curves at various magnetic fields for a quasi-2D organic conductor (BEDT-TTF)$_2$Br(DIA). **c** Angular dependence of interlayer resistance at different temperatures for θ-(BEDT-TTF)$_2$I$_3$. The peak width $2\Delta\theta_{\text{peak}}$ is independent of magnetic field strength and temperature

As discussed in Sect. 4.5, the coherence peak should not be observed in the weakly incoherent regime. In Fig. 4.14a, the interlayer resistance of (BEDT-TTF)$_2$NH$_4$Hg(SCN)$_4$ with long τ shows the AMRO but no coherence peak at $\theta = 90°$. The results show that the interlayer transport of this sample is in the weakly incoherent regime. Note that the coherence peak is observed in quasi-2D systems only when the interlayer transport is coherent.

4.6.3 Observation Condition of Coherence Peak

The uncertainty of the energy due to scattering is $\hbar/\tau$, and the width of the Fermi edge due to finite temperature is given by k_BT. Therefore, to clearly observe the coherence peak, the transfer integral t_c, which corresponds to the transition amplitude between the layers, should be larger than these quantities;

$$t_c > \hbar/\tau, \quad t_c > k_BT. \tag{4.41}$$

In Fig. 4.15a, the electron at point A undergoes cyclotron motion on the side of the Fermi surface with the frequency Ω_c in Eq. (4.33). Taking $\Omega_c\tau = 1$, we can define the threshold magnetic field as

$$B_c = \frac{\hbar}{ec\tau}\sqrt{\frac{m_c}{2t_c}}, \tag{4.42}$$

above which the coherence peak is observed. From the relation $\Delta\theta_{\text{peak}} \approx t_c k_F c/E_F$ in Eq. (4.40), the Fermi wave number k_F can be estimated from the peak width. Figure 4.18a presents the closed orbits on the side of the Fermi surface for θ-(BEDT-TTF)$_2$I$_3$. One type of closed orbit (solid circle) is formed on the convex surface around point X in the magnetic field along the k_x-axis ($B \| k_x$), while two types of closed orbits (solid and dotted circles) are formed on the convex surface around points Y and on the concave surface around R in $B \| k_y$, respectively. The larger closed orbit (solid curve) has a larger contribution to the conduction in $B \| k_y$. The experimental results in Fig. 4.18b show a wider peak width for the $\varphi = 0°$ rotation plane than for $\varphi = 90°$. Since $\Delta\theta_{\text{peak}} \propto k_F$ from Eq. (4.40), the result is consistent with the fact $k_a > k_b$, as seen in Fig. 4.11a. The detailed analyses of the peak width are reported in the literature [19].

4.6.4 Weak Localization Effect

The organic conductor κ_H-(DMEDO-TSeF)$_2$[Au(CN)$_4$](THF), where DMEDO-TSeF is dimethyl(ethylenedioxy)tetraselenafulvalene and THF is tetrahydrofuran, has a very thick insulating layer between the conducting DMEDO-TSeF layers, and thus has an extremely 2D electronic state. The Fermi surface consists of quasi-1D and quasi-2D parts, identical to those in Fig. 4.18a. As shown in Fig. 4.19a,

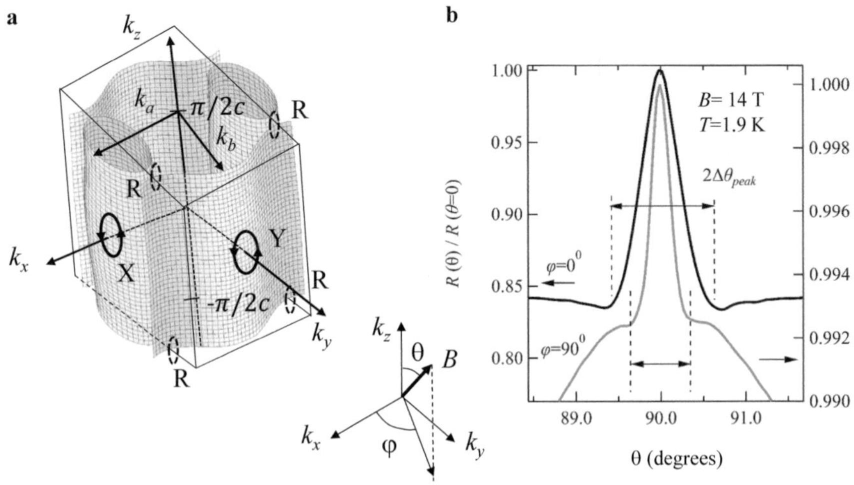

Fig. 4.18 **a** Fermi surface and closed orbits on the side for θ-(BEDT-TTF)$_2$I$_3$, **b** Coherence peaks for magnetic field rotations in $\varphi = 0°$ (k_xk_z) plane and $\varphi = 90°$ (k_yk_z) plane

the interlayer resistance has a broad minimum around $\theta = 90°$ [20], which cannot be explained in the framework of the Chambers formula as discussed in Sect. 4.5. In addition, a small peak is observed at $\theta = 90°$. Figure 4.19b shows the close-up of the peaks as a function of the perpendicular component of the magnetic field $B\cos(\theta)$. The peak width in this plot is independent of the magnetic field, which is inconsistent with the coherence peak picture.

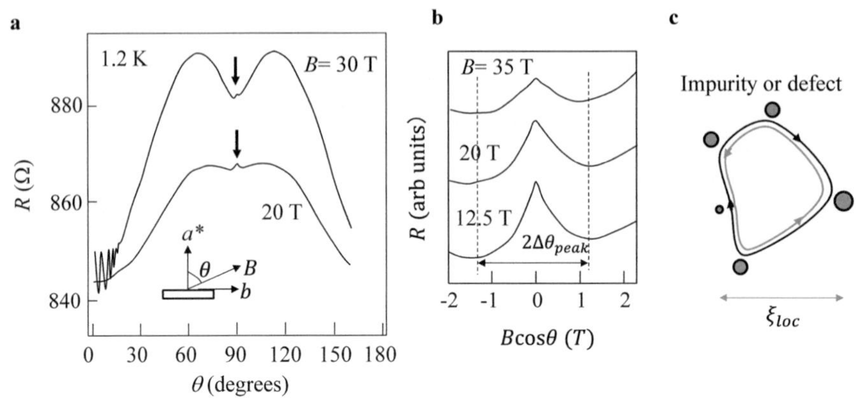

Fig. 4.19 **a** Interlayer resistance as a function of magnetic field angle θ for extremely 2D organic conductor, κ_H-(DMEDO-TSeF)$_2$[Au(CN)$_4$](THF) (reproduced with permission from [20]). Arrows indicate small peaks at $\theta = 90°$. **b** Close-up of the peaks near $\theta = 90°$. Peak width is scaled by the perpendicular magnetic field $B\cos(\theta)$. **c** Schematic of quantum interference loop causing weak localization in 2D system. $\xi_{\rm loc}$ is the localization length

In general, electronic states tend to localize at low temperatures in the presence of inhomogeneities (impurities or defects). This phenomenon is called weak localization and is more pronounced in lower dimensional systems. The weak localization is ascribed to quantum interference between two closed paths in which an electron can propagate clockwise and counterclockwise around a loop as schematically shown in Fig. 4.19c. The quantum interference forms a standing wave in the loop, which increases the net resistance. A magnetic field perpendicular to the orbital plane of the loop causes a phase difference between the clockwise and counterclockwise propagation and resultantly breaks the quantum interference. Therefore, the resistance decreases with increasing perpendicular magnetic field in the weak localization regime. In κ_H-(DMEDO-TSeF)$_2$[Au(CN)$_4$](THF), the electronic state is probably in the weak localization regime due to disorders in the [Au(CN)$_4$](THF) anion layers. The broad minimum around $\theta = 90°$ in Fig. 4.19a is likely due to the weak localization in the conduction layers. Since the weak localization is suppressed by the perpendicular magnetic field to the layers, the weak localization effect can cause a resistance peak at $\theta = 90°$, whose width is scaled by the perpendicular magnetic field $B\cos(\theta)$ as shown in Fig. 4.19b. This weak localization is another origin of the peak in the parallel magnetic fields, different from the coherence peak discussed above. On the other hand, quantum oscillation in the resistance is observed at large perpendicular magnetic fields for κ_H-(DMEDO-TSeF)$_2$[Au(CN)$_4$](THF). The fact shows that the weak localization length ξ_{loc} is much longer than the cyclotron orbit radius. The detailed theoretical discussion is made in the literature [21]. In this way, there are two mechanisms of the resistance peak in the parallel magnetic field, the coherence peak or weak localization peak. The origin of the peak can be discussed from the magnetic field dependence of the peak.

4.7 Effect of Open Orbit on Side of Quasi-2D Fermi Surface

The electrons at points B and C in Fig. 4.15a have open orbits on the Fermi surface in magnetic fields. In this section, we see the effects of these open orbits on the electrical conduction. In a parallel magnetic field $(B_x, 0, 0)$, the velocity v_z of the electrons on the open orbit is given by Eq. (4.36). Substituting it into the Chambers formula Eq. (4.15), we obtain the following simple solution [22],

$$\sigma_{zz}(B_x) = \sigma_{zz}(0)\frac{1}{\sqrt{1 + (k_F c \omega_c \tau)^2}}, \tag{4.43}$$

where the cyclotron frequency is defined as $\omega_c = eB_x/m_c$ and $\sigma_{zz}(0)$ is given in Eq. (4.22). This is derived in Appendix 4.4. The resistivity is given by

$$\rho_{zz}(B_x) = 1/\sigma_{zz}(B_x) = \rho_{zz}(0)\sqrt{1 + (k_F c \omega_c \tau)^2}. \tag{4.44}$$

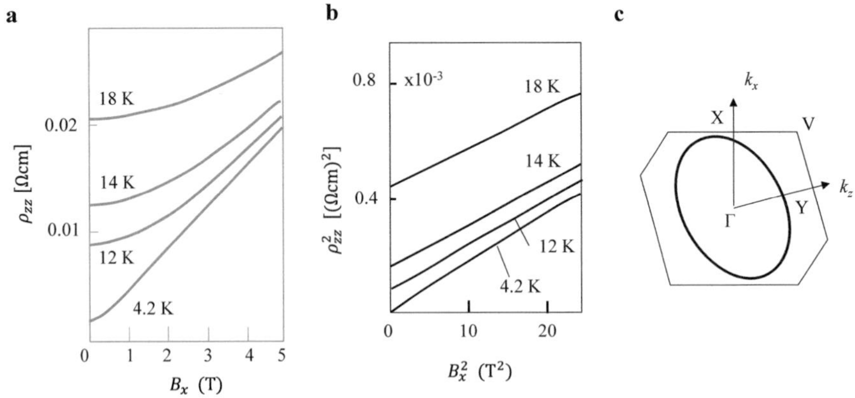

Fig. 4.20 **a** Magnetic field dependence of the interlayer resistance $\rho_{zz}(B_x)$ for the quasi-2D organic conductor (BEDT-TTF)$_3$Cl(DFBIB) (reproduced with permission from [22]), **b** $\rho_{zz}^2(B_x)$ versus B_x^2 plot, **c** Calculated 2D Fermi surface of (BEDT-TTF)$_3$Cl(DFBIB) [23]

This equation can be applied to the entire range of magnetic fields, from weak to strong. It is approximated as $\rho_{zz}(B_x) - \rho_{zz}(0) \propto B_x^2$ at low magnetic fields and $\rho_{zz}(B_x) \propto B_x$ at high magnetic fields. Squaring both sides of Eq. (4.44), we have

$$\rho_{zz}^2(B_x) = \rho_{zz}^2(0) + \alpha^2 B_x^2, \quad \alpha = \frac{\pi \hbar^4 k_F}{2em_c^2 t_c^2}. \tag{4.45}$$

Therefore, in the plot of $\rho_{zz}^2(B_x)$ versus B_x^2, the intercept gives $\rho_{zz}^2(0)$ and the slope gives α^2, which is independent of temperature and scattering time. Figure 4.20a and b respectively presents the magnetic field dependence of the interlayer resistance $\rho_{zz}(B_x)$ and $\rho_{zz}^2(B_x)$ at different temperatures for a quasi-2D organic conductor (BEDT-TTF)$_3$Cl(DFBIB), where DFBIB stands for 1,4-difluoro-2,5-bis(iodoethynyl)benzene. This conductor has a triclinic structure, whose calculated Fermi surface is depicted in Fig. 4.20c [23]. From the temperature-independent slope we obtain the material-dependent parameter $k_F/m_c^2 t_c^2$.

4.8 Case for Quasi-1D Fermi Surface

So far, we have focused our discussion only on quasi-2D Fermi surfaces. In this section, we will consider the orbital motion of the electrons on quasi-1D Fermi surfaces. No quantum oscillations arise from quasi-1D Fermi surfaces due to the absence of extremal cross sections. However, as in the case of AMRO for quasi-2D Fermi surfaces, a characteristic angle-dependent magnetoresistance can also be observed for quasi-1D Fermi surfaces. From the results, we can obtain the information on the quasi-1D Fermi surfaces.

4.8.1 Orbital Motion

We consider a simple energy band forming a quasi-1D Fermi surface with the orthorhombic lattice constants a, b, and c,

$$E(\boldsymbol{k}) = \hbar v_F(|k_x| - k_F) - 2t_b \cos(k_y b) - 2t_c \cos(k_z c), \quad (v_F > 0), \tag{4.46}$$

where $\hbar v_F k_F > 2t_b \gg 2t_c$. The energy dispersion in the k_x-axis is approximated by a linear function, which is reasonable in a narrow energy range. The energy band and Fermi surface are depicted in Fig. 4.21a and b, respectively. For $t_b = t_c = 0$, we note that $E_F = 0$ at $|k_x| = k_F$. This band gives the velocity in real space,

$$\begin{aligned} v_x &= \frac{\partial E}{\hbar \partial k_x} = v_F \text{ for } k_x > 0, \\ &= -v_F \text{ for } k_x < 0. \end{aligned} \tag{4.47}$$

Here we define the magnetic field direction (θ, φ) as shown in the inset of Fig. 4.21b. In the magnetic field, the orbital motion of the electrons is indicated by the dotted curves. From $\hbar \dot{\boldsymbol{k}} = \boldsymbol{F} = -e\boldsymbol{v} \times \boldsymbol{B}$, we have

$$\hbar \dot{k}_y = e(v_x B_z - v_z B_x) \approx e v_x B_z, \tag{4.48}$$

$$\hbar \dot{k}_z = e(v_y B_x - v_x B_y), \tag{4.49}$$

where $\dot{k}_y(\dot{k}_z)$ is the velocity along the $k_y(k_z)$-axis in k-space. The time to pass the Brillouin zone in the k_y-axis is $T_y = (2\pi/b)/|\dot{k}_y|$, which is defined as the period of the orbital motion of the electrons. The angular frequency of the motion ω_z has

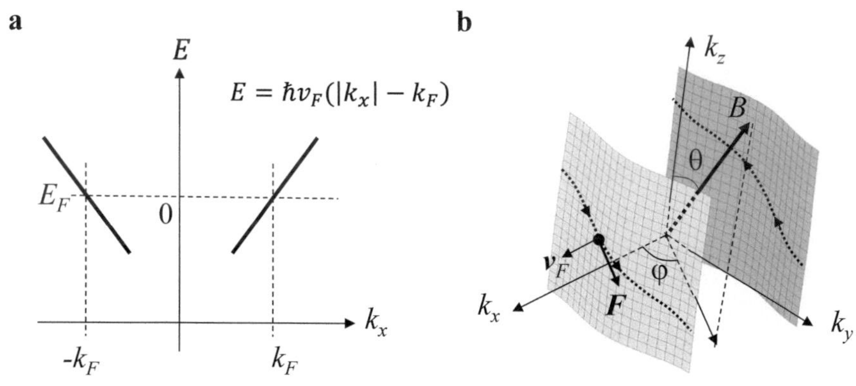

Fig. 4.21 **a** 1D energy bands given by Eq. (4.46), **b** Quasi-1D Fermi surface formed by the energy bands and definition of the magnetic field angles. Dotted curves show the electron orbits in magnetic field

the relation $T_y = 2\pi/\omega_z$. Similarly, the period of the orbital motion in the k_z-axis is defined by $T_z = (2\pi/c)/|\dot{k}_z| = 2\pi/\omega_y$. Taking $\hbar\dot{k}_z \approx -ev_xB_y$ in Eq. (4.49) and $|v_x| = v_F$ from Eq. (4.47), we have

$$\omega_z = \frac{v_F beB_z}{\hbar}, \quad \omega_y = \frac{v_F ceB_y}{\hbar}. \tag{4.50}$$

These values are defined as the angular frequencies of the periodic motions on the 1D Fermi surface. By integrating Eq. (4.48) over time, we have $k_yb = \omega_zt + k_{y0}b$ for $k_x > 0$. The y-axis velocity is

$$v_y(t) = \frac{\partial E}{\hbar\partial k_y} = \frac{2t_bb}{\hbar}\sin(k_yb) = \frac{2t_bb}{\hbar}\sin(\omega_zt + k_{y0}b). \tag{4.51}$$

Substituting Eq. (4.51) into Eq. (4.49), we have

$$\hbar\dot{k}_z = \frac{2t_bb}{\hbar}eB_x\sin(\omega_zt + k_{y0}b) - ev_FB_y. \tag{4.52}$$

Integrating Eq. (4.52) over time, we obtain

$$k_zc = -\gamma\cos(\omega_zt + k_{y0}b) - \omega_yt + k_{z0}c + \gamma\cos(k_{y0}b), \tag{4.53}$$

where

$$\gamma = \frac{2t_bc}{\hbar v_F}\frac{B_x}{B_z}. \tag{4.54}$$

Note $k_z = k_{z0}$ at $t = 0$. Therefore, the z-axis velocity is written as

$$\begin{aligned} v_z(\boldsymbol{k}(t)) &= \frac{2t_cc}{\hbar}\sin(k_zc) \\ &= \frac{2t_cc}{\hbar}\sin[-\gamma\cos(\omega_zt + k_{y0}b) - \omega_yt + k_{z0}c + \gamma\cos(k_{y0}b)]. \end{aligned} \tag{4.55}$$

For $k_x < 0$, we have $\omega_y \to -\omega_y, \omega_z \to -\omega_z, \gamma \to -\gamma$ in Eqs. (4.51) and (4.55) because $v_x = -v_F$. In this way, we obtain all the velocity components, v_x, v_y, v_z and thus can calculate the conductivity tensor according to the Chambers formula.

4.8.2 AMRO for Quasi-1D Fermi Surface

First, we obtain the AMRO formula at general magnetic field angles (θ, φ) as defined in Fig. 4.21b. By substituting Eq. (4.55) into Eq. (4.15), the z-axis conductivity is calculated as [17]

$$\begin{aligned}\sigma_{zz}^{1D}(B) &= \sigma_{zz}^{1D}(0) \sum_{\pm,n=-\infty}^{\infty} \frac{J_n^2(\gamma)/2}{1+\left(n\omega_z \pm \omega_y\right)^2\tau^2} \\ &= \sigma_{zz}^{1D}(0) \sum_{\pm,n=-\infty}^{\infty} \frac{J_n^2(\gamma)/2}{1+\left(\frac{v_F eB}{\hbar}\right)^2 [nb\cos(\theta) \pm c\sin(\theta)\sin(\varphi)]^2\tau^2},\end{aligned} \tag{4.56}$$

$$\sigma_{zz}^{1D}(0) = 2D(E_F)\left(\frac{et_c c}{\hbar}\right)^2 \tau,\ D(E_F) = \frac{2}{\pi bc\hbar v_F}, \tag{4.57}$$

where $D(E_F)$ is the density of states including the spin degree of freedom per unit volume. The sign $\pm$ in Σ means to take the sum of both signs. The above formula is derived in Appendix 4.5. Because of the Bessel function $J_n^2(\gamma)$ and the term $[nb\cos(\theta) \pm c\sin(\theta)\sin(\varphi)]^2$ in the denominator, Eq. (4.56) produces a complicated angular dependence as seen below.

As discussed in Sect. 4.5, even in the case of weakly incoherent transport for the quasi-2D Fermi surface, the same AMRO formula, Eq. (4.21) is obtained. For the quasi-1D Fermi surface, the AMRO is also calculated based on a quantum tunneling picture in the whole angle region [24]. Surprisingly, as long as B_x is not in the strong limit, the same AMRO formula as Eq. (4.56) is obtained even in the incoherent limit ($t_c \ll \hbar/\tau$). Therefore, Eq. (4.56) is a general formula available in a wide range of the scattering time. At a strong magnetic field in the x-axis, new Aharonov-Bohm-type oscillations are discussed [17, 24, 25]

By substituting Eq. (4.47) into Eq. (4.15), we similarly obtain

$$\sigma_{xx}^{1D}(B) = D(E_F)(ev_F)^2\tau = D(E_F)E_F\frac{e^2\tau}{m_c} = \frac{N_e e^2\tau}{m_c} = \sigma_{xx}^{1D}(0), \tag{4.58}$$

where $E_F = \hbar v_F k_F = m_c v_F^2$ and $N_e = D(E_F)E_F$ is the number of electrons per unit volume. By substituting Eq. (4.51) into Eq. (4.15), the other components are obtained as

$$\begin{aligned}\sigma_{yy}^{1D}(B) &= 2D(E_F)\left(\frac{et_b b}{\hbar}\right)^2 \frac{\tau}{1+(\omega_z\tau)^2} = \sigma_{yy}^{1D}(0)\frac{1}{1+(\omega_z\tau)^2}, \\ \sigma_{yy}^{1D}(0) &= 2D(E_F)\left(\frac{et_b b}{\hbar}\right)^2 \tau.\end{aligned} \tag{4.59}$$

It is easily verified that $\sigma_{xy}^{1D} = \sigma_{yz}^{1D} = \sigma_{zx}^{1D} = 0$. In this way, we obtain the conductivity tensor,

$$\boldsymbol{\sigma}^{1D} = \begin{pmatrix} \sigma_{xx}^{1D} & 0 & 0 \\ 0 & \sigma_{yy}^{1D} & 0 \\ 0 & 0 & \sigma_{zz}^{1D} \end{pmatrix} = \begin{pmatrix} \sigma_{xx}^{1D}(0) & 0 & 0 \\ 0 & \sigma_{yy}^{1D}(0)\frac{1}{1+(\omega_z\tau)^2} & 0 \\ 0 & 0 & \sigma_{zz}^{1D}(0)\sum\limits_{\pm,n=-\infty}^{\infty}\frac{J_n^2(\gamma)/2}{1+(n\omega_z\pm\omega_y)^2\tau^2} \end{pmatrix}. \tag{4.60}$$

The term σ_{xx}^{1D} is independent of the magnetic field, and all off-diagonal components are zero, reflecting the fact that all electrons can move almost exclusively in the k_x direction. Since the tensor $\boldsymbol{\sigma}^{1D}$ has only diagonal components, the resistivity tensor $\boldsymbol{\rho}^{1D}$ is simply given by

$$\boldsymbol{\rho}^{1D} = \begin{pmatrix} \rho_{xx}^{1D} & 0 & 0 \\ 0 & \rho_{yy}^{1D} & 0 \\ 0 & 0 & \rho_{zz}^{1D} \end{pmatrix} = \begin{pmatrix} 1/\sigma_{xx}^{1D} & 0 & 0 \\ 0 & 1/\sigma_{yy}^{1D} & 0 \\ 0 & 0 & 1/\sigma_{zz}^{1D} \end{pmatrix}. \tag{4.61}$$

Figure 4.22 shows the calculated results of $\rho_{zz}^{1D}(\theta)/\rho_{zz}^{1D}(0) = \sigma_{zz}^{1D}(0)/\sigma_{zz}^{1D}(\theta)$ with the parameters $\hbar v_F/\hbar v_F = 0.24$, $(v_F eBb/\hbar)\tau = 15$ and $c/b = 1$, where $\rho_{zz}^{1D}(0) = 1/\sigma_{zz}^{1D}(0)$. In the angle range $5° < \varphi < 90°$, dips appear as indicated by the arrows in Fig. 4.22. At lower φ, sharp peaks are evident as indicated by the arrows in Fig. 4.22b. The origins of these behaviors are discussed below.

Next, we see what happens at the magnetic field in the xz plane $(B_x, 0, B_z)$. The averaged velocity in the z-axis $v_z(t)$ is given by

$$\begin{aligned} v_z(t) &= \frac{1}{T_y}\frac{2t_c c}{\hbar}\int_0^{T_y}\sin\left[-\gamma\cos\left(\omega_z t + k_{y0}b\right) + k_{z0}c + \gamma\cos\left(k_{y0}b\right)\right]\mathrm{d}t \\ &= \frac{1}{2\pi}\frac{2t_c c}{\hbar}\sin\left[k_{z0}c + \gamma\cos\left(k_{y0}b\right)\right]\int_0^{2\pi}\cos\left\{-\gamma\cos\left(\alpha + k_{y0}b\right)\right\}\mathrm{d}\alpha \\ &= \frac{2t_c c}{\hbar}\sin\left[k_{z0}c + \gamma\cos\left(k_{y0}b\right)\right]J_0(\gamma), \end{aligned} \tag{4.62}$$

where $\omega_z t = \alpha$ and $T_y = 2\pi/\omega_z$. Since $v_z(t)$ is expressed by the Bessel function $J_0(\gamma)$, $v_z(t)$ vanishes when

$$\gamma = \frac{2t_b c}{\hbar v_F}\frac{B_x}{B_z} = \frac{2t_b c}{\hbar v_F}\tan\theta_n = \pm\pi\left(n - \frac{1}{4}\right),\ n = 1, 2, 3, \ldots. \tag{4.63}$$

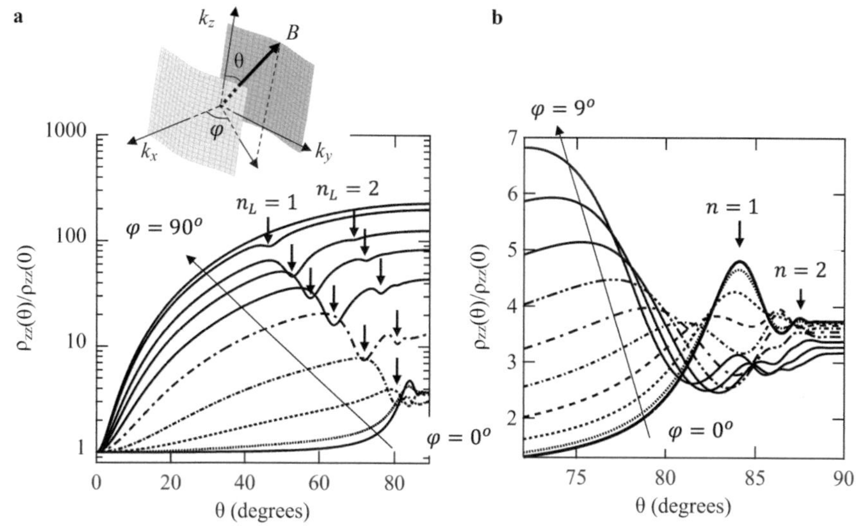

Fig. 4.22 Calculated results of $\rho_{zz}^{1D}(\theta)/\rho_{zz}^{1D}(0)$ by Eq. (4.56). Parameters used are $\hbar v_F/\hbar v_F = 0.24$, $(v_F eBb/\hbar)\tau = 15$ and $c/b = 1$. **a** Results for $\varphi = 0°, 2°, 5°, 10°, 20°, 30°, 40°, 50°, 70°, 90°$. Arrows indicate dips with $n_L = 1$ or 2 in Eq. (4.73). **b** Calculated results for $\varphi = 0° \sim 9°$ with 1° step. Arrows indicate peaks with $n = 1$ or 2 in Eq. (4.63) for $\varphi = 0°$

Therefore, the resistance in the z-axis has periodic peaks as a function of $\tan\theta$, which is very similar to the case for the quasi-2D Fermi surface, Eq. (4.9).

In the magnetic field $(B_x, 0, B_z)$, we have $\omega_y = 0$, and thus Eq. (4.56) is written as

$$\sigma_{zz}^{1D}(B) = \sigma_{zz}^{1D}(0) \sum_{n=-\infty}^{\infty} \frac{J_n^2\left(\frac{2t_b c}{\hbar v_F}\tan\theta\right)}{1 + (n\omega_z\tau)^2}. \tag{4.64}$$

Note that this functional form is almost identical to the AMRO for the quasi-2D Fermi surface, Eq. (4.21). Figure 4.23a shows the AMRO observed in the quasi-1D organic conductor $(TMTSF)_2ClO_4$ [26]. This conductor has four sheets (two pairs) of quasi-1D Fermi surface because the superlattice potential is formed due to the ClO_4 anion order at 24 K as illustrated in Fig. 1.10f: the b-axis is doubled ($b' = 2b$). Nevertheless, each pair of the bands can be approximated by Eq. (4.46) and thus Eq. (4.64) is available to the case of the four sheets. The calculated results $\rho_{zz}^{1D}(\theta)/\rho_{zz}^{1D}(0) = \sigma_{zz}^{1D}(0)/\sigma_{zz}^{1D}(\theta)$ by Eq. (4.64) are presented in Fig. 4.23b. For simplicity, the same parameters $\hbar v_F/\hbar v_F = 0.24$ and $c/b = 1$ are used as in Fig. 4.22. As $(v_F eBb/\hbar)\tau$ increases, the peaks indicated by the arrows become evident. For $(v_F eBb/\hbar)\tau = 15$, the overall features are consistent with the experimental result for $B = 8$ T in Fig. 4.23a. However, the peak at $\theta = 90°$ is

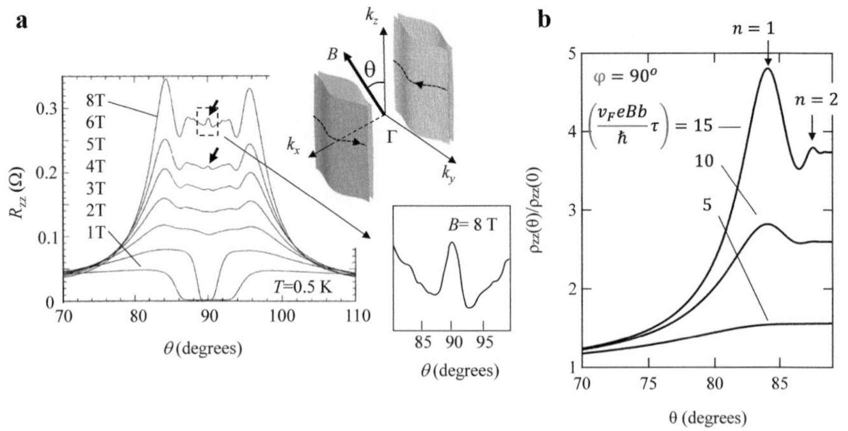

Fig. 4.23 **a** AMRO observed in quasi-1D organic conductor $(TMTSF)_2ClO_4$ (reproduced with permission from [26]). The Fermi surface consists of four sheets (two pairs) as depicted in the upper inset. The magnetic field is rotated in the $k_z k_x$ plane. The zero resistance near $\theta = 90°$ in a low magnetic field range arises from the superconductivity. Coherence peaks (indicated by arrows) are observed at $\theta = 90°$ in a high magnetic field range. The lower inset is the close-up of the coherence peak for $B = 8$ T. **b** Calculated results of $\rho_{zz}^{1D}(\theta)/\rho_{zz}^{1D}(0) = \sigma_{zz}^{1D}(0)/\sigma_{zz}^{1D}(\theta)$ by Eq. (4.64) with the parameters, $\hbar v_F/\hbar v_F = 0.24$ and $c/b = 1$. Arrows indicate the AMRO peaks with $n = 1$ and 2

not reproduced by Eq. (4.64) since the effect of the small closed orbits on the side of the Fermi surface is not included in the calculation as discussed below.

4.8.3 Coherence Peak

As in the case of the quasi-2D Fermi surface, closed orbits on the convex surface around $(\sim k_F, 0, 0)$ and on the concave surface around $(\sim k_F, \pm\pi/b, \pm\pi/c)$ appear on the side of the Fermi surface in a magnetic field parallel to the k_x-axis for the quasi-1D Fermi surface, as shown in Fig. 4.24a. In the magnetic field $(B_x, 0, 0)$, we have $\hbar\dot{k}_y = -ev_z B_x$ and $\hbar\dot{k}_z = ev_y B_x$ from $\hbar\dot{\boldsymbol{k}} = \boldsymbol{F} = -e\boldsymbol{v} \times \boldsymbol{B}$. Using $v_i = (1/\hbar)\partial E/\partial k_i$, the velocities are given by

$$v_y = \frac{2t_b b}{\hbar}\sin(k_y b),\ v_z = \frac{2t_c c}{\hbar}\sin(k_z c), \tag{4.65}$$

and thus

$$\begin{aligned}\ddot{k}_y &= -\frac{e\dot{v}_z B_x}{\hbar} = -\frac{eB_x}{\hbar}\frac{2t_c c^2}{\hbar}\cos(k_z c)\dot{k}_z = -\frac{eB_x}{\hbar}\frac{2t_c c^2}{\hbar}\cos(k_z c)\frac{ev_y B_x}{\hbar} \\ &= -\left(\frac{eB_x}{\hbar}\right)^2\frac{2t_c c^2}{\hbar}\cos(k_z c)\frac{2t_b b}{\hbar}\sin(k_y b).\end{aligned} \tag{4.66}$$

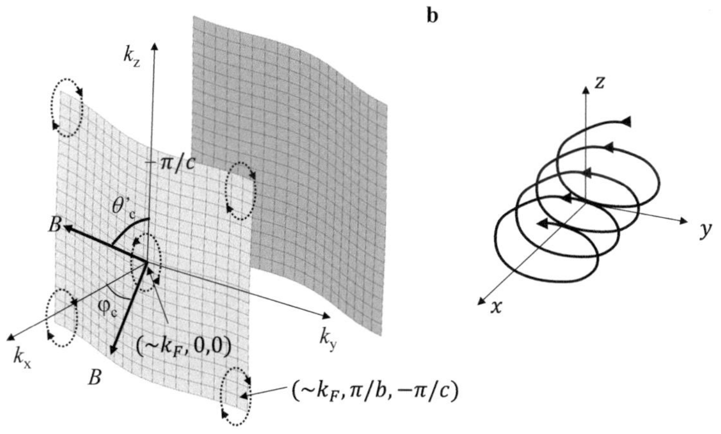

Fig. 4.24 **a** Quasi-1D Fermi surfaces and closed orbits for $B \| k_x$. Dotted curves show cyclotron orbits on the convex surface around $(\sim k_F, 0, 0)$ and on the concave surface around $(\sim k_F, \pm\pi/b, \pm\pi/c)$ in magnetic field parallel to the k_x-axis. The closed orbits disappear when the magnetic field is tilted toward the k_z-axis or the k_y-axis, whose critical angles are θ'_c and φ_c, respectively. **b** Orbital (helical) motion in real space corresponding to the cyclotron orbit around $(\sim k_F, 0, 0)$

Similarly, we have

$$\ddot{k}_z = -\left(\frac{eB_x}{\hbar}\right)^2 \frac{2t_b b^2}{\hbar} \cos(k_y b) \frac{2t_c c}{\hbar} \sin(k_z c).$$

For the closed orbit around $(\sim k_F, 0, 0)$, by approximating $\cos(k_z c) \approx 1$, $\sin(k_y b) \approx k_y b$, $\cos(k_y c) \approx 1$, and $\sin(k_z c) \approx k_z c$, we obtain the equations of the motion on the Fermi surface,

$$\ddot{k}_y = -\Omega_c'^2 k_y, \quad \ddot{k}_z = -\Omega_c'^2 k_z. \tag{4.67}$$

The cyclotron frequency is given by

$$\Omega'_c = \frac{2eB_x bc}{\hbar^2} \sqrt{t_b t_c}. \tag{4.68}$$

In real space, the electron undergoes a helical motion in the x-axis as shown in Fig. 4.24b.

The cyclotron orbit around $(\sim k_F, \pm\pi/b, \pm\pi/c)$ has the same cyclotron frequency as that around $(\sim k_F, 0, 0)$, but the rotation direction is opposite. In both cases, the cyclotron motion on the Fermi surface without being scattered leads to the coherence motion even in the least conducting z-axis.

From Eq. (4.46), we have

$$\left|\frac{\partial k_z}{\partial k_x}\right| = \frac{\hbar v_F}{2t_c c\,|\sin(k_z c)|}. \tag{4.69}$$

This function has an inflection point at $k_z = \pi/2c$. As seen in Fig. 4.16b, when the magnetic field is tilted from the k_x-axis to the k_z-axis, the closed orbits on the convex and concave surfaces disappear at the inflection point, which is defined as the critical angle θ'_c,

$$\frac{\pi}{2} - \theta'_c \equiv \Delta\theta'_{\text{peak}} = \left|\frac{\partial k_x}{\partial k_z}\right|_{k_z=\pi/2c} = \frac{2t_c c}{\hbar v_F} = \frac{2t_c}{E_F} k_F c. \tag{4.70}$$

Here the Fermi energy is defined as $E_F = \hbar v_F k_F$. In the small angle range $\pi/2 - \theta'_c < \theta < \pi/2 + \theta'_c$, the closed orbits on the Fermi surface are formed. The result is identical to Eq. (4.40) except for the factor 2. As discussed in Sect. 4.6.2, the closed orbit does not contribute to the z-axis conductivity; the z-axis velocity averaged over the closed orbits vanishes, $\langle v_z(t)\rangle = 0$. As the magnetic field is tilted from the x-axis, the closed orbits disappear and then the z-axis conductivity is enhanced; the resistance shows the peak at $\theta = 90°$ as indicated by the arrows in Fig. 4.23a. The peak at $\theta = 90°$ is verified in numerical calculations by the Chambers formula for the energy band given by Eq. (4.46) [26], which includes the effect of the small closed orbits on the side of the Fermi surface. This peak at $\theta = 90°$ is also recognized as the coherence peak as discussed for the quasi-2D Fermi surface in Sect. 4.6.

Another model describing the quasi-1D energy band is given by [27]

$$E(\boldsymbol{k}) = \varepsilon - 2t_a\cos(k_x a) - 2t_b\cos(k_y b) - 2t_c\cos(k_z c), \tag{4.71}$$

where the tight binding approximation is used in all directions. Unlike Eq. (4.46), this energy band makes higher order corrugations of the Fermi surface. Taking $k_x = k_F$, and $k_z = \pi/2c$, we can define the peak width

$$\Delta\theta'_{\text{peak}} = \left|\frac{\partial k_x}{\partial k_z}\right|_{k_z=\pi/2c} = \frac{t_c c}{t_a a}\left|\frac{\sin(k_z c)}{\sin(k_x a)}\right| = \frac{t_c c}{t_a a \sin(k_F a)}. \tag{4.72}$$

If the energy band is half-filled ($k_F = \pi/2a$), the volume of the states occupied by the electrons is a half of the first Brillouin zone. It gives $\Delta\theta'_{\text{peak}} = t_c c/t_a a$. For the quasi-1D organic conductor $(TMTSF)_2ClO_4$, substituting the reasonable values, $a = 0.73$ nm, $c = 1.35$ nm, $t_a = 200$ meV, and $t_c = 2$ meV, we obtain $\Delta\theta'_{\text{peak}} = 1°$. This coherence peak width is roughly consistent with the result in Fig. 4.23.

4.8.4 Lebed Magic Angle

Returning to Eq. (4.56), we see the conductivity $\sigma_{zz}^{1D}(B)$ for the $\varphi \neq 0°$ rotation. The AMRO in this rotation is different from that for the quasi-2D Fermi surface, Eq. (4.21). Equation (4.56) shows that the conductivity (resistance) has a peak (dip) when the magnetic field angle satisfies the relation $n\omega_z \pm \omega_y = 0$. Substituting Eq. (4.50) into the relation, we define the angles θ_L as

$$\tan\theta_L = \frac{B_y}{B_z} = \frac{n_L b}{c} \quad (n_L = \pm1, \pm2, \pm3, \ldots), \tag{4.73}$$

which are called Lebed magic angles [28]. In particular, for the $\varphi = 90°$ rotation ($k_y k_z$ plane) in Fig. 4.21b, we have $\gamma = 0$. Since $J_0(0) = 1$ and $J_{n>0}(0) = 0$, the conductivity is given by

$$\sigma_{zz}^{1D}(B) = \sigma_{zz}^{1D}(0)\frac{1}{1+\omega_y^2\tau^2} = \sigma_{zz}^{1D}(0)\frac{1}{1+\omega_{y0}^2\tau^2\sin^2(\theta)}, \tag{4.74}$$

where $\omega_y = v_F ceB_y/\hbar = (v_F ce/\hbar)B\sin\theta = \omega_{y0}\sin\theta$. Figure 4.25a shows the calculated results of $\rho_{zz}^{1D}(\theta) = 1/\sigma_{zz}^{1D}(\theta) = \rho_{zz}^{1D}(0)\left[1+\omega_{y0}^2\tau^2\sin^2(\theta)\right]$ at different $\omega_{y0}\tau$ values. This rotation produces smooth sinusoidal curves without any particular structure.

Figure 4.25b shows the z-axis resistance for the quasi-1D organic conductor $(TMTSF)_2PF_6$ [29], whose Fermi surfaces are identical to the inset of **a**. When the magnetic field is rotated in the $k_y k_z$ plane, the interlayer resistance R_{zz} shows characteristic behavior. At low magnetic fields, R_{zz} has a sinusoidal anisotropy with broad maxima at $\theta = \pm 90°$, which is consistent with the calculated results in Fig. 4.25a. At high magnetic fields, however, R_{zz} shows the opposite anisotropy

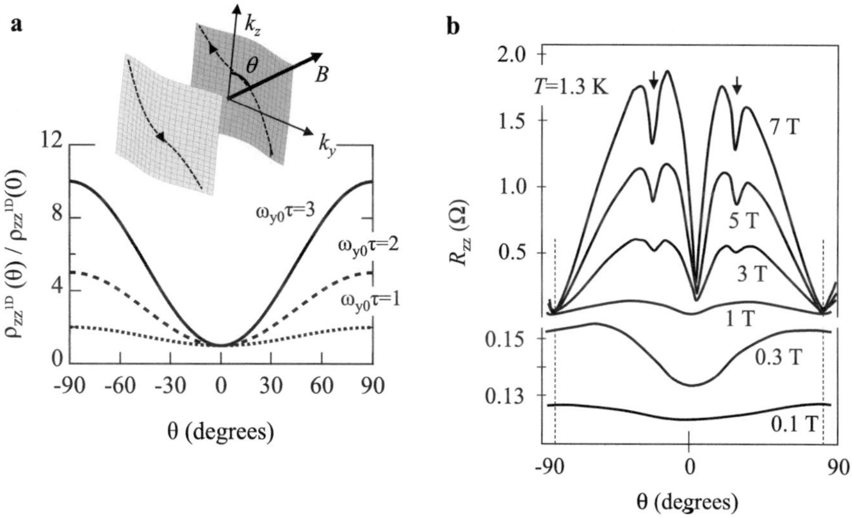

Fig. 4.25 **a** Calculated results of the z-axis resistivity by Eq. (4.74) for magnetic field rotation in $k_y k_z$ plane. Inset: quasi-1D Fermi surface. **b** Resistance for the quasi-1D organic conductor $(TMTSF)_2PF_6$ (reproduced with permission from [29]). Arrows indicate the dips at θ_L with $n_L = \pm1$.

(minima at $\theta = \pm 90°$) accompanied by sharp dips at the angles indicated by the arrows. The angles of the sharp dips correspond to the Lebed magic angles θ_L with $n_L = \pm 1$ ($\omega_z = \pm\omega_y$) in Eq. (4.73). The sharp resistance dips at these angles can be qualitatively explained by the numerical calculations based on the energy band in Eq. (4.71) [27], but not by the band in Eq. (4.46). The fact shows that the higher order corrugation of the Fermi surface is required to reproduce the sharp dips. For the $\varphi \neq 0 (\gamma \neq 0)$ rotation, the term $[nb\cos(\theta) \pm c\sin(\theta)\sin(\varphi)]^2$ of the denominator in Eq. (4.56) leads to the dips in the resistance at θ_L as already seen in Fig. 4.22a.

A significant point is that the numerical calculation cannot reproduce the background resistance with the opposite anisotropy observed at the high magnetic fields (minima at $\theta = \pm 90°$). This suggests the presence of another mechanism of the interlayer conduction, which is not included in the Chambers formula. As a possible cause, magnetic-field-induced incoherent interlayer conduction based on the geometry of the Fermi surfaces has been discussed [30–32]. Although the mechanism of the opposite anisotropy remains controversial, the interlayer resistivity dips at the Lebed magic angles in Fig. 4.25b can be ascribed to the periodic orbital motion on the quasi-1D Fermi surface.

4.8.5 Third Angular Effect

Let us see what happens when the magnetic field is rotated in the $k_x k_y$ plane. In this rotation ($B_z = 0$), Eq. (4.56) cannot be used because $\gamma \to \infty$. We take $\hbar \dot{k}_z = ev_y B_x - ev_F B_y$ in Eq. (4.49) and $v_y = (2t_b b/\hbar)\sin(k_y b) \to (2t_b b/\hbar)\sin(k_{y0} b)$ in Eq. (4.65) for simplicity. By integrating $\dot{k}_z$ over time, we have

$$\begin{aligned} k_z c &= -\frac{v_F ceB_y}{\hbar}t + k_{z0}c + \frac{v_y ceB_x}{\hbar} \\ &= -\frac{v_F ceB_y}{\hbar}t + k_{z0}c + \frac{ceB_x}{\hbar}\frac{2t_b b}{\hbar}\sin(k_{y0}b)t \\ &= k_{z0}c - \left[\sin(\varphi) - \eta\cos(\varphi)\sin(k_{y0}b)\right]\omega_{y0}t \\ &= k_{z0}c - \Omega_{1D}t, \end{aligned} \tag{4.75}$$

where $B_y = B\sin(\varphi), B_x = B\cos(\varphi), \omega_{y0} = (v_F ce/\hbar)B$ and $\Omega_{1D} = \left[\sin(\varphi) - \eta\cos(\varphi)\sin(k_{y0}b)\right]\omega_{y0}$. The parameter

$$\eta = \frac{2t_b b}{\hbar v_F} = \frac{2t_b}{E_F}k_F b \tag{4.76}$$

shows the relative strength of the Fermi surface corrugation $2t_b$ to the Fermi energy E_F. The velocity is given by

$$v_z(\boldsymbol{k}(t)) = \frac{2t_c c}{\hbar}\sin(k_z c) = \frac{2t_c c}{\hbar}\sin(k_{z0}c - \Omega_{1D}t). \tag{4.77}$$

By substituting Eq. (4.75) into Eq. (4.15), σ_{zz}^{1D} is calculated as [33],

$$\sigma_{zz}^{1D}(B) = \sigma_{zz}^{1D}(0)\frac{\sqrt{X+\sqrt{X^2+4Y^2}}}{\sqrt{2}\sqrt{X^2+4Y^2}},\ \sigma_{zz}^{1D}(0) = 2D(E_F)\left(\frac{et_c c}{\hbar}\right)^2\tau, \tag{4.78}$$

where $X = 1-(Y^2 - Z^2)$, $Y = \omega_{y0}\tau\sin(\varphi)$, and $Z = \eta\omega_{y0}\tau\cos(\varphi)$. This is derived in Appendix 4.6.

Taking $\eta = 0 (Z = 0)$, we obtain $\sigma_{zz}^{1D}(B) = \sigma_{zz}^{1D}(0)/\left(1+\omega_y^2\tau^2\right)$, consistent with Eq. (4.74). It is easy to calculate the other components of the conductivity tensor in this approximation,

$$\sigma_{xx}^{1D}(B) = D(E_F)(ev_F)^2\tau = \sigma_{xx}^{1D}(0),\ \sigma_{yy}^{1D}(B) = 2D(E_F)\left(\frac{et_b b}{\hbar}\right)^2\tau = \sigma_{yy}^{1D}(0). \tag{4.79}$$

All off-diagonal components vanish. The conductivity tensor in the magnetic field $(B\cos(\varphi), B\sin(\varphi), 0)$ is summarized as

$$\boldsymbol{\sigma}^{1D} = \begin{pmatrix} \sigma_{xx}^{1D} & 0 & 0 \\ 0 & \sigma_{yy}^{1D} & 0 \\ 0 & 0 & \sigma_{zz}^{1D} \end{pmatrix} = \begin{pmatrix} \sigma_{xx}^{1D}(0) & 0 & 0 \\ 0 & \sigma_{yy}^{1D}(0) & 0 \\ 0 & 0 & \sigma_{zz}^{1D}(0)\frac{\sqrt{X+\sqrt{X^2+4Y^2}}}{\sqrt{2}\sqrt{X^2+4Y^2}} \end{pmatrix}. \tag{4.80}$$

Note that only $\sigma_{zz}^{1D}(B)$ depends on the magnetic field. The z-axis resistivity is written as $\rho_{zz}^{1D}(B) = 1/\sigma_{zz}^{1D}(B)$.

The results of the z-axis resistance for the quasi-1D conductor $(TMTSF)_2ClO_4$ are presented in Fig. 4.26a at different magnetic fields. At low magnetic fields, R_{zz} shows a sinusoidal curve with a minimum at $\theta = 0$. As the magnetic field increases, R_{zz} becomes large accompanied by kinks or dips near $\theta = 0$ as indicated by arrows. The kinks or dips are called the third angular effect [34], which are recognized as the effect of the small closed orbits formed on the Fermi surface. As seen in Fig. 4.24a, the small closed orbits are formed on the convex and concave surfaces for $B\|k_x$. When the magnetic field is tilted toward the k_y-axis, the closed orbits disappear at a critical angle φ_c. As discussed in Sect. 4.8.3, the critical angle corresponds to the inflection point ($k_y = \pi/2b$) of the Fermi surface in the k_xk_y plane,

$$\varphi_c = \left|\frac{\partial k_x}{\partial k_y}\right|_{k_y=\pi/2b} = \frac{2t_b b}{\hbar v_F} = \frac{2t_b}{E_F}k_F b. \tag{4.81}$$

This quantity is identical to η in Eq. (4.76). For the energy band given by Eq. (4.71), taking $k_x = k_F = \pi/2a$, and $k_y = \pi/2b$, we can define the critical angle,

$$\varphi_c = \left|\frac{\partial k_x}{\partial k_y}\right|_{k_y=\pi/2b} = \frac{t_b b}{t_a a}. \tag{4.82}$$

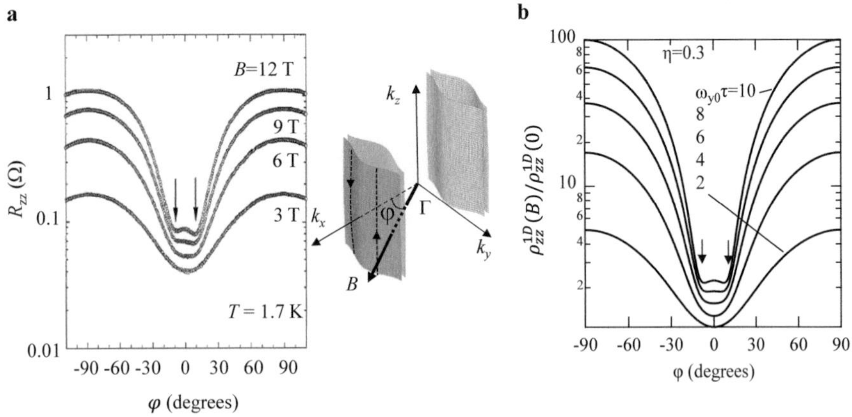

Fig. 4.26 **a** Resistance of the quasi-1D organic conductor $(TMTSF)_2ClO_4$ for the magnetic field rotation in the k_xk_y plane (reproduced with permission from [34]). Inset: four sheets of the quasi-1D Fermi surface in $(TMTSF)_2ClO_4$. **b** Calculated results of the z-axis resistivity $\rho_{zz}^{1D}(B) = 1/\sigma_{zz}^{1D}(B)$ for $\eta = 0.3$ by Eq. (4.78), where $\rho_{zz}^{1D}(0) = 1/\sigma_{zz}^{1D}(0)$

The b-axis needs to be doubled due to the ClO_4 anion order at low temperatures for $(TMTSF)_2ClO_4$. Using the appropriate values, $b \rightarrow 2b = 1.54\,\text{nm}$, $2t_b/E_F \approx 0.09$, and $k_F \approx \pi/2a$ ($a = 0.73$ nm), the resistivity $\sigma_{zz}^{1D}(\varphi) = 1/\sigma_{zz}^{1D}(\varphi)$ for $\eta = 0.3$ is calculated at different $\omega_{y0}\tau$ values in Fig. 4.26b. We obtain $\varphi_c = (2t_b/E_F)k_F b' = \eta \approx 0.3$ rad $\approx 17°$, which is in agreement with the experimental results. Detailed numerical calculations of the time-averaged velocity $\langle v_z \rangle$ have been made on the quasi-1D Fermi surface at a magnetic field in the k_yk_z plane [35]. The results consistently show that the third angular effect arises from the incomplete cancelation of v_z over the closed orbits. Another picture to explain the third angular effect [36, 37] is the presence of "effective carriers" with open orbits near the inflection points. As the magnetic field direction approaches φ_c, more electrons near the inflection points contribute to the conductivity, resulting in the kinks or dips in the z-axis resistance. Although the detailed interpretation is still controversial, it is clear that the kinks or dips are related to the orbital motion of the electrons near the inflection points of the Fermi surface.

If the kink or dip is related to the small closed orbit on the side of the Fermi surface, the threshold magnetic field B_c' can be defined as $\Omega_c'\tau = 1$ in Eq. (4.68),

$$B_c' = \frac{\hbar^2}{2ebc\tau\sqrt{t_b t_c}}. \tag{4.83}$$

The kink or dip will be observed for $B > B_c'(\Omega_c'\tau > 1)$. Using the reasonable values, $a = 0.73$ nm, $c = 1.35$ nm, $t_b = 10$ meV, and $t_c = 2$ meV, and $\tau = 10^{-11}$ s, we obtain $B_c' = 2.3$ T. This value is in reasonable agreement with the experimental results of Fig. 4.26a. As described above, a variety of behaviors

Table. 4.1 Features of z-axis resistance for quasi-1D Fermi surface

Orbital motion in magnetic field	Features of z-axis resistance
B in $k_x k_z$ plane, $\boldsymbol{B}(B_x, 0, B_z)$ 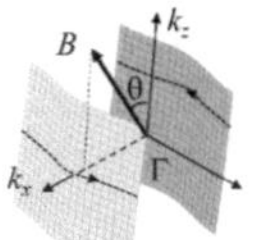	Periodic peaks at θ_n, $\frac{2t_b c}{\hbar v_F}\tan\theta_n = \pm\pi(n - 1/4)$, $(n = 1, 2, 3, \ldots)$
B in $k_y k_z$ plane, $\boldsymbol{B}\left(0, B_y, B_z\right)$ 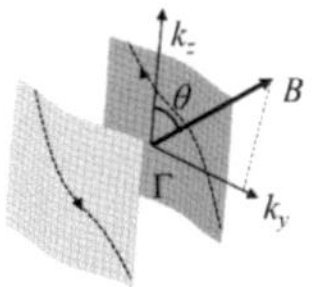	Sharp dips at Lebed magic angles θ_L,$\tan\theta_L = \frac{B_y}{B_z} = \frac{n_L b}{c}$ $(n_L = \pm 1, \pm 2, \pm 3, \ldots)$
B in $k_x k_y$ plane, $\boldsymbol{B}\left(B_x, B_y, 0\right)$ 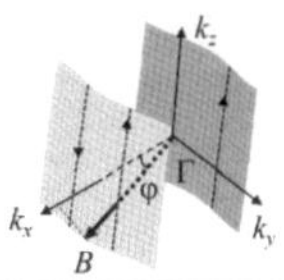	Kink or dip at φ_c,$\varphi_c = \frac{2t_b}{E_F}k_F b$ or $\frac{t_b b}{t_a a}$

appear in the z-axis resistance for the quasi-1D Fermi surface. The results are summarized in Table 4.1.

4.9 Summary

We summarize the Table 4.2 physical quantities obtained from the angular dependence of the z-axis resistance in low-dimensional electron systems. Although we cannot determine each physical quantity from a single-independent measurement, we can obtain various microscopic information about the Fermi surface from the complementary measurements Table 4.3.

4.9.1 Quasi-2D Fermi Surface

See Table 4.2.

4.9.2 Quasi-1D Fermi Surface

Table 4.3.

Table. 4.2 Summary of physical quantities obtained from the angle dependence of the z-axis resistance for quasi-2D Fermi surface

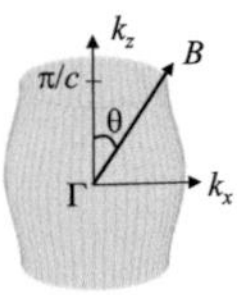	Energy band of quasi-2D Fermi surface: $E = \frac{\hbar^2}{2m_c}\left(k_x^2 + k_y^2\right) - 2t_c\cos(k_z c)$
Physical quantity	Method of observation and related expression
k_F	AMRO period (periodic peaks): $\Delta(\tan\theta) = \frac{\pi}{k_F c}$.
τ, k_F, m_c	AMRO amplitude: $\rho_{zz}(B) = \frac{1}{\sigma_{zz}(B)}$, $\sigma_{zz}(B) = \sigma_{zz}(0)\sum_{n=-\infty}^{\infty}\frac{J_n^2(k_F c\tan\theta)}{1+(n\omega_c\tau)^2}$, $\omega_c = \frac{eB_z}{m_c}$.
t_c, E_F, k_F	Coherence peak width: $\Delta\theta_{\text{peak}} = \frac{2m_c t_c c}{\hbar^2 k_F} = \frac{t_c}{E_F}k_F c$.
τ, m_c, t_c	Threshold magnetic field of coherence peak: $B_c = \frac{\hbar}{ec\tau}\sqrt{\frac{m_c}{2t_c}}$.
τ, k_F, m_c	Magnetic field dependence for $B \perp z$ ($B \parallel$ layer): $\rho_{zz}(B) = \rho_{zz}(0)\sqrt{1+(k_F c\omega_c\tau)^2}$, $\omega_c = eB_x/m_c$.

Table. 4.3 Summary of physical quantities obtained from the angle dependence of the z-axis resistance for quasi-1D Fermi surface

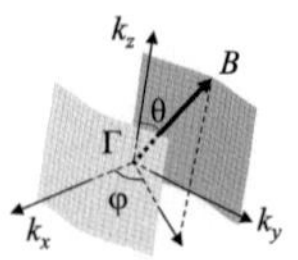	Energy band of quasi-1D Fermi surface: $E(\boldsymbol{k}) = \hbar v_F(\lvert k_x\rvert - k_F) - 2t_b\cos(k_y b) - 2t_c\cos(k_z c)$ or $E(\boldsymbol{k}) = \varepsilon - 2t_a\cos(k_x a) - 2t_b\cos(k_y b) - 2t_c\cos(k_z c)$
Physical quantity	Method of observation and related expression
t_b, v_F, E_F, k_F	AMRO period (periodic peaks) for $k_x k_z$ rotation: $\frac{2t_b c}{\hbar v_F}\tan\theta_n = \frac{2t_b k_F c}{E_F}\tan\theta_n = \pm\pi\left(n - \frac{1}{4}\right)$, $n = 1, 2, 3, \ldots$
t_c, v_F, E_F, k_F	Coherence peak width for $k_x k_z$ rotation: $\Delta\theta'_{\text{peak}} = \frac{2t_c c}{\hbar v_F} = \frac{2t_c k_F c}{E_F}$ or $\frac{t_c c}{t_a a\,\sin(k_F a)}$.
t_b, v_F, E_F, k_F	Kink or dip angle for $k_x k_y$ rotation: $\varphi_c = \frac{2t_b b}{\hbar v_F} = \frac{2t_b k_F b}{E_F}$ or $\frac{t_b b}{t_a a\sin(k_F a)}$.
t_b, t_c, τ	Threshold magnetic field of coherence peak or third angular effect: $B'_c = \frac{\hbar^2}{2ebc\tau\sqrt{t_b t_c}}$.

Appendix 4.1

The Chambers formula can be derived as follows. Let $f_k(\boldsymbol{r}, t)$ be the local electron density with a wave number vector $\boldsymbol{k}$ at a time t. Assume that the temperature is constant and homogeneous, the number of electrons is conserved, and the electronic state is in a steady state. It means that the number of electrons entering and leaving a microscopic space is equal. Since $\boldsymbol{k}$ in $f_k(\boldsymbol{r}, t)$ is time-dependent, the time variation of the local electron density in an electric field $\boldsymbol{E}$ is given by

$$\frac{\partial f_k(\boldsymbol{r}, t)}{\partial t} = \dot{\boldsymbol{k}} \cdot \frac{\partial f_k(\boldsymbol{r}, t)}{\partial \boldsymbol{k}} = -\frac{e\boldsymbol{E}}{\hbar} \cdot \frac{\partial f_k(\boldsymbol{r}, t)}{\partial \boldsymbol{k}}. \tag{4.84}$$

Here the equation of motion $\boldsymbol{F} = \hbar\dot{\boldsymbol{k}} = -e\boldsymbol{E}$ is used. Since the density $f_k(\boldsymbol{r}, t)$ is regarded as the Fermi distribution function $f_k(\boldsymbol{r}, t) = f(E_k)$,

$$\frac{\partial f_k(\boldsymbol{r}, t)}{\partial \boldsymbol{k}} = \frac{\partial f(E_k)}{\partial E_k}\frac{\partial E_k}{\partial \boldsymbol{k}} = \hbar \boldsymbol{v}_k(t)\frac{\partial f(E_k)}{\partial E_k}, \tag{4.85}$$

and thus

$$\frac{\partial f_k(\boldsymbol{r}, t)}{\partial t} = -e\boldsymbol{v}_k(t) \cdot \boldsymbol{E}\frac{\partial f(E_k)}{\partial E_k}. \tag{4.86}$$

Assuming that the electric field is sufficiently small and the steady-state density deviates slightly from the thermal equilibrium density $f_k^{(0)}(\boldsymbol{r}, t)$, we define the deviation as

$$g_k(\boldsymbol{r}, t) = f_k^{(0)}(\boldsymbol{r}, t) - f_k(\boldsymbol{r}, t).$$

Next, we consider the scattering effect on the steady state. The electrons are accelerated by the electric field but the acceleration is interrupted by the scattering. This scattering effect will be written as

$$\frac{\partial g_k(t)}{\partial t} = -\frac{g_k(t)}{\tau}. \tag{4.87}$$

If the electric field becomes zero at $t = 0$, we have the solution $g_k(t) = g_k(0)\exp(-t/\tau)$, which means that the electron density comes back to the thermal equilibrium state, $g_k(t) \to 0$ for $t \to \infty$. The deviation under a constant electric field is generally given by

$$g_k(t) = -\frac{1}{\tau}\int_{-\infty}^{t} \mathrm{d}t' g_k(t')\exp(t'/\tau). \tag{4.88}$$

Differentiating both sides by t and $t \to 0$, we find that Eq. (4.87) is satisfied. Note that the origin of time can be chosen arbitrarily without loss of generality, so the upper limit of this integral can be set to $t = 0$. This equation implies that the deviation from the thermal equilibrium density at time t is influenced by the previous deviation in the range of $\sim \tau$. In the steady state, the sum of the variation by the electric field, Eq. (4.86) and that by the scattering, Eq. (4.87) must be zero,

$$\frac{\partial f_{\boldsymbol{k}}(\boldsymbol{r}, t)}{\partial t} + \frac{\partial g_{\boldsymbol{k}}(t)}{\partial t} = 0,$$

and thus we obtain

$$-\frac{g_{\boldsymbol{k}}(t)}{\tau} - e\boldsymbol{v}_{\boldsymbol{k}}(t) \cdot \boldsymbol{E}\frac{\partial f(E_{\boldsymbol{k}})}{\partial E_{\boldsymbol{k}}} = 0. \tag{4.89}$$

Substituting Eq. (4.89) into the right side of Eq. (4.88), we have,

$$\begin{aligned} g_{\boldsymbol{k}}(0) &= -\frac{1}{\tau}\int_{-\infty}^{0} \mathrm{d}t' g_{\boldsymbol{k}}(t') \exp(t'/\tau) \\ &= \frac{\partial f(E_{\boldsymbol{k}})}{\partial E_{\boldsymbol{k}}}\int_{-\infty}^{0} \mathrm{d}t' e\boldsymbol{v}_{\boldsymbol{k}}(t') \cdot \boldsymbol{E} \exp(t'/\tau). \end{aligned} \tag{4.90}$$

The electric current is given by the sum over all $\boldsymbol{k}(0)$ states,

$$\begin{aligned} \boldsymbol{j} &= -2\sum_{\boldsymbol{k}(0)} e\boldsymbol{v}_{\boldsymbol{k}}(0) g_{\boldsymbol{k}}(0) \\ &= -2e^2 \sum_{\boldsymbol{k}(0)} \boldsymbol{v}_{\boldsymbol{k}}(0) \frac{\partial f(E_{\boldsymbol{k}})}{\partial E_{\boldsymbol{k}}} \int_{-\infty}^{0} \mathrm{d}t' \boldsymbol{v}_{\boldsymbol{k}}(t') \cdot \boldsymbol{E} \exp(t'/\tau). \end{aligned} \tag{4.91}$$

The current was doubled by taking the spin degree of freedom into account. In Eq. (4.91), $\boldsymbol{v}_{\boldsymbol{k}}(0)$ is oriented in the current direction and only $\boldsymbol{v}_{\boldsymbol{k}}(t')$ parallel to the electric field contributes the integral. The components σ_{ij} of the conductivity tensor $\boldsymbol{\sigma}$ are defined for $\boldsymbol{j} \parallel i$-axis and $\boldsymbol{E} \parallel j$-axis. By comparing Eq. (4.91) with $\boldsymbol{j} = \boldsymbol{\sigma}\boldsymbol{E}$, we obtain the Chambers formula,

$$\sigma_{ij} = 2e^2 \sum_{\boldsymbol{k}(0)} \left(-\frac{\mathrm{d}f(E)}{\mathrm{d}E}\right) v_i(\boldsymbol{k}(0)) \int_{-\infty}^{0} \mathrm{d}t\ v_j(\boldsymbol{k}(t)) \exp(t/\tau),\ i, j = x, y, z.$$

Appendix 4.2

We derive the analytical solution of the AMRO, Eq. (4.21). In the Chambers formula,

$$\sigma_{zz}(B) = 2e^2 \sum_{k(0)} \left(-\frac{\mathrm{d}f}{\mathrm{d}E}\right) v_z(k(0)) \int_{-\infty}^{0} \mathrm{d}t \; v_z(k(t)) \exp(t/\tau),$$

we can write $-\mathrm{d}f/\mathrm{d}E = \delta(E_F - E) = \left(m_c/\hbar^2 k_F\right)\delta(k_F - k_\perp)$ from the relation $E = \hbar^2 k_\perp^2/2m_c$. The sum of $\boldsymbol{k}(0)$ is written as

$$\sum_{k(0)} \left(-\frac{\mathrm{d}f}{\mathrm{d}E}\right) \to \frac{1}{(2\pi)^3} \int_{-\pi/c}^{\pi/c} \mathrm{d}k_{z0} \int_{0}^{\infty} \mathrm{d}k_\perp \int_{-\pi}^{\pi} k_\perp \mathrm{d}\varphi_0 \left(\frac{m_c}{\hbar^2 k_F}\right) \delta(k_F - k_\perp). \quad (4.92)$$

By taking $Z = k_{z0}c + k_\perp c \tan\theta \cos\varphi_0$ in Eq. (4.20), the velocity is expressed as

$$\begin{aligned} v_z(\boldsymbol{k}(t)) &= \left(\frac{2t_c c}{\hbar}\right) \sin[Z - k_\perp c \tan\theta \cos(\omega_c t + \varphi_0)] \\ &= v_{z1}(\boldsymbol{k}(t)) + v_{z2}(\boldsymbol{k}(t)), \\ v_{z1}(\boldsymbol{k}(t)) &= \left(\frac{2t_c c}{\hbar}\right) \frac{1}{2i} e^{i\{Z - k_\perp c \tan\theta \cos(\omega_c t + \varphi_0)\}}, \\ v_{z2}(\boldsymbol{k}(t)) &= -\left(\frac{2t_c c}{\hbar}\right) \frac{1}{2i} e^{-i\{Z - k_\perp c \tan\theta \cos(\omega_c t + \varphi_0)\}}. \end{aligned} \quad (4.93)$$

Using the definitions of the nth order Bessel function $J_n(z)$,

$$e^{iz \sin\alpha} = \sum_{n=-\infty}^{\infty} J_n(z) e^{in\alpha}, \; e^{iz\cos\alpha} = \sum_{n=-\infty}^{\infty} e^{in\pi/2} J_n(z) e^{in\alpha}, \quad (4.94)$$

we have the exponential function of $v_{z1}(\boldsymbol{k}(t))$,

$$\begin{aligned} e^{-ik_\perp c \tan\theta \cos(\omega_c t + \varphi_0)} &= \sum_{n=-\infty}^{\infty} e^{in\pi/2} J_n(-k_\perp c \tan\theta) e^{in(\omega_c t + \varphi_0)} \\ &= \sum_{n=-\infty}^{\infty} e^{-in\pi/2} J_n(k_\perp c \tan\theta) e^{-in(\omega_c t + \varphi_0)}. \end{aligned} \quad (4.95)$$

Here we used the relation,$J_n(-z) = J_{-n}(z)$ and then replaces $-n$ with n. Therefore, we obtain

$$v_{z1}(\boldsymbol{k}(t)) = \left(\frac{2t_c c}{\hbar}\right) \frac{1}{2i} e^{iZ} \sum_{n=-\infty}^{\infty} e^{-in\pi/2} J_n(k_\perp c \tan\theta) e^{-in(\omega_c t + \varphi_0)}. \quad (4.96)$$

Using the relation,

$$v_z(\boldsymbol{k}(0)) = \left(\frac{2t_c c}{\hbar}\right)\sin(k_{z0}c) = \left(\frac{2t_c c}{\hbar}\right)\frac{1}{2i}\left(e^{ik_{z0}c} - e^{-ik_{z0}c}\right),$$

we can calculate the conductivity σ_{zz1} arising from $v_{z1}(\boldsymbol{k}(t))$,

$$\begin{aligned}
\sigma_{zz1} &= 2e^2\sum_{\boldsymbol{k}(0)}\left(-\frac{\mathrm{d}f}{\mathrm{d}E}\right)v_z(\boldsymbol{k}(0))\int_{-\infty}^{0}\mathrm{d}t\ v_{z1}(\boldsymbol{k}(t))\exp(t/\tau) \\
&= \frac{2e^2}{(2\pi)^3}\int_{-\pi/c}^{\pi/c}\mathrm{d}k_{z0}\int_0^{\infty}\mathrm{d}k_{\perp} \\
&\quad\times\int_{-\pi}^{\pi}k_{\perp}\mathrm{d}\varphi_0\frac{m_c}{\hbar^2 k_F}\delta(k_F - k_{\perp})v_z(\boldsymbol{k}(0))\int_{-\infty}^{0}\mathrm{d}t\ v_{z1}(\boldsymbol{k}(t))\exp(t/\tau) \\
&= \frac{2e^2}{(2\pi)^3}\frac{m_c}{\hbar^2}\left(\frac{2t_c c}{\hbar}\right)^2\left(\frac{-1}{4}\right)\int_{-\pi/c}^{\pi/c}\mathrm{d}k_{z0}\int_{-\pi}^{\pi}\mathrm{d}\varphi_0 \\
&\quad\times\int_{-\infty}^{0}dt\left(e^{ik_{z0}c} - e^{-ik_{z0}c}\right)e^{i(k_{zo}c + k_F c\tan\theta\cos\varphi_0)} \\
&\quad\times\sum_{n=-\infty}^{\infty}e^{-in\pi/2}J_n(k_F c\tan\theta)e^{-in(\omega_c t+\varphi_0)}e^{t/\tau} \\
&= \frac{2e^2}{(2\pi)^3}\frac{m_c}{\hbar^2}\left(\frac{2t_c c}{\hbar}\right)^2\left(\frac{2\pi}{4c}\right)\int_{-\pi}^{\pi}\mathrm{d}\varphi_0 \\
&\quad\times\int_{-\infty}^{0}\mathrm{d}te^{ik_F c\tan\theta\cos\varphi_0}\sum_{n=-\infty}^{\infty}e^{-in\pi/2}J_n(k_F c\tan\theta)e^{-in(\omega_c t+\varphi_0)}e^{t/\tau},
\end{aligned} \tag{4.97}$$

where

$$\begin{aligned}
\int_{-\pi}^{\pi}\mathrm{d}\varphi_0 e^{ik_F c\tan\theta\cos\varphi_0}e^{-\frac{in\pi}{2}}e^{-in(\omega_c t+\varphi_0)} &= e^{-in\omega_c t}\int_{-\pi}^{\pi}\mathrm{d}\varphi_0 e^{ik_F c\tan\theta\cos\varphi_0}e^{-in(\varphi_0+\frac{\pi}{2})} \\
&= e^{-in\omega_c t}\int_{-\pi}^{\pi}\mathrm{d}\varphi' e^{i(k_F c\tan\theta\sin\varphi' - n\varphi')} \\
&= 2\pi J_n(k_F c\tan\theta)e^{-in\omega_c t}.
\end{aligned}$$

In the above equation, we put $\varphi' = \varphi_0 + \pi/2$ and changed the integration range since the function to be integrated is a periodic function in the range $[-\pi, \pi]$. In the last equality, we used the integral formula of the Bessel function,

$$J_n(z) = \frac{1}{2\pi} \int_{-\pi}^{\pi} e^{i(z \sin\alpha - n\alpha)} \mathrm{d}\alpha. \tag{4.98}$$

In this way, we obtain

$$\sigma_{zz1}(B) = \frac{e^2 m_c t_c^2 c}{\pi \hbar^4} \sum_{n=-\infty}^{\infty} [J_n(k_F c \tan\theta)]^2 \int_{-\infty}^{0} \mathrm{d}t\, e^{-in\omega_c t + t/\tau}. \tag{4.99}$$

Similarly,

$$\sigma_{zz2}(B) = \frac{e^2 m_c t_c^2 c}{\pi \hbar^4} \sum_{n=-\infty}^{\infty} [J_n(k_F c \tan\theta)]^2 \int_{-\infty}^{0} \mathrm{d}t\, e^{in\omega_c t + t/\tau}. \tag{4.100}$$

By integrating and summing $\sigma_{zz1}(B)$ and $\sigma_{zz2}(B)$, we obtain the AMRO formula,

$$\sigma_{zz}(B) = \sigma_{zz1}(B) + \sigma_{zz2}(B) = \frac{2e^2 m_c t_c^2 c\tau}{\pi \hbar^4} \sum_{n=-\infty}^{\infty} \frac{J_n^2(k_F c \tan\theta)}{1 + (n\omega_c \tau)^2}.$$

Appendix 4.3

We calculate $\sigma_{xy}(B)$ by using the Chambers formula. From Eq. (4.23), we have

$$v_x(0) = \frac{\hbar k_x}{m_c} = \left(\frac{\hbar k_\perp}{m_c}\right) \cos\varphi_0,\ v_y(t) = \frac{\hbar k_y}{m_c} = \left(\frac{\hbar k_\perp}{m_c}\right) \sin(\omega_c t + \varphi_0).$$

By substituting these into Eq. (4.15), the Hall conductivity is calculated as

$$\begin{aligned}
\sigma_{xy}(B) &= 2e^2 \sum_{\boldsymbol{k}(0)} \left(-\frac{\mathrm{d}f}{\mathrm{d}E}\right) v_x(\boldsymbol{k}(0)) \int_{-\infty}^{0} \mathrm{d}t\, v_y(\boldsymbol{k}(t)) \exp(t/\tau) \\
&= \frac{2e^2}{(2\pi)^3} \int_{-\pi/c}^{\pi/c} \mathrm{d}k_{z0} \int_{0}^{\infty} \mathrm{d}k_\perp \int_{-\pi}^{\pi} k_\perp \mathrm{d}\varphi_0 \left(\frac{m_c}{\hbar^2 k_F}\right) \delta(k_F - k_\perp)
\end{aligned}$$

$$\times \int_{-\infty}^{0} dt\exp(t/\tau)\left(\frac{\hbar k_\perp}{m_c}\right)^2 \cos\varphi_0 \sin(\omega_c t + \varphi_0).$$

Using the relation

$$\cos\varphi_0 \sin(\omega_c t + \varphi_0) = \frac{1}{4i}\left(e^{i\varphi_0} + e^{-i\varphi_0}\right)\left(e^{i(\omega_c t+\varphi_0)} - e^{-i(\omega_c t+\varphi_0)}\right),$$

we integrate $\sigma_{\mathrm{xy}}(B)$ over $k_{z0}, k_\perp$ and φ_0 and then obtain

$$\begin{aligned}\sigma_{xy}(B) &= \frac{2e^2}{(2\pi)^3}\left(\frac{2\pi}{c}\right)\left(\frac{m_c}{\hbar^2}\right)\left(\frac{\hbar k_F}{m_c}\right)^2 \frac{2\pi}{4i}\int_{-\infty}^{0} dte^{t/\tau}\left(e^{i\omega_c t} - e^{-i\omega_c t}\right)\\ &= \left(\frac{m_c}{\pi c\hbar^2}\right)\left(\frac{\hbar^2 k_F^2}{2m_c}\right)\frac{e^2}{m_c}\frac{-\omega_c\tau^2}{1+(\omega_c\tau)^2}\\ &= -D(E_F)E_F\frac{e^2\tau}{m_c}\frac{\omega_c\tau}{1+(\omega_c\tau)^2}\\ &= -\left(\frac{N_e e^2\tau}{m_c}\right)\frac{\omega_c\tau}{1+(\omega_c\tau)^2}.\end{aligned}$$

Here again,

$$D(E_F) = \frac{m_c}{\pi c\hbar^2},\ E_F = \frac{\hbar^2 k_F^2}{2m_c},\ \text{and}\ N_e = \frac{k_F^2}{2\pi c}.$$

By using Eq. (4.23), we can easily obtain $\sigma_{xx}(B)$ and $\sigma_{yy}(B)$ in Eq. (4.24).

Appendix 4.4

In a parallel magnetic field $(B_x, 0, 0)$, most of the electrons on the quasi-2D Fermi surface have open orbits as shown in Fig. 4.27. Using the relation $m_c v_y = \hbar k_y = \hbar k_\perp \sin\varphi_0$ in Eq. (4.36), we have the velocity in the z-axis,

$$\begin{aligned}v_z(k(t)) &= \left(\frac{2t_c c}{\hbar}\right)\sin(k_z c)\\ &= \left(\frac{2t_c c}{\hbar}\right)\sin[k_{z0}c + k_\perp c\omega_c t\sin\varphi_0],\\ \omega_c &= \frac{eB_x}{m_c}.\end{aligned} \tag{4.101}$$

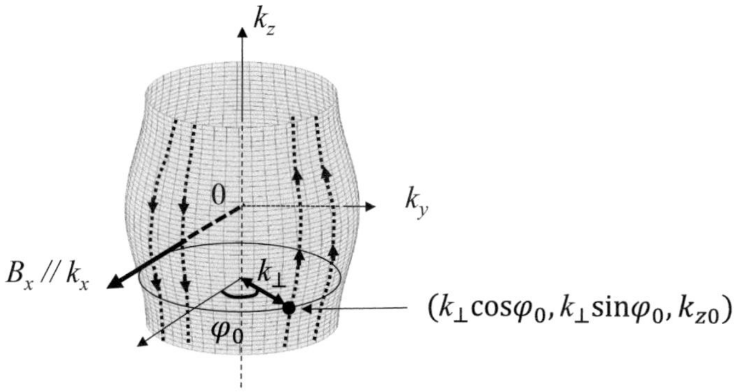

Fig. 4.27 Quasi-2D Fermi surface and electron orbits (dotted curves) in parallel magnetic field

The Chambers formula is written as

$$\begin{aligned}\sigma_{zz}(B_x) &= 2e^2 \sum_{\boldsymbol{k}(0)} \left(-\frac{\mathrm{d}f}{\mathrm{d}E}\right) v_z(\boldsymbol{k}(0)) \int_{-\infty}^{0} \mathrm{d}t\ v_z(\boldsymbol{k}(t)) \exp(t/\tau) \\ &= \frac{2e^2}{(2\pi)^3} \int_{-\pi/c}^{\pi/c} \mathrm{d}k_{z0} \int_0^{\infty} \mathrm{d}k_\perp \int_{-\pi}^{\pi} k_\perp \mathrm{d}\varphi_0 \frac{m_c}{\hbar^2 k_F} \delta(k_F - k_\perp) v_z(\boldsymbol{k}(0)) \\ &\quad \times \int_{-\infty}^{0} \mathrm{d}t\ v_z(\boldsymbol{k}(t)) \exp(t/\tau) \\ &= \frac{2e^2}{(2\pi)^3} \frac{m_c}{\hbar^2} \left(\frac{2t_c c}{\hbar}\right)^2 \int_{-\pi/c}^{\pi/c} \mathrm{d}k_{z0} \int_{-\pi}^{\pi} \mathrm{d}\varphi_0 \\ &\quad \times \int_{-\infty}^{0} \mathrm{d}t\ e^{t/\tau} \sin(k_{z0}c) \sin[k_{z0}c + k_F c\omega_c t \sin\varphi_0].\end{aligned}$$

Using the relation

$$\begin{aligned}&\sin(k_{z0}c) \sin[k_{z0}c + k_F c\omega_c t \sin\varphi_0] \\ &\quad = -\frac{1}{4}\left[e^{ik_{z0}c} - e^{-ik_{z0}c}\right] \\ &\quad \times \left[e^{i(k_{z0}c + k_F c\omega_c t\ \sin\varphi_0)} - e^{-i(k_{z0}c + k_F c\omega_c t\ \sin\varphi_0)}\right],\end{aligned}$$

the conductivity is written as

$$\sigma_{zz}(B_x) = \frac{2e^2}{(2\pi)^3}\frac{m_c}{\hbar^2}\left(\frac{2t_c c}{\hbar}\right)^2\frac{2\pi}{c}\left(\frac{1}{4}\right)\int_{-\pi}^{\pi}\mathrm{d}\varphi_0$$

$$\times\int_{-\infty}^{0}\mathrm{d}t\ e^{t/\tau}\left[e^{ik_F c\omega_c t\sin\varphi_0} + e^{-ik_F c\omega_c t\sin\varphi_0}\right]$$

$$= \frac{2e^2}{(2\pi)^3}\frac{m_c}{\hbar^2}\left(\frac{2t_c c}{\hbar}\right)^2\left(\frac{2\pi}{4c}\right)\int_{-\pi}^{\pi}\mathrm{d}\varphi_0\frac{2\tau}{1+(k_F c\omega_c\tau\sin\varphi_0)^2}. \quad (4.102)$$

Using the following integral formula,

$$\int_{-\pi}^{\pi}\mathrm{d}\varphi_0\frac{1}{1+(\alpha\sin\varphi_0)^2} = \frac{2\pi}{\sqrt{1+\alpha^2}}, \quad (4.103)$$

we obtain Eq. (4.43),

$$\sigma_{zz}(B_x) = \sigma_{zz}(0)\frac{1}{\sqrt{1+(k_F c\omega_c\tau)^2}},\ \sigma_{zz}(0) = \frac{2e^2 m_c t_c^2 c\tau}{\pi\hbar^4}.$$

Similarly, we can calculate the other components,

$$\sigma_{xx} = \sigma_{yy} = \frac{N_e e^2\tau}{m_c},\ N_e = \frac{k_F^2}{2\pi c},\ \sigma_{xy} = \sigma_{yz} = \sigma_{zx} = 0.$$

In this calculation, the effect of the closed orbits on the side of the Fermi surface is not included. The results are consistent with Eqs. (4.24) and (4.25) for $\omega_c = eB_z/m_c = 0$.

Appendix 4.5

Here we drive Eqs. (4.56) and (4.57). From Eq. (4.55), the velocity is given by

$$v_z(k(t)) = \frac{2t_c c}{\hbar}\sin(k_z c)$$

$$= \frac{2t_c c}{\hbar}\sin\left[-\gamma\cos\left(\omega_z t + k_{y0}b\right) - \omega_y t + k_{z0}c + \gamma\cos\left(k_{y0}b\right)\right].$$

Taking $Z = -\omega_y t + k_{z0}c + \gamma\cos\left(k_{y0}b\right)$, we define $v_{z1}(\boldsymbol{k}(t))$ and $v_{z2}(\boldsymbol{k}(t))$ as

$$v_z(\boldsymbol{k}(t)) = \left(\frac{2t_c c}{\hbar}\right)\frac{1}{2i}\left[e^{i\{Z-\gamma\cos(\omega_z t+k_{y0}b)\}} - e^{-i\{Z-\gamma\cos(\omega_z t+k_{y0}b)\}}\right]$$

$$= v_{z1}(\boldsymbol{k}(t)) + v_{z2}(\boldsymbol{k}(t)). \tag{4.104}$$

From the formula of the nth order Bessel function $J_n(z)$ in Eq. (4.94),

$$\begin{aligned} e^{-i\gamma \cos(\omega_z t + k_{y0} b)} &= \sum_{n=-\infty}^{\infty} e^{in\pi/2} J_n(-\gamma) e^{in(\omega_z t + k_{y0} b)} \\ &= \sum_{n=-\infty}^{\infty} e^{-in\pi/2} J_n(\gamma) e^{-in(\omega_z t + k_{y0} b)}, \end{aligned} \tag{4.105}$$

where we used the relation $J_n(-z) = J_{-n}(z)$ and then replaced $-n$ with n. The velocity is written as

$$v_{z1}(\boldsymbol{k}(t)) = \left(\frac{2t_c c}{\hbar}\right) \frac{1}{2i} e^{iZ} \sum_{n=-\infty}^{\infty} e^{-in\pi/2} J_n(\gamma) e^{-in(\omega_z t + k_{y0} b)}. \tag{4.106}$$

Similarly, we obtain

$$v_{z2}(\boldsymbol{k}(t)) = -\left(\frac{2t_c c}{\hbar}\right) \frac{1}{2i} e^{-iZ} \sum_{n=-\infty}^{\infty} e^{in\pi/2} J_n(\gamma) e^{in(\omega_z t + k_{y0} b)}. \tag{4.107}$$

For $t = 0$, the velocity is

$$v_z(\boldsymbol{k}(0)) = \left(\frac{2t_c c}{\hbar}\right) \sin[k_{z0} c] = \left(\frac{2t_c c}{\hbar}\right) \frac{1}{2i} \left[e^{ik_{z0} c} - e^{-ik_{z0} c}\right].$$

From the relation $\mathrm{d}f/\mathrm{d}E = -\delta(E_F - E) = (1/\hbar v_F)\delta(k_F - |k_x|)$, the sum is written as

$$\sum_{k(0)} \left(-\frac{\mathrm{d}f}{\mathrm{d}E}\right) \to \frac{1}{(2\pi)^3} \int_{-\pi/c}^{\pi/c} \mathrm{d}k_{z0} \int_{-\pi/b}^{\pi/b} \mathrm{d}k_{y0} \int_{-\pi/a}^{\pi/a} \mathrm{d}k_x \frac{1}{\hbar v_F} \delta(k_F - |k_x|). \tag{4.108}$$

The conductivity σ_{zz1}^{1D} arising from $v_{z1}(\boldsymbol{k}(t))$ is

$$\begin{aligned} \sigma_{zz1}^{1D} &= 2e^2 \sum_{\boldsymbol{k}(0)} \left(-\frac{\mathrm{d}f}{\mathrm{d}E}\right) v_z(\boldsymbol{k}(0)) \int_{-\infty}^{0} \mathrm{d}t\ v_{z1}(\boldsymbol{k}(t)) \exp(t/\tau) \\ &= \frac{2e^2}{(2\pi)^3} \int_{-\pi/c}^{\pi/c} \mathrm{d}k_{z0} \int_{-\pi/b}^{\pi/b} \mathrm{d}k_{y0} \end{aligned}$$

$$\times \int_{-\pi/a}^{\pi/a} \mathrm{d}k_x \frac{1}{\hbar v_F} \delta(k_F - |k_x|) v_z(\boldsymbol{k}(0)) \int_{-\infty}^{0} \mathrm{d}t\ v_{z1}(\boldsymbol{k}(t)) \exp(t/\tau)$$

$$= \frac{2e^2}{(2\pi)^3} \left(\frac{2t_c c}{\hbar}\right)^2 \left(\frac{-1}{4}\right) \int_{-\pi/c}^{\pi/c} \mathrm{d}k_{z0} \int_{-\pi/b}^{\pi/b} \mathrm{d}k_{y0} \int_{-\pi/a}^{\pi/a} \mathrm{d}k_x \frac{1}{\hbar v_F} \delta(k_F - |k_x|)$$

$$\times \int_{-\infty}^{0} \mathrm{d}t \left[e^{ik_{z0}c} - e^{-ik_{z0}c}\right] e^{iZ} \sum_{n=-\infty}^{\infty} e^{-\frac{in\pi}{2}} J_n(\gamma) e^{-in(\omega_z t + k_{y0} b)} \exp(t/\tau).$$

Substituting $Z = -\omega_y t + k_{z0}c + \gamma \cos(k_{y0}b)$ into the above equation, we have

$$\sigma_{zz1}^{1D} = \frac{2e^2}{(2\pi)^3} \left(\frac{2t_c c}{\hbar}\right)^2 \left(\frac{2\pi}{4c}\right)$$
$$\times \int_{-\pi/b}^{\pi/b} \mathrm{d}k_{y0} \int_{-\pi/a}^{\pi/a} \mathrm{d}k_x \frac{1}{\hbar v_F} \delta(k_F - |k_x|)$$
$$\times \int_{-\infty}^{0} \mathrm{d}t \sum_{n=-\infty}^{\infty} e^{-i[\omega_y t - \gamma \cos(k_{y0}b)]} e^{-\frac{in\pi}{2}} J_n(\gamma) e^{-in(\omega_z t + k_{y0} b)} \exp(t/\tau). \quad (4.109)$$

Here we divide the k_x integral into two parts,

$$\int_{-\pi/a}^{\pi/a} \mathrm{d}k_x = \int_{0}^{\pi/a} \mathrm{d}k_x + \int_{-\pi/a}^{0} \mathrm{d}k_x.$$

For $k_x < 0$, we have to change $\omega_y \to -\omega_y$, $\omega_z \to -\omega_z$, $\gamma \to -\gamma$ and thus

$$\sigma_{zz1}^{1D} = \frac{2e^2}{(2\pi)^3} \frac{1}{\hbar v_F} \left(\frac{2t_c c}{\hbar}\right)^2 \left(\frac{2\pi}{4c}\right) \int_{-\pi/b}^{\pi/b} \mathrm{d}k_{y0}$$
$$\times \int_{-\infty}^{0} \mathrm{d}t \sum_{\pm, n=-\infty}^{\infty} e^{\pm i\gamma \cos(k_{y0}b)} e^{-\frac{in\pi}{2}} J_n(\pm\gamma) e^{-ink_{y0}b} e^{\mp i(n\omega_z + \omega_y)t} e^{t/\tau}.$$

The sign $\pm$ in Σ means to take the sum of both signs. Using $J_n(-z) = J_{-n}(z)$ and then replacing $-n$ with n, the above equation is written as

$$\sigma_{zz1}^{1D} = \frac{2e^2}{(2\pi)^3} \frac{1}{\hbar v_F} \left(\frac{2t_c c}{\hbar}\right)^2 \left(\frac{2\pi}{4c}\right) \int_{-\pi/b}^{\pi/b} \mathrm{d}k_{y0}$$

$$\times \int_{-\infty}^{0} \mathrm{d}t \sum_{\pm,n=-\infty}^{\infty} e^{\pm i\gamma\cos(k_{y0}b)} e^{\mp\frac{in\pi}{2}} J_n(\gamma) e^{\mp ink_{y0}b} e^{-i(n\omega_z \pm \omega_y)t} e^{t/\tau}. \quad (4.110)$$

Similarly, we calculate σ_{zz2}^{1D},

$$\sigma_{zz2}^{1D} = \frac{2e^2}{(2\pi)^3} \frac{1}{\hbar v_F} \left(\frac{2t_c c}{\hbar}\right)^2 \left(\frac{2\pi}{4c}\right) \int_{-\pi/b}^{\pi/b} \mathrm{d}k_{y0}$$

$$\times \int_{-\infty}^{0} \mathrm{d}t \sum_{\pm,n=-\infty}^{\infty} e^{\mp i\gamma\cos(k_{y0}b)} e^{\pm\frac{in\pi}{2}} J_n(\gamma) e^{\pm ink_{y0}b} e^{i(n\omega_z \pm \omega_y)t} e^{t/\tau}. \quad (4.111)$$

Taking $X = k_{y0}b + \pi/2$, we have the relations,

$$\int_{-\pi/b}^{\pi/b} \mathrm{d}k_{y0} e^{i\gamma\cos(k_{y0}b)} e^{-in(k_{y0}b+\pi/2)} = \frac{1}{b}\int_{-\pi}^{\pi} \mathrm{d}X e^{i\gamma\sin X} e^{-inX} = \frac{2\pi}{b} J_n(\gamma),$$

$$\int_{-\pi/b}^{\pi/b} \mathrm{d}k_{y0} e^{-i\gamma\cos(k_{y0}b)} e^{in(k_{y0}b+\pi/2)} = \frac{1}{b}\int_{-\pi}^{\pi} \mathrm{d}X e^{-i\gamma\sin X} e^{inX} = \frac{2\pi}{b} J_n(\gamma),$$

and then σ_{zz1}^{1D} and σ_{zz2}^{1D} are respectively given by

$$\sigma_{zz1}^{1D} = \frac{1}{\pi bc\hbar v_F} \left(\frac{et_c c}{\hbar}\right)^2 \sum_{\pm,n=-\infty}^{\infty} J_n^2(\gamma) \int_{-\infty}^{0} \mathrm{d}t e^{t/\tau} e^{-i(n\omega_z \pm \omega_y)t}, \quad (4.112)$$

$$\sigma_{zz2}^{1D} = \frac{1}{\pi bc\hbar v_F} \left(\frac{et_c c}{\hbar}\right)^2 \sum_{\pm,n=-\infty}^{\infty} J_n^2(\gamma) \int_{-\infty}^{0} \mathrm{d}t e^{t/\tau} e^{i(n\omega_z \pm \omega_y)t}. \quad (4.113)$$

By integrating them, we obtain Eq. (4.56),

$$\sigma_{zz}^{1D}(B) = \sigma_{zz1}^{1D}(B) + \sigma_{zz2}^{1D}(B)$$
$$= \sigma_{zz}^{1D}(0) \sum_{\pm,n=-\infty}^{\infty} \frac{J_n^2(\gamma)/2}{1 + (n\omega_z \pm \omega_y)^2 \tau^2},$$
$$\sigma_{zz}^{1D}(0) = 2D(E_F)\left(\frac{et_c c}{\hbar}\right)^2 \tau,$$

where $D(E_F) = 2/(\pi bc\hbar v_F)$ is the density of states per unit volume.

Appendix 4.6

We derive Eq. (4.78). Using $v_z(\boldsymbol{k}(t))$ in Eq. (4.77), the conductivity is written as

$$
\begin{aligned}
\sigma_{zz}^{1D} &= 2e^2 \sum_{\boldsymbol{k}(0)} \left(-\frac{\mathrm{d}f}{\mathrm{d}E}\right) v_z(\boldsymbol{k}(0)) \int_{-\infty}^{0} \mathrm{d}t\; v_z(\boldsymbol{k}(t)) \exp(t/\tau) \\
&= \frac{2e^2}{(2\pi)^3} \int_{-\pi/c}^{\pi/c} \mathrm{d}k_{z0} \int_{-\pi/b}^{\pi/b} \mathrm{d}k_{y0} \\
&\quad \times \int_{-\pi/a}^{\pi/a} \mathrm{d}k_x \frac{1}{\hbar v_F} \delta(k_F - |k_x|) v_z(\boldsymbol{k}(0)) \int_{-\infty}^{0} \mathrm{d}t\; v_z(\boldsymbol{k}(t)) \exp(t/\tau) \\
&= \frac{2e^2}{(2\pi)^3} \left(\frac{2t_c c}{\hbar}\right)^2 \int_{-\pi/c}^{\pi/c} \mathrm{d}k_{z0} \int_{-\pi/b}^{\pi/b} \mathrm{d}k_{y0} \int_{-\pi/a}^{\pi/a} \mathrm{d}k_x \frac{1}{\hbar v_F} \delta(k_F - |k_x|) \\
&\quad \times \int_{-\infty}^{0} \mathrm{d}t\; \sin(k_{z0}c) \sin(k_{z0}c - \Omega_{1D}t) \exp(t/\tau).
\end{aligned}
\tag{4.114}
$$

Since there are two sheets of the quasi-1D Fermi surface, the k_x integral is written as

$$
\int_{-\pi/a}^{\pi/a} \mathrm{d}k_x \frac{1}{\hbar v_F} \delta(k_F - |k_x|) = \frac{2}{\hbar v_F}. \tag{4.115}
$$

By rewriting

$$
\begin{aligned}
\sin(k_{z0}c)\sin(k_{z0}c - \Omega_{1D}t) &= \sin^2(k_{z0}c)\cos(\Omega_{1D}t) - \sin(k_{z0}c)\cos(k_{z0}c)\sin(\Omega_{1D}t) \\
&= \frac{1-\cos(2k_{z0}c)}{2}\cos(\Omega_{1D}t) - \frac{\sin(2k_{z0}c)}{2}\sin(\Omega_{1D}t),
\end{aligned}
$$

we calculate the integral,

$$
\begin{aligned}
&\int_{-\pi/c}^{\pi/c} \mathrm{d}k_{z0} \int_{-\infty}^{0} \mathrm{d}t \sin(k_{z0}c)\sin(k_{z0}c - \Omega_{1D}t)\exp(t/\tau) \\
&\qquad = \frac{\pi}{c} \int_{-\infty}^{0} \mathrm{d}t \cos(\Omega_{1D}t)\exp(t/\tau) = \frac{\pi}{c} \frac{\tau}{1 + (\Omega_{1D}\tau)^2}.
\end{aligned}
\tag{4.116}
$$

Therefore, we obtain

$$\sigma_{zz}^{1D} = \frac{2e^2}{(2\pi)^3}\left(\frac{2t_c c}{\hbar}\right)^2\left(\frac{2}{\hbar v_F}\right)\left(\frac{\pi}{c}\right)\int_{-\pi/b}^{\pi/b} \mathrm{d}k_{y0}\frac{\tau}{1+(\Omega_{1D}\tau)^2}. \tag{4.117}$$

Substituting $\Omega_{1D} = \left[\sin(\varphi) - \eta\cos(\varphi)\sin\left(k_{y0}b\right)\right]\omega_{y0}$ into Eq. (4.117), we obtain

$$\begin{aligned}
\sigma_{zz}^{1D}(B) &= \sigma_{zz}^{1D}(0)\frac{\sqrt{X+\sqrt{X^2+4Y^2}}}{\sqrt{2}\sqrt{X^2+4Y^2}},\\
\sigma_{zz}^{1D}(0) &= 2D(E_F)\left(\frac{et_c c}{\hbar}\right)^2\tau,\\
D(E_F) &= 2/\pi bc\hbar v_F,\\
X = 1-\left(Y^2-Z^2\right),\ Y &= \omega_{y0}\tau\sin(\varphi),\ Z = \eta\omega_{y0}\tau\cos(\varphi).
\end{aligned}$$

In this calculation, we used the integral formula,

$$\int_{-\pi}^{\pi}\mathrm{d}x\frac{1}{1+2b\sin(x)+c^2\sin^2(x)} = \frac{\pi c^2}{\sqrt{b^2-c^2}}\times\left(\frac{1}{\sqrt{-2b\sqrt{b^2-c^2}+2b^2-c^2\left(c^2+1\right)}} - \frac{1}{\sqrt{2b\sqrt{b^2-c^2}+2b^2-c^2\left(c^2+1\right)}}\right).$$

References

1. Yamaji, K.: On the Angle dependence of the magnetoresistance in quasi-two-dimensional organic superconductors. J. Phys. Soc. Jpn. **58**, 1520 (1989). https://doi.org/10.1143/JPSJ.58.1520
2. Kajita, K., Nishio, Y., Takahashi, T., Sasaki, W., Kato, R., Kobayashi, H., Kobayashi, A., Iye, Y.: A new type oscillatory phenomenon in the magneto-transport of θ-(BEDT-TTF)$_2$I$_3$. Solid State Commun. **70**, 1189–1193 (1989). https://doi.org/10.1016/0038-1098(89)90761-8
3. Kartsovnik, M.V., Kononovich, P.A., Laukhin, V.N., Shchegolev, I.F.: Anisotropy of magnetoresistance and the Shubnikov-de Haas oscillations in the organic metal β-(ET)2IBr$_2$. Pis'ma Zh. Eksp. Teor. Fiz. **48**, 498–501 (1988)
4. Sugawara, S., Tamura, M.: Anisotropic mobility and carrier dynamics in the β-type BEDT-TTF salts as studied by inter-layer transverse magneto-resistance. Sci. Technol. Adv. Mater. **14**, 045004 (2013). https://doi.org/10.1088/1468-6996/14/4/045004
5. Yagi, R., Iye, Y., Osada, T., Kagoshima, S.: Semiclassical interpretation of the angular-dependent oscillatory magnetoresistance in quasi-two-dimensional systems. J. Phys. Soc. Jpn. **59**, 3069 (1990). https://doi.org/10.1143/JPSJ.59.3069

6. Klepper, S.J., Athas, G.J., Brooks, J.S., Tokumoto, M., Kinoshita, T., Tamura, N., Kinoshita, M.: Angular magnetoresistance oscillations and magnetic breakdown in θ-(BEDT-TTF)$_2$I$_3$. Synth. Met. **70**, 835–836 (1995). https://doi.org/10.1016/0379-6779(94)02670-T
7. House, A.A., Harrison, N., Blundell, S.J., Deckers, I., Singleton, J., Herlach, F., Hayes, W., Perenboom, J.A.A.J., Kurmoo, M., Day, P.: Oscillatory magnetoresistance in the charge-transfer salt β"-(BEDT−TTF)2AuBr 2 in magnetic fields up to 60 T: evidence for field-induced Fermi-surface re-construction. Phys. Rev. B **53**, 9127 (1996)
8. Terashima, T., Uji, S., Aoki, H., Tamura, M., Kinoshita, M., Tokumoto, M.: Reexamination of angle dependent magnetoresistance oscillation in θ-(BEDT-TTF)2I3. Synth. Met. **70**, 845–846 (1995). https://doi.org/10.1016/0379-6779(94)02675-O
9. Kimata, M., Terashima, T., Kurita, N., Satsukawa, H., Harada, A., Kodama, K., Sato, A., Imai, M., Kihou, K., Lee, C.H., Kito, H., Eisaki, H., Iyo, A., Saito, T., Fukazawa, H., Kohori, Y., Harima, H., Uji, S.: Quasi-two-dimensional Fermi surfaces and coherent interlayer transport in KFe$_2$As$_2$. Phys. Rev. Lett. **105**, 246403 (2010). https://doi.org/10.1103/PhysRevLett.105.246403
10. Ohmichi, E., Adachi, H., Mori, Y., Maeno, Y., Ishiguro, T., Oguchi, T.: An-gle-dependent magnetoresistance oscillation in the layered perovskite Sr$_2$RuO$_4$. Phys. Rev. B **59**, 7263 (1999). https://doi.org/10.1103/PhysRevB.59.7263
11. Enomoto, K., Uji, S., Yamaguchi, T., Terashima, T., Konoike, T., Nishimura, M., Enoki, T., Suzuki, M., Suzuki, I.S.: Fermi surface and interlayer transport in high-stage MoCl$_5$ graphite intercalation compounds. Phys. Rev. B **73**, 045115 (2006). https://doi.org/10.1103/PhysRevB.73.045115
12. Kawamura, M., Endo, A., Hirasawa, M., Katsumoto, S., Iye, Y.: Angular dependent magnetoresistance oscillation in GaAs/AlGaAs superlattice. Phys. B **249–251**, 882 (1998). https://doi.org/10.1016/S0921-4526(98)00335-4
13. Hussey, N.E., Abdel-Jawad, M., Carrington, A., Mackenzie, A.P., Balicas, L.: A coherent three-dimensional Fermi surface in a high-transition-temperature superconductor. Nature **425**, 814 (2003). https://doi.org/10.1038/nature01981
14. Lewin, S.K., Analytis, J.G.: Angle-dependent magnetoresistance as a probe of Fermi surface warping in HgBa$_2$CuO$_{4+\delta}$. Phys. Rev. B **92**, 195130 (2015). https://doi.org/10.1103/PhysRevB.98.075116
15. Yagi, R., Iye, Y.: On the peak inversion of the angular dependent magnetoresistance oscillation of cylindrical Fermi surfaces with different corrugation symmetries. Solid State Commun. **89**, 275 (1994). https://doi.org/10.1016/0038-1098(94)90697-1
16. Moses, P., Mckenzie, R.H.: Periodic orbit resonances in layered metals in tilted magnetic fields. Phys. Rev. B **60**, 7998 (1999). https://doi.org/10.1103/PhysRevB.60.R11241
17. Osada, T., Ohmichi, E.: Magnetotransport and magnetic-field-induced density waves in low-dimensional layered conductors. J. Phys. Soc. Jpn. **75**, 051006 (2006). https://doi.org/10.1143/JPSJ.75.051006
18. Hanasaki, N., Kagoshima, S., Hasegawa, T., Osada, T., Miura, N.: Contribution of small closed orbits to magnetoresistance in quasi-two-dimensional conductors. Phys. Rev. B **57**, 1336 (1998). https://doi.org/10.1103/PhysRevB.57.1336
19. Singleton, J., Goddard, P.A., Ardavan, A., Harrison, N., Blundell, S.J., Schlueter, A., Kini, A.M.: Test for interlayer coherence in a quasi-two-dimensional superconductor. Phys. Rev. Lett. **88**, 037001 (2002). https://doi.org/10.1103/PhysRevLett.88.037001
20. Kawamoto, T., Mori, T., Graf, D., Brooks, J.S., Takahide, Y., Uji, S., Shi-rahata, T., Imakubo, T.: Interlayer charge disproportionation in the layered organic superconductor κ_H−(DMEDO−TSeF)$_2$[Au(CN)$_4$](THF) with polar dielectric insulating layers. Phys. Rev. Lett. **109**, 147005 (2012). https://doi.org/10.1103/PhysRevLett.109.147005
21. Kennett, M.P., McKenzie, R.H.: Quantum interference and weak localization effects in the interlayer magnetoresistance of layered metals. Phys. Rev. B **78**, 024506 (2008). https://doi.org/10.1103/PhysRevB.78.024506

22. Jindo, R., Sugawara, S., Tajima, N., Yamamoto, H.M., Kato, R., Nishio, Y., Kajita, K.: Out-of-plane resistance of quasi-two dimensional metal (BEDT-TTF)$_3$Cl(DFBIB) in transverse magnetic fields. J. Phys. Soc. Jpn. **75**, 013705 (2006)
23. Yamamoto, H.M., Yamaura, J., Kato, R.: Multicomponent molecular conductors with supramolecular assembly: Iodine-containing neutral molecules as building blocks. J. Am. Chem. Soc. **120**, 5905 (1998). https://doi.org/10.1021/ja980024u
24. Osada, T., Kurauchi, M.: General quantum picture for magnetoresistance angular effects in quasi-one-dimensional conductors. Synth. Met. **133–134**, 75 (2003). https://doi.org/10.1016/S0379-6779(02)00279-5
25. Cooper, B.K., Yakovenko, V.M.: Interlayer Aharonov–Bohm interference in tilted magnetic fields in quasi-one-dimensional organic conductors. Phys. Rev. Lett. **96**, 037001 (2006). https://doi.org/10.1103/PhysRevLett.96.037001
26. Danner, G.M., Kang, W., Chaikin, P.M.: Measuring the Fermi surface of quasi-one-dimensional metals. Phys. Rev. Lett. **72**, 3714 (1994). https://doi.org/10.1103/PhysRevLett.72.3714
27. Osada, T., Kami, N., Kondo, R., Kagoshima, S.: Semiclassical and non-classical angular effects of magnetoresistance in (TMTSF)$_2$X. Synth. Met. **103**, 2024–2027 (1999). https://doi.org/10.1016/S0379-6779(98)00289-6
28. Lebed, A.G.: Anisotropy of an instability for a spin density wave induced by a magnetic field in a Q1D conductor. Pis'ma Zh. Eksp. Teor. Fiz. **43**, 137–139 (1986)
29. Chashechkina, E.I., Chaikin, P.M.: Magic angles and the ground states in (TMTSF)$_2$PF$_6$. Phys. Rev. Lett. **80**, 2181 (1998). https://doi.org/10.1103/PhysRevLett.80.2181
30. Strong, S.P., Clarke, D.G., Anderson, P.W.: Magnetic field induced confinement in strongly correlated anisotropic materials. Phys. Rev. Lett. **73**, 1007 (1994). https://doi.org/10.1103/PhysRevLett.73.1007
31. Danner, G.M., Chaikin, P.M.: Non-Fermi-liquid behavior in transport in (TMTSF)$_2$PF$_6$. Phys. Rev. Lett. **75**, 4690 (1995). https://doi.org/10.1103/PhysRevLett.75.4690
32. Clarke, D.G., Strong, S.P., Chaikin, P.M., Chashechkina, E.I.: The Quantum-Classical Metal. Science **279**, 2071 (1998). https://doi.org/10.1126/science.279.5359.2071
33. Sugawara, S., Ueno, T., Kawasugi, Y., Tajima, N., Nishio, Y., Kajita, K.: Out-of-plane resistance under transverse magnetic field in quasi-one-dimensional layered metals. J. Phys. Soc. Jpn. **75**, 053704 (2006). https://doi.org/10.1143/JPSJ.75.053704
34. Osada, T., Kagoshima, S., Miura, N.: Third angular effect of magneto-resistance in quasi-one-dimensional conductors. Phys. Rev. Lett. **77**, 5261 (1996). https://doi.org/10.1103/PhysRevLett.77.5261
35. Yoshino, H., Murata, K.: Origin of the third angular effect of magneto-resistance of quasi-one-dimensional metals. J. Phys. Soc. Jpn. **68**, 3027 (1999). https://doi.org/10.1143/JPSJ.68.3027
36. Lebed, A.G., Bagmet, N.N.: Nonanalytical magnetoresistance, the third angular effect, and a method to investigate Fermi surfaces in quasi-two-dimensional conductors. Phys. Rev. B **55**, R8654 (1997). https://doi.org/10.1103/PhysRevB.55.R8654
37. Lee, I.J., Naughton, M.: Effective electrons and angular oscillations in quasi-one-dimensional conductors. Phys. Rev. B **57**, 7423 (1998). https://doi.org/10.1103/PhysRevB.57.7423

Index

S. Uji, *Fermiology*, Lecture Notes in Physics 1042,
https://doi.org/10.1007/978-981-95-0820-4

MIX
Papier aus verantwortungsvollen Quellen
Paper from responsible sources
FSC® C105338

If you have any concerns about our products,
you can contact us on
ProductSafety@springernature.com

In case Publisher is established outside the EU,
the EU authorized representative is:
Springer Nature Customer Service Center GmbH
Europaplatz 3, 69115 Heidelberg, Germany

Printed by Libri Plureos GmbH
in Hamburg, Germany